Fundamentals of the Physical-Chemistry of Pulverized Coal Combustion

NATO ASI Series

Advanced Science Institutes Series

A Series presenting the results of activities sponsored by the NATO Science Committee, which aims at the dissemination of advanced scientific and technological knowledge, with a view to strengthening links between scientific communities.

The Series is published by an international board of publishers in conjunction with the NATO Scientific Affairs Division

A	Life Sciences	Plenum Publishing Corporation
B	Physics	London and New York
C	Mathematical and Physical Sciences	D. Reidel Publishing Company Dordrecht, Boston, Lancaster and Tokyo
D	Behavioural and Social Sciences	Martinus Nijhoff Publishers Dordrecht, Boston and Lancaster
E	Applied Sciences	
F	Computer and Systems Sciences	Springer-Verlag Berlin, Heidelberg, New York
G	Ecological Sciences	London, Paris and Tokyo
H	Cell Biology	

Series E: Applied Sciences – No. 137

Fundamentals of the Physical-Chemistry of Pulverized Coal Combustion

edited by

J. Lahaye and G. Prado

Centre de Recherches sur la Physico-Chimie des Surfaces Solides – C.N.R.S.
24, Avenue du Président Kennedy
68200 Mulhouse
France

1987 **Martinus Nijhoff Publishers**
Dordrecht / Boston / Lancaster
Published in cooperation with NATO Scientific Affairs Division

Proceedings of the NATO Advanced Research Workshop on 'Fundamentals of the Physical-Chemistry of Pulverized Coal Combustion', Les Arcs, France, July 28 – August 1, 1986

Library of Congress Cataloging in Publication Data

NATO Advanced Research Workshop on "Fundamentals of the
    Physical-Chemistry of Pulverized Coal Combustion"
    (1986 : Les Arcs, Bourg-Saint-Maurice, France)
    Fundamentals of the physical-chemistry of pulverized
coal combustion.

    (NATO ASI series. Series E, Applied sciences ; no.
137)
    Includes index.
    1. Coal, Pulverized--Congresses. 2. Coal--Combustion
--Congresses. I. Lahaye, J., 1937-   . II. Prado, G.
III. Title. IV. Series.
TP328.N35 1986   621.402'3   87-15250
ISBN 90-247-3573-4

ISBN 90-247-3573-4 (this volume)
ISBN 90-247-2689-1 (series)

Distributors for the United States and Canada: Kluwer Academic Publishers, P.O. Box 358, Accord-Station, Hingham, MA 02018-0358, USA

Distributors for the UK and Ireland: Kluwer Academic Publishers, MTP Press Ltd, Falcon House, Queen Square, Lancaster LA1 1RN, UK

Distributors for all other countries: Kluwer Academic Publishers Group, Distribution Center, P.O. Box 322, 3300 AH Dordrecht, The Netherlands

Printed in The Netherlands

PREFACE

The study of coal for the production of energy is certainly not a new area of research. Many research works were carried out to improve the efficiency of industrial and domestic facilities. In the sixties, however, because of the availability and low cost of petroleum, coal consumption decreased and the research effort in this area was minimum.

Meanwhile, the situation has totally changed.Considering the reserves of oil and the instability of regions where they are located, it is becoming absolutely necessary to develop other sources of energy.The major alternative to oil appears to be coal, at least for the near future. Indeed, the reserves known today represent several centuries of energy consumption.It is therefore becoming urgent to develop efficient and non polluting technologies to produce energy from coal. The main possibilities are :

. liquefaction
. gasification
. directed combustion.

Research and development efforts on liquefaction have been considerably reduced because of high cost of technologies involved and poor prospects for the next two decades. Research works on gasification are progressing ; it is a promising approach.

However, direct combustion either in pulverized coal furnaces or in fluidized beds is the more promising way of expanding rapidly the utilization of coal. These techniques are already used in some facilities but many environmental problems remain, slowing down their development.

It is of primary importance to be able to achieve a proper physical description with accurate models of the different processes occurring in industrial facilities,in order to develop efficient and non polluting equipments.

Beginning in the early eighties, many laboratories have been involved in coal sciences. Several computer codes have been developed, describing flow aerodynamics with some coupling with chemical reactions. In order to avoid the manipulation of empirical parameters which make these models inappropriate for solving practical problems, it is necessary to introduce data relative to the fundamental mechanisms of the combustion of a coal particle. It is also important to take into account the different types of coal and for a given coal the heterogeneity of maceral composition and size of the particles.

At the Twentieth Symposium (International) on combustion held at Ann Arbor (Michigan) in 1984, a fairly large number of papers were concerned with the combustion of coal. We noticed that, unfortunately, it was difficult to do an acceptable synthesis of the results published by the different laboratories. We decided at that time, encouraged by our colleagues from M.I.T. involved in coal combustion science, to organize a workshop to put together for a week the specialists of the domain in order to establish the state of the art and formulate recommendations for future work.

The final programme was divided into six sessions. Eight main lectures were completed by twelve communications.

In the first session a review of coal properties known to be important to coal combustion behaviour was presented.

When coal is subjected to high temperature in oxidative or not oxidative atmosphere, devolatilization occurs. In the second session, three main lectures were devoted to the kinetics of devolatilization, the role of volatiles in coal combustion and the morphological transformation of coal during pyrolysis.

Heterogeneous combustion of the char obtained by devolatilization of coal was reviewed in the third session.

The largest obstacle to the public acceptance of increased coal use is the perception that coal combustion is polluting. The fourth session addressed this key problem.

Much of the recent progress in the understanding of coal combustion is attributable to new advanced diagnostics. The fifth session reviewed the techniques for in situ measurements of coal flames.

The sixth session was concerned with the transfer of fundamental results to the modeling of burners and boilers.

The last part of the meeting was devoted to a synthesis of the workshop by three sub-committees.

The question, comments, and answers submitted in a written form have been edited in the present volume.

Special thanks are addressed to the Scientific Committee, Profs. J.B. Howard, H. Jüntgen and F.E. Lockwood, Dr R.E. Mitchell, Profs. A.F. Sarofim and T.F. Wall, for their efficient collaboration in the preparation of the meeting.

We are grateful to C. Denninger, F. Muller, P. and S. Wagner for their technical assistance before and during the workshop.

J. LAHAYE

G. PRADO

PARTICIPANTS

1.   G. De Soete
2. R.H. Essenhigh
3.   M. Morgan
4. D.R. Hardesty
5. H.J. Mühlen
6.   M. Hertzberg
7. K.H. Van Heek
8.   P. Anglesio
9. J.F. Muller
10. E.M. Suuberg
11.   R. Cyprès
12.   F. Beretta
13. I.W. Smith
14.   A. Williams
15.   G. Flament
16. B.S. Haynes
17. J.L. Roth
18.   F. KAPTEIJN
19.   I. Gulyurtlu
20. J.H. Pohl
21.   G. Leyendecker
22.   E. Saatdjian
23.   T. Outassourt
24. H.P. Odenthal
25.

26.   W. Zinser
27. P.R. Solomon
28.   Z. Habib
29.   O. Charon
30. J.R. Richard
31.   P. Roberts
32.   A. Garo
33.   P. Wagner
34.   S. Wagner
35. J.G. Smith
36. J.M. Beer
37.   H. Jüntgen
38. T.F. Wall
39.   J. Lahaye
40.   G. Prado
41. R.E. Mitchell
42. A.F. Sarofim
43. J.B. Howard
44.   W. Thielen
45. J.M. Vleeskens
46. F.C. Lockwood

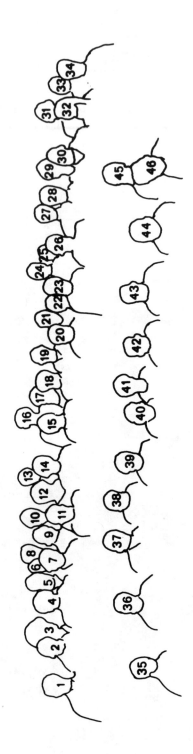

C O N T E N T S

INTRODUCTION SESSION

COAL CHARACTERIZATION IN RELATION TO COAL COMBUSTION

Harald Jüntgen
Professor Dr.rer.nat.

1. <u>INTRODUCTION</u>

Most coals are used worldwide for combustion today. Generally all kinds of coals are applicable for combustion.
The major methods of burning are fixed bed firing, fluidized
bed firing and suspension firing. The last technique is
used in big industrial plants for electric power generation.
In fixed bed firing heating rate is slow (a few K/s),
and combustion time ranges from minutes to hours, depending
on particle size. In fluidized bed combustion heating
rates can be estimated at 10^3 to 10^4 K/s, burning temperatures are between 800 and 900° C, and burning times
are in the order of minutes. In suspension firing heating
rates also are high, temperatures are more than 1500° C,
and burning times are in the order of seconds. In these
firing systems characteristic and very different particle
sizes are used: Lump coal in a range from 5 to 50 mm in
fixed bed firing, crushed coal in a size range from 1
to 5 mm in fluidized bed firing, and pulverized coal with
sizes of below 100 μ in suspension firing.

It is generally agreed that particle size of coal is the
most defining parameter with respect to the dominant reaction mechanism of combustion /1/. Beyond that coal properties also influence the combustion behaviour. As combustion
of coal is a complex process consisting of several parallel
or consecutive reactions, which are partly governed by
heat and mass transfer phenomena, it can not be expected
that there are simple correlations between parameters
of physical and chemical coal structure on the one hand
and overall characterization parameters of coal combustion
on the other hand. There is the fundamental need to have
a clear picture of the influence of coal properties on
combustion behaviour, because coal properties are very
significant for practical questions of design and operation
of combustion technology. There is the need to design
boilers for a much wider range of coals, for equipment
both smaller and larger than previously used, and for
more exact operation conditions /2/.

The relationship between coal properties and combustion
behaviour can be considered under various aspects. First
of all it has to be taken into account that coal properties
systematically change with coal rank. Some chemical, physical and calorique properties of coal are needed for achieving the heat and mass balance for boiler design.

As previously shown the particle size is a fundamental
question of using different burning technologies, and
hence it follows that design of crushing equipment is
a precondition of boiler performance. Therefore mechanical
properties of coal are important for the characterization
of crushing behaviour and grain size distribution.

An especially important task is the prediction of the
combustion behaviour. That is only possible, if the mechanism of the basic steps of combustion is identified, and
if a relationship between the characteristic coal properties and important parameters of each basic step has been
established. Then there is the hope that the modelling
of single steps of combustion and their coaction can also
be performed.

To meet the demand for minimizing the discharge of sulfur,
nitrogen oxides, and particulates the correlation between
the occurence of nitrogen and sulfur in coal and the formation of correspondent gaseous compounds during combustion
must be understood better. Also coal properties by which
boiler slagging and fouling are affected are of great
interest.

The aim of this paper is mainly to discuss the relationships
between coal properties and performance of basic steps
of combustion like pyrolysis, ignition and combustion
of pyrolysis products, and of ignition and combustion
of residual char. Another important point is the description
of coal nitrogen in respect to NO_x formation.

2. CLASSIFICATION OF COAL ACCORDING TO RANK

Coal is a chemically and physically heterogeneous rock,
mainly containing organic matter. It principally consists
ofcarbon, hydrogen and oxygen and on less amounts of sulfur
and nitrogen. Coal also contains ash forming inorganic
components distributed as discrete particles of mineral
matter throughout the coal substance.

Coals originated through the accumulation of plant debris
that were later covered, compacted and changed into the
organic rock we find today. The conversion - called "coalification" - is based on biological reactions in the first
stage, followed by a second phase in which geochemical
reactions take place. This phase can be described by a
very slow pyrolysis reaction under specially low rates
of heating /3/4/. This transformation successively leads
to peat (in the case of humic coals) lignite, bituminous
coal and anthracite. The progress in this coalification
scale (figure 1) is called the rank of coal and is suitable
for coal classification. Coal properties determining rank
like moisture content, carbon content, content of volatile matter and mean reflectance of vitrinite are listed
in table 1. There the relationship to the rank is also
shown.

Typical ultimate analysis of coals of different rank are
given in figure 2. From there it can be seen, that with
increasing rank the H/C ratio and the O/C ratio are decrea-
sing.That can be interpreted - as will be shown in the
following - by a loss of functional oxygen containing
groups and by an increase of the aromaticity of the coal-
macromolecule.

The properties of coal are not only influenced by their
rank but also in a certain extent by the kind of precursor
material and the kind of the environments during the coalification process. So humic coals are developed in mostly
oxidative environments and pass through a peat stage in
which the so called huminification takes place. Sapropelic
coals which are relatively rare are formed in more reductive environments and do not pass a peat stage.

The petrographic components of coals are defined as macerals and can be characterized by microscopy. They are
derived from different precursor materials. Single maceral are summerized in maceral groups, those precursor
materials, H/C ratio and aromaticity values are listed
in figure 3. The most frequent maceral group in humic
coals is that of vitrinites. Therefore the determination
of the rank is based on the mean reflectance of the vitrinite.

The development of the different macerals can be best
demonstrated in the H/C versus O/C diagram shown in figure 4
D.W. van Krevelen has given an extensive explanation in
his book /5/.

A diagram of more practical purpose is that of figure 5
in which experimentally derived relationships between
volatile matter, carbon content, hydrogen content, H/C
ratio and mean reflectance are shown /6/. A correlation
between the maximum vitrinite reflectance and the content
of vitrinite on C, S, H and N is given in figure 6 /7/.

3. COAL PROPERTIES NEEDED FOR ESTABLISHING HEAT AND MASS
   BALANCE FOR COAL COMBUSTION

The standard coal data needed for setting up the mass
balance, their relationship to coal rank and to other
parameters are listed in figure 7 /8/. Most data, with
the exception of ash content, mineral matter, S-content,
N- content and specific heat of coke are related to rank
parameters as volatiles, mean reflectance or ultimate
analysis.

In the next figures some examples of correlations of these
data are given for special coal basins. Figure 8 shows
the C-, H-, and O-content of coals from the Ruhr district
as a function of volatiles /9/. In figure 9 the correlation
between volatiles and C-content is presented for lignites

and hard coals /10/. Here data of coals from mostly Poland
are compared with those from Bulgaria, CSSR, FRG, France
and USSR. It follows from this figure that the scattering
of data for the different coals basins is especially large
in the range of volatiles between 35 and 50 %. Figures
10 and 11 are related to a broad band of solid fuels with
ranks between peat and anthracite from Bulgaria /11/.
Figure 10 gives relationships between C-, H-, O-content
and volatiles, and figure 11 presents a correlation between
the sum of O- and S-content and the C-content of coals.
This strong correlation shows that sulfur and oxygen are
exchangable to a certain extent which means that in coal
genesis organic sulfur can be formed by a chemical reaction
with oxygen containing functional groups.

As to the physical and caloric properties of coals true
density and porosity as a function of volatiles are given
in figure 12 /12/. The true specific heats of bituminous
coals can be correlated to the atomic ratio H_{ar} + 1.3 0/C,
where H_{ar} is the aromatic H-content, and to the temperature (figure 13) /13/. The true specific heat of coke
is only dependent on temperature. A correspondent correlation for blast furnace coke in the temperature range up
to 1000° C is given into /14/:

$$C_p = 0.837 + 1.54 \cdot 10^{-3}T - 5.4 \cdot 10^{-7} T^2 \text{ (kJ/kg K)}.$$

Figure 14 shows the relationship between net caloric value
and volatiles for coals from the Ruhr district /9/. There
are also a number of different formula describing the
caloric value as a function of the ultimate analysis of
coal and the heats of combustion of the elements, e.g.

$$Q \text{ (cal/g)} = 80,30 \text{ C} + 339 \text{ H} + 22.5 \text{ S} - 37.70 \text{ /15/}$$
or
$$H_u \text{ (MJ/kg)} = 33.91 + 121,42 \left(H - \frac{O}{8}\right) + 104,67 \text{ /16/}$$

in which C, H, S, O mean the content of C, H, S and O
in % in water- and ashfree coal.

4. COAL PROPERTIES NEEDED FOR CHARACTERIZATION OF GRAIN SIZE AND OF CRUSHING BEHAVIOUR

As mentioned in chapter 1 the particle size of coal is the
most defining parameter with respect to the dominant mechanism of combustion. Therefore the characterization of particle
size distribution is very important for systematic combustion
studies in research and for the design and operation of combustion equipment used for the different firing systems fixed
bed firing, fluidized bed firinbg and suspension firing.

It is not the task of this paper to describe methods for
the determination of particle size distribution. However
it shall be stressed that particle size of coal can change
during the course of combustion. First changes can occur

during pyrolysis of high volatile bituminous coals containing large parts of vitrain (litotype consisting on
distinct maceral groups). Here different effects of swelling
combined with grain size expansion or of destruction of
grains combined with grain size reduction have been observed. Parameters governing the behaviour of different
coals during pyrolysis are on the one hand process parameters as heating rate, on the other hand coal properties
as the rank and the maceral composition of coals. This
may also be one of the possible explanations for some
observations that the combustion behaviour of macerals
is different /21/. Some more information of change of
particle size and density are given in chapter 5.

Particle size of char is also influenced by combustion
at high temperatures. As in this temperature region film
diffusion is rate determining and the particle burns at
outer surface. The particle size decreases during combustion, which is also confirmed by experiments /17/.

From the point of designing crushing equipment the crushing
behaviour of coals is important. An empirical test for
the characterization of the grindability of coal has been
developed by Hardgrove. 50 g of an original grain size
between 0,59 and 1,19 mm is ground in a ball mill. After
sixty revolutions the coal is sieved over a sieve of 75 µ
and the grain fraction passing this mesh - MP - is determined. From there the Hardgrove Index °H is calculated
as follows

$$°H = 13 + 6.93 \ MP$$

A high Hardgrove index therefore means a high grindability.
The grindability of coals in a wide range of rank is presented in figure 15 /9/10/. The picture above shows that
there are two maxima of grindability, the first in the
region of lignites, the second in the range of hard coals.
The behaviour of hard coals is given more distinctly in
the picture below together with a mathematical correlation
for both areas on this side and beyond the maximum. It
is interesting to see, that the maximum of grindability
is at the same volatile content, which also characterizes
the minimum of porosity (figure 12).

5. COAL PROPERTIES NEEDED FOR THE PREDICTION OF COMBUSTION BEHAVIOUR AND MODELLING OF SINGLE STEPS OF COMBUSTION

5.1 Previous Results on Combustion Parameters and Chemical Structure

Many attempts have been made to establish empirical correlations between parameters of combustion behaviour of
coals like ignition temperature or burning time and coal
properties corresponding with coal rank /17/18/19/20/21/22/.
In many cases the correlations found are quite poor or
the scattering of the measured points is very large. A
general result of these tests is that ignition temperature
is as lower and burning time is as shorter (using corresponding particle sizes) as lower the rank of coal is.

A fundamental discussion of concepts of coal structure
in relation to combustion behaviour is given by P.H. Given
/23/. He comes to the following conclusions: Two aspects

of structure are in principle quite distinct: structural aspects determining mass transport within coal particles and structure in the sense of organic chemistry. He comes to the conclusion that the pre-existing pore structure in coals does not seem to be relevant for mass transport during pyrolysis and combustion because the pore structure is changed very much in the course of degradation of these coal macro-molecule.

He further stresses the importance of the mobile phase in coals for fast formation and ignition of volatile matter and also elucidates the different behaviour of macerals: so exinites are of high hydrogen and aliphatic hydrogen contents and of high volatile matter yield. Presumably, therefore, they should readily ignite and show rapid burn out. In contrast burn out is much less complete with coals having substantial contents of fusinite and semifusinite (belonging to the maceral group inertinite) /24/. According to his conclusions on the role of various induvidual structure features the aromaticity will tend to decrease pyrolysis and combustion rates and retard burn-out. The reverse is true for hydroaromatic and other aliphatic structures. The presence of OH functional groups will increase pyrolysis rates.

If the aim is to confirm and to extend these general conclusions in more detail a precondition of better understanding the coaction of coal structure and its combustion behaviour is a fundamental knowledge of the combustion mechanism of single coal particles. Therefore in the following a short survey shall be given on the description of single reaction steps in relation to relevant coal properties. These are the devolatilization as a function of chemical coal structure, the ignition and burning of volatiles depending on amount and chemical composition of volatiles, the formation of char considering the swelling behaviour of coals and the change of particle size and porosity during devolatilization, and the ignition and combustion of chars depending on reactivity and particle size.

5.2 Mechanism of combustion of individual coal particles

A survey on experimental methods and results of Bergbau-Forschung GmbH of the course of combustion of individual coals particles is given by Jüntgen /25/26/27/28/29/30/. Finely ground coal particles are heated up with a definite rate between 150 and 10^4 K/s using an electrically heated wire cloth. The reaction behaviour can be analysed by rapid gas analysis using a fast operating mass spectrometer and by optical observation using a high speed camera. As it is shematically shown in figure 16 (above) the combustion of the coal particle takes place in the three steps: heating up and formation of volatiles, ignition and combustion of volatile matter, and ignition and burn of of the char. This can be derived from the measured CO_2-formation as given in the left hand picture below of figure 16 for the combustion of coals of different

rank but of the same particle size in an air atmosphere
when heated up with a rate of 10^4 K per minute. Two clearly
seperated peaks can be distinguished on the graph. The
first one is ascriable to the burning of the volatiles,
and the second one to the combustion of the char. It can
be seen clearly, that the intensity of the first peak
is increasing with the content of volatile matter, and
that the ignition temperature of volatiles shifts from
430° C to 600° C with decreasing content of volatiles
from 39.5 % to 10 %. The second peak is studied in more
detail by burning the correspondent coal chars made by
pyrolysis in oxygen free atmosphere at the same heating
rate. Results are shown in the right hand picture of figure 16
below. The most important results are that chars ignite
at higher temperatures compared with the volatiles, that
ignition temperatures of chars increase from 590° C to
860° C, if volatiles of precursor coals decrease from
59,5 % to 10 %, and that burning times of chars also increase with decreasing volatiles of precursor coals.

More detailed studies have shown that the mechanism is
more complex and depends on grain size, rate of heating
and amount of volatiles. Results with a high volatile
coal are shown in figure 17. In the left hand picture
the three observed mechanisms are explained: 1) particle
ignition without seperate ignition of volatiles; 2) successive ignition of volatiles and char particle; and 3)
simultaneous ignition of volatiles and char particle.
The right hand picture shows under which conditions the
different mechanisms take place: mechanism 1) in the case
of very small particles in a wide range of heating rates,
mechanism 2) for larger grains and relatively low heating
rates, and mechanism 3) for larger grains and high heating
rates.

5.3 Some Aspects of Chemical Coal Structure

In the foregoing chapter the significance of devolatilization for combustion of coal particles has been shown.
Now correlations between devolatilization and chemical
structure of coal shall be discussed in more detail. Therefore first of all some aspects of chemical structure have
to be taken in mind.

Coal seems to be a two component system consisting of
a macro-molecule also called coal matrix and a so called
mobile phase of single molecules with relative small molecular weight which are trapped in the holes of the macro-
molecular network. Many efforts are made to elucidate
the nature of the macromolecule, whereby methods of oxidative or reductive degradation of the macro-molecule
are used with the identification of the soluble degradation
products /5/31/. Besides this kind of investigations also
spectroscopic and other physical methods as ^{1}NMR- and
^{13}C NMR-methods are used to analyse the overall distribution
of aliphatic, hydroaromatic and aromatic hydrocarbon structure and to identify functional oxygen and sulfur containing
groups. All knowledge of the chemical structure of the
macro-molecule can be condensed and schematically demonstrated in a model structure.

The concept of mobile phase is discussed recently in great
detail /33/. It appears that there are two (or possibly
three) populations of protons in coals. Mobile protons
could be in relatively small molecules which are enclosed
within the macro-molecular network. From some other observations e.g. a comparison of the amount of small molecules
received from extraction, pyrolysis and hydrogenation
it could be concluded that the content of relatively small
molecules is larger than expected from the previous results
of extraction measurements alone and is equal to the content
of substances containing mobile protons. However, there
is a disagreement on the admissibility of extraction experiments. An other argument for small molecules in mobile
phase is the fact that many molecules received from extraction, pyrolysis and hydrogenation experiments are of biological origin and so received without chemical alteration
during the preparation methods for their recovery.

As to the chemical constitution of macro-molecule only
some characteristic results shall be discussed here. Figure 18 shows the mean content of different oxygen functionalities in coals in a wide range of rank /34/. Oxygen
functionalities are determined by different chemical methods
and infrared spectra. Lignites contain high oxygen contents
and a great variety of functionalities like phenolic hydroxyl, carboxylic acid groups, carbonyl groups, ether groups
and methoxy groups. In hard coals there occur only phenolic
hydroxyl, carboxylic acid groups and to a small extent
carbonyl groups. Ether groups have an especial significance
for the behaviour during pyrolysis. Figure 19 gives a
survey on the distribution of carbon and hydrogen in vitrain
coals /35/. It can be seen that aromaticity of carbon
and hydrogen is increasing with rank while non aromatic
structures are decreasing. The significance of the different
structures for the pyrolysis behaviour of coals will be
discussed in chapter 5.4. An example for a coal model
molecule /33/ is given in figure 20. It can simultaneously
summarize and illustrate the main chemical structural
features of coal. Nowadays its construction is possible
by computer assisted analyses of main results of functional
groups and C- and H-distributions in coals. By construction
of a space filling model it can be determined whether
the proposed chemical structures are sterically accessible
/36/.

The coal model reflects the macro-molecule of a coal with
83 wf % C. It consists of polynuclear aromatic-hydroaromatic ring systems linked either by bridges of methylen,
ethylen or ethers. Oxygen is also present in the form
of phenolic OH groups and in heterocyclic bondings. The
distribution of S and N is similar. This picture represents
only a part of the macro-molecule and is, in turn, linked
to the rest of the macro-molecule by other bridges.

12

To propose a relation between coal structure and its behaviour during pyrolysis a discussion of the different
specific bonding energies between the aromatic units and
the hydrocarbon bridges or ether bridges linking them
is fruitful. From general knowledge on bonding energies
in organic molecules as shown in <u>figure 21</u> it can be concluded that the C-C bond energies in the coal macro-molecule increase in the following sense: C-C bonds in bridges
(dependent on size of the attached aromatic structures),
C-C bonds between alkyl groups and aromatic structures,
C-C bonds in aromatic rings. It may be expected that in
pyrolysis the sequence of cracking of C-C bonds as a function of temperature is in the same sense as the strength
of bond dissociation energies of the different bond-types.
In this way the first reaction will be the cracking of
weak bonds in the aliphatic hydrocarbon bridges resulting
in the formation of alkyl aromatics followed by the cleavage of alkyl groups and the formation of alkyl radicals
and aromatic ring structures. The strong aromatic C-C
bonds lead to the fact that the aromatic ring system stays
relative stable during pyrolysis. Corresponding consideration can also be made related to the cleavage of ether
groups which bond dissociation energies in general are
somewhat greater than those of aliphatic hydrocarbon bridges. These considerations are very important for the mechanism of pyrolysis which is discussed in more detail in
chapter 5.4.

As to the mobile phase, considerations on the biological
origin of small molecules in coal can give some more eluci-
dation of this problem.

The formation of biomarker compounds - also called geochemical fossils - from biological precursors is strongly
related to the kind of precursor material and to the geochemical history of coalification. Most biomarker hydrocarbons are directly formed by the so called lipids (fats,
waxes, vegetable oils, terpenoids, steroids) /37/. The
transformation of diterpenoids e.g. dehydroabietane in
phenanthren during coalification from lignite to anthracite is proven by identification of correspondent intermediates and products in coal extracts /38/. Also biomarker
compounds like pristane, phytane and farnesane are identified in coal extracts /39/. As biological precursor
molecule for hopane derivatives, found in coal extracts,
bacteriohopanetetrol, a cell membrane compound of special
bacteria, could be identified /40/. Different biomarker
compounds like straight chain alkanes are also found in
supercritical coal extracts /41/ and in the light oil
of coal hydrogenation /33/42/. Recent experiments performed
at Bergbau-Forschung have shown, that most of biomarker
compounds found by extraction can be found without or
with only small chemical alterations in the products of
mild coal degradation, using hydrogen donor solvents,
in those of pyrolysis and of hydropyrolysis /43/.

According to the origin of these compounds of lipids it
is to assume, that exinites are most rich on mobile phase

and on biomarker molecules. This recently has been proven
by the comparison of fast pyrolysis results of a humic
and a sapropelic coal. An example is shown in <u>figure 22</u>.
Here the yield of straight chain alkanes with C-numbers
between 15 and 33 after 10 s shock heating in a Curie-Point
apparatus at 1 bar H_2 is plotted against the C-number
for two different coals at temperatures of 600, 700 and
800° C. It can be seen that the sapropelic coal Lady Victoria (45 % Exinites) has a five times greater yield on
C_{15} to C_{25} hydrocarbons than the humic coal Westerholt
(12 % Exinites). It is also interesting to see that the
yield is only little influenced by the temperature of
shock heating. So it can be concluded that even hydrocarbons with high carbon numbers can destillate out of
the coal particle of a size between 0.063 and 0.1 mm during
shock heating without remarkable decomposition. In the
study it could furtheron be shown that high pressure and
high particle size lead to secondary reactions within
the coal particle with decomposition of these hydrocarbons by cracking reactions /44/.

5.4 Devolatilization of Coal in Relation to its Chemical Structure

The overall reaction of pyrolysis takes place by heating
up coal in an inert atmosphere and leads to the products:
char, tar and gas. For the burning behaviour tar and gas
formation is the most important point of view. Pyrolysis
is studied very intensively during long time /5/26/45/46/
/47/48/49/50/51/52/. Parameters influencing kinetics and
mechanism are kind and particle size of coals, heating
rate, final temperature, kind of gas atmosphere and gas
pressure. The kinetics of pyrolysis are complex. It can
be assumed that many secondary reactions take place in
parallel reactions which are of first order related to
the special pyrolysis products formed by the correspondent
reaction /53/. According to the residence time of primary
products in the particle or in the hot gas atmosphere
surrounding the particle consecutive reactions between
the products of primary reactions can take place. These
are preferred in cases of high residence times especially
in large coal particles and under the conditions of high
gas pressure. The primary reactions often consist of
cracking reactions mostly with the consequence of forming
coke into the coal particle and of reducing the yield
of liquid and gaseous products. In combustion systems
with fine coal particles and high rate of heating a minimum
of secondary reactions and a maximum of yield of the liquid
and gaseous pyrolysis products can be expected.

The mechanism of pyrolysis /54/55/ has to be understood,
if correlations between pyrolysis behaviour of coal and
its chemical structure are to be derived. Here the following
points shall be discussed in more detail: the destillation of mobile phase, the formation of tar and gas molecules by degradation of the macro-molecule and the formation
of char by condensation of big aromatic clusters within
the solid particle. From this conclusions can be drawn
to the correlation between the CH-distribution in the
coal macro-molecule and the evolution of pyrolysis products.

As shown in figure 22 the destillation of products from
the mobile phase can be studied by performing pyrolysis
experiments and identifying molecules formed by lipids
precursor molecules. Additionally the study of kinetics
of the formation of tar and methane during pyrolysis can
give some evidence on the behaviour of molecules of mobile
phase. It can be assumed that a molecule of mobile phase
is trapped in holes of the coal matrix because transport
processes are very slow at room temperature due to the
fact that the molecule has to pass bottle-neck pores.
It has been shown that activation energy of activated
diffusion - the transport mechanism in bottle-neck pores -
is sharply increasing with molecular size /56/. The transport, however, can be accelerated by increasing temperature.
So it can be expected that in the temperature range of
350° C the diffusion is not more inhibited and the molecules of mobile phase can evapourate and diffuse through
the holes of the solid out of the grain. This process
is not combined with a formation of gases, because these
can desorb at much lower temperatures and also the diffusion
out of the particle can take place at lower temperatures
due to their smaller molecular size.

Other liquid hydrocarbons can be formed by the degradation
of the coal macro-molecule. However this formation is
combined with a gas formation as will be shown in the
following. Therefore the destillation of molecules of
mobile phase can be proven by the comparison of the formation of tar, which contains both products from mobile
phase and from the degradation of the macro-molecule,
and the formation of methane which is only a byproduct
of degradation of the macro-molecule.

Figure 23 (left hand picture) shows that tar formation
during pyrolysis of a high volatile coal at low heating
rates begins at about 250° C in 1 bar nitrogen. With increasing gas pressure the begin of tar evolution is shifted
to higher temperatures and the amount decreases. In contrast
to this, methane formation begins at 400° C and is independent of gas pressure (right hand picture) /57/. From
these findings it can be concluded that tar formation
begins at lower temperatures than methane formation and
is pressure dependent. That means that the destillation
of mobile phase begins earlier during pyrolysis and the
first fraction of formed tar consists of molecules of
mobile phase. Tar compounds of degradation of macro-molecule
are only formed at higher temperatures when the methane
formation begins, which is not pressure dependent.

The significance in relation to combustion is that aliphatic
hydrocarbons can evolve very early during pyrolysis, if
coal contains a great content of mobile phase as e.g.
exinites rich coals. It can be assumed that also the ignition temperature of these compounds is especially low.

The degradation of the macro-molecule can be explained
by taking as an example the degradation of an ethen molecule
which is substituted by phenanthren and by tetralin as
shown in figure 24. This is a very simple but typical
model molecule for the macro-molecule of coal. The first
step consists of a hydrogen donor reaction by aromatization
the tetralin system in naphthalene.It follows the cleavage
of the weakest bond between the carbon atoms of the ethen-
bridge, lateron also the alkyl residues can be cleavaged.
In the presence of exessive hydrogen from the hydrogen
donor reaction the radicals can be saturated with hydrogen
forming naphthalene, phenanthrene and methane. These products can leave the solid particle by suitable transport
processes. If not enough hydrogen is present radicals
can combine to bigger aromatic structures which cannot
leave the grain anymore because transport processes are
inhibited. These aromatic structures can condensate at
higher temperatures to big aromatic clusters under formation
of hydrogen.

This simple picture, which does not consider the role
of ether groups during degradation clearly shows the different role of CH-distribution: hydroaromatic hydrogen is
necessary to saturate radicals which are formed by cleavage
of weak C-C bonds.

Figure 25 summarizes the main reactions of pyrolysis /55/
consisting of desorption of water and gases at about $120°$ C,
destillation of mobile phase, beginning at temperatures
up to $250°$ C, and forming an aliphatic tar, degradation
of macro-molecule under formation of an aromatic tar and
gases beginning up to $400°$ C, and the condensation of
big aromatic structures to char, combined with formation
of H_2, CO and N_2 which begins at about $600°$ C.

In figure 26 an overall picture is given describing the
coaction of main reaction steps in a coal macro-molecule
which is composed ofa variety of aromatic-hydroaromatic
structural units linked with different bridge elements
and containing in its holes a great variety of molecules
of mobile phase. These can destillate and leave the particle by diffusion, if temperature is high enough to allow
free diffusion. The cleavage of bridges and saturation
of radicals is dependent firstly on the bond energy of
C-C or C-O bonds in the bridges and secondly on the amount
of hydroaromatic structures which make the hydrogen available for the saturation of radicals. According to the
size of bond energies the degradation may take place in
several steps. Figure 27 gives a summary of the significance
of the different CH-distributions for the product formation
during pyrolysis. CH_3, CH_2 and CH in side chains or bridges
result in the formation of C_1 to C_3 hydrocarbons. Hydroaromatic CH_2 makes the hydrogen available for saturation
of radicals and in this way has an important function
to decide how much radicals can be saturated to molecules
leaving the solid. CH-aromatics lead to tar and coke depending on the number of aromatic rings and on hydrogen available.

5.5 Amount and Chemical Composition of Volatiles

As pointed out in chapter 5.2 an important step of coal
combustion is the ignition and burning of volatiles. Figure 16 shows that ignition of volatiles is dependent
on amount of volatiles and of rank of the parent coal.
In more detail it must be expected that there exist correlations between quality of volatiles and their ignition
behaviour. Unfortunately there are no exact correlations
available. Since tar is the main volatile product of pyrolysis, the amount of tar evolved and its characterization
dependent on rank of coal may be interesting to interprete
the ignition behaviour. Figure 28 shows (above) the tar
formation during pyrolysis at 3 K/min and 200 K/s depending
on gas pressure for anthracite and hard coals of different
rank /57/. It can be seen that the amount is increasing
with volatiles, though it is not proportional to volatiles,
and that the tar amount is greater, if higher heating
rates are used. The tar quality is characterized below
by the C/H ratio and the sum of CH (aromatic plus aliphatic)
and phenolic OH. It can be seen that the C/H ratio is
decreasing with increasing volatiles. That means, that
the aromaticity is decreasing. The same conclusion also
follows from the sum of CH + OH. Summerizing it can be
concluded that the lower ignition temperature of volatiles
with increasing volatiles can be explained by the increasing
amount of tar and by the decreasing aromaticity of tars.

5.6 Change of Particle Size and Porosity by Pyrolysis

As concluded in chapter 5.3 the pore structure of coal
is not relevant for combustion, since the pore structure
is very much changed by pyrolysis. This development is
dependent on kind of coal, rate of heating and final temperature of pyrolysis. Also internal surfaces are changing
during char formation. In general internal surface increases
with increasing temperature of pyrolysis, goes through
a maximum and then decreases. The position and height
of the maximum is highly dependent on kind of coal and
method used for determining internal surface /58/. Details
of this and other studies shall not further be discussed
here. However an other fact arising with char formation
is so significant for the burn-out of residual char especially at high temperatures that it shall be mentioned
here. For swelling coals the particles in general change
their size and as a consequence also its apparent density
and porosity, as reported in /17/. Here the statement
is made that chars from nonswelling and swelling coals
behave differently during combustion. While for chars
from non swelling coals mean diameter and density of a
sample decrease after partial combustion, for chars from
swelling coals the mean diameter of a sample decreases
whereas its density rises after partial combustion. This
rise is attributed to the more rapid combustion of the
very-low-density particles.

A more detailed study of the change of vitrain and durain
particles from high volatile bituminous coals pyrolysed
in a wire net apparatus and filmed with a high speed camera
shows the very different behaviour of the individual coal
lithotypes /59/. Some results are summerized in figure 29.

The left hand picture above shows the principle development
of particle size during fast pyrolysis under helium pressure.
The grain first swells very much, reaches a maximum of
particle size at 800° C, and then begins to shrink. The
relative extension of the particle can be correlated to
the tar formation as shown in the left hand picture below.
The right hand figure shows the different behaviour of
vitrain and durain particles for two different coals.
By addition of these curves the average change of mean
particle volume and density can be predicted. This result
is very important for the combustion behaviour of char
particles of swelling coals, however, these findings elucidate only the principle of possible effects. Further
experiments are underway at Bergbau-Forschung at pressures
of 1 bar and in the presence of oxygen.

5.7 Combustion of Char

According to temperatures different mechanisms are responsible for the combustion of char: the chemical reaction at low temperatures,the film diffusion at high temperatures,and a transition range at temperatures in which
pore diffusion is rate determing. The reaction rate in
the temperature region, where chemical reaction is rate
determining is governed by the chemical reactivity of
the char, while in the region of film diffusion the reaction
rate mainly depends on particle size. In the transition
region reactivity and pore effectiveness factor are relevant parameters determining reaction rate.

Figure 30 shows the temperature dependence of the effective
reaction rate in the three regions. It is determined by
the activation energy in the region of chemical reaction
by the half of this activation energy in the region of
pore diffusion , and is very low in the region of film
diffusion /60/. In figure 31 intrinsic rate data of carbons
and chars per unit area of internal surface are given
as a function of temperature /61/. Corrections are made
to extrapolate the literature data to an oxygen pressure
of 1 bar, taking account reaction orders related to oxygen
between 0,3 and 1, and to an effectiveness factor of 1
to avoid influences of transport phenomena in pores. The
average activation energy is 179 kJ/mole, the spread for
individual data lies between 124 and 290 kJ/mol. The value
for pure carbons is 250 kJ/mol. The intrinsic rates depend
on the internal surface area (which in the figure is extrapolated to 1), the degree of graphitization and catalytic
effects ensuing from the presence of ash compounds. It
may be, that the degree of graphitization and the internal
surface of chars is a function of the rank of the precursor
coals as it can be supposed from figure 16. Other results

of kinetics of heterogeneous reactions are reported in
/82/.

Figure 32 shows that at very high combustion temperatures
of 1500 and 2000 K particle size is the determining parameter. So burning times decrease with decreasing particle
diameter between 10^3 and 10^2 μ from about 6 s to 0,1 s.
Burning times can be calculated by the Nusselt square
law equation in which burning time is a function of the
square of the initial diameter.

6.. CHARACTERIZATION OF COAL IN RELATION TO ENVIRONMENTAL
    PROBLEMS OF COMBUSTION

6.1 Significance of Coal Sulfur

As mentioned in chapter 1 the demand for minimizing the
discharge of S- and N-compounds and the trace elements
is an important task to meet environmental impact of coal
combustion. Related to this problem the characterization
of S, N and other compounds in coal leading to the discharge
of HCl, HF and trace elements is neccessary.

S is general converted to SO_2 with only traces of SO_3 in coal
combustion, the latter and parts of SO_2 can be absorbed in
flue ash, if these contains suitable mineral Ca compounds.
Sulfur in coal mainly exists in inorganic form, mostly as pyrite,
and as organic sulfur, bound directly in substances of mobile
phase or in the macro-molecule of coal. S-analytic is important
from the point of its removal before burning. Pyritic sulfur
in principle can be removed by mechanical methods of coal preparation, if size of pyrite crystals is not to small. Aslo
the microbiological conversion of pyritic sulfur into sulfuric
acid at pH of about two using Thiobacillus ferro oxidans /63/64/
65/ is under development to remove pyritic-sulfur distributed
very finely in the coal particle. The removal of organic sulfur
is much more difficult, reductive chemical treatment leads
only to low degrees of removal with undesired side reactions,
changing the coal substance. Therefore great research effort
is made also to convert organic coal sulfur in sulfuric acid
by microbiological processes. As model compound mostly dibenzo-
thiophene is used which in principle can be degradated by
oxidative biochemical reactions /66/67/. It could be found,
that also the analogous heterocyclic nitrogen compound, the
carbazole is attached by bacteria /68/. A new organism was
developed from a mixed culture of naturally occuring soil micro-
organisms by mutagenic alteration of the microbial DMA. The
microorganism is capable of oxididzing the sulfur in dibenzo-
thiophene, but does not break the carbon ring structure of
the molecule. This organism is also able to remove up to 47
% of the organic sulfur from coal /69/.

6.2 Significance of Coal Nitrogen for the NO_x-Formation
    in Flames

It is well known that Nitrogen oxides are formed according
to three different mechanisms during coal combustion: at high
temperatures above 1200° C mainly from oxygen and nitrogen

and to a small extent by the reaction of radicals like OH with
nitrogen and at lower temperatures above 750°C by reactions
of coal-nitrogen with oxygen. This latter reaction is especially
important for coal combustion processes at low temperatures
like fluidized bed combustion, but also plays a significant
role in coal combustion in general. The fate of coal nitrogen
during coal combustion is studied by several authors /70/71/
/72/73 /83/84/ with the result that there is no simple relation
between coal nitrogen and nitric oxide formed by combustion.
Also the behaviour of various nitrogen containing species as
NH_3, amines or pyridin in relation to nitric oxid formation
has been investigated in gas flames /74/75/. A further object
of research has been the formation of nitrogen containing species
during pyrolysis /76/77/78/. As a result of these studies it
can be stated, that the following nitrogen containing compounds
are formed during pyrolysis: the gaseous compounds N_2, NH_3,
and HCN, various liquid compounds as amines and heterocycles,
and nitrogen bond in the graphitic lattice of the formed char.
The oxidation behaviour of the gaseous and volatile nitrogen
containing species seems to be different from that of the nitrogen in char. On the one hand the latter seems to react mostly
to NO, on the other hand this primarily formed NO can also
be reduced to nitrogen by the carbon content of the char.

To find correlations between nitrogen content and occurrence
in coal and NO_x formations during combustion these findings
may be discussed in more detail. Typical results of nitrogen
distributions in volatiles are shown in figure 33 after vacuum
devolatilization /78/. From this it follows that percentage
of parent-coal nitrogen retained in char is dependent on temperature and on kind of coal. The nitrogen in char decreases
with increasing temperatures, and as shown in /77/, temperatures of about 1900° C are needed to release the nitrogen
completely from char. In the temperature range up to 1200° C
the main amount of nitrogen is found in tar. As to the
nitrogen in volatiles the amount in tar is in principle
proportional to tar formation and therefore a function
of coal quality (see figure 28), and decreasing with rank.

Model combustion studies /79/ show that the behaviour
of nitrogen containing hydrocarbons is very different
during combustion as a function of oxygen content (figure 34). So pyrrolidin forms only a small amount of NO_x,
main product is HCN, while pyrrol and methylpyridin form
increasing parts of NO: From there it follows that NO_x
formation during combustion is a function of chemical constitution of the substance burnt and especially of the kind of
nitrogen bonding.

Therefore it seems to be very important to study the chemical
constitution of N containing compounds in tar. A first result
investigating a pyrolysis tar (Fischer-carbonization) of a
high volatile coal from the Ruhr district is shown in figure 35. The tar was seperated in fractions of different basity
and different N-content by ion exchange (right hand picture)
Also the molecular weight distribution of the tar fractions
has been determined by gel permeation chromatography (left
hand picture). It can be seen that in general the part of compounds with low basity and high potential for NO_x-formation

is high. However the total N-content and the molecular weight
distribution of the different fractions are different. Fractions
4 and 5 (bases 2 and bases 3) have a high mean molecular weight
and at the same time a high total content of nitrogen, while
the fractions with a relatively low molecular weight have partly a lower nitrogen content. That means, that pyrolysis tary
compounds with low N content have been envolved firstly and
substances with higher N content later on. This result must
be confirmed also for other tars. Surely a modelling of homogeneous reactions during burning of these substances will be
possible in principle, if the pyrolysis rate of these fractions
is known and if the behaviour of individual N-containing substances during combustion is understood better.

A next question is, whether the N distribution on char
and on volatiles is influenced by the heating rate. Some
results of Bergbau-Forschung /81/ performed in a wire
net apparatus comparing the nitrogen remaining in the
char at a final temperature of 950° C dependent on particle
size and heating rate are presented in figure 36. It can
be seen that the heating rate between 200 and 1000 K/s
has no influence on the nitrogen content of char, while
the nitrogen content increases with increasing particle
size. It can be assumed that nitrogen containing tar of
high molecular weight is cracked into coke and gas during
pyrolysis of grains with greater particle sizes, therefore
their N content is higher.

The next figure 37 shows a typical result of the NO formation during combustion of the residual char in the wire-
net apparatus, whose wire-net is flown through by a helium-
oxygen-mixture with an oxygen content of 9.9 %. It can
be seen, that the ratio of $NO/(CO + CO_2)$ is constant during
the combustion and that a part between 80 and 100 % of
the char-nitrogen is converted to NO. Therefore the kinetics
of NO formation can be established, if the kinetics of
the CO and CO_2-formation are known.

Also the problem of NO reduction by a char has been investigated at Bergbau-Forschung /81/. It could be observed
that the reaction rate is very high at the beginning of
each experiment, however after about 80 minutes, an inhibition of reaction takes place (figure 38). Further experiments to establish the kinetics of reduction have been
performed with this char after reaching the steady state.
It is assumed that the inhibition is caused by the intermediate formation of surface oxide complexes. Activation
energies for the chemical reaction of NO reduction have
been determined to be 120 kJ/mol. The transition temperature from chemical kinetics to pore diffusion as rate
determining step depends on particle size and NO concentration. Above this temperature reaction rates become
not only dependent on concentration of NO, but also on
particle size as shown in figure 39. The concentration
dependence follows a typical Langmuiur-Hinshelwood kinetics.

The last findings were used to estimate the NO-reduction
effect by the char in a fluidized bed (figure 40) taking
into account the parameters listed in figure 40 and two
different carbon concentrations of 0.7 and 2.1 g/cm^3 which
are typical for fluidized bed combustion of a high volatile A bituminous coal and an anthracite. Due to the higher
carbon content of the fluidized bed during combustion
of anthracite a higher degree of NO reduction is to be
expected.

From this short presentation of some newer results it
follows that the knowledge of N content and molecular
weight distributions of fractions with different basity
is very important for modelling the NO$_x$-formation during
combustion of volatiles. Correspondent data depend on
the kind of coal. Furtheron the extent of nitrogen in
char, which depends on grain-size and temperature of carbonization, has significance for flame modelling because
this part of fuel nitrogen is converted into NO nearly
completelyduring burning. The kinetics of NO formation
is strongly correlated to that of CO and CO$_2$ formation.
Also the reduction of NO by the char is of importance.
It can lead to a remarkable reduction of NO content dependent on temperature, particle size, NO concentration,
and carbon content of the bed.

7. CONCLUSION

For establishing heat and mass balance for coal combustion
moisture content, ash content, ultimate analysis, true
and apparent density, calorific value, specific heat of
coal and specific heat of coke are needed. These data
are available for many coals and mostly related to volatiles
or other rank parameters. Also the grindability of coal
can be characacterized to design crushing equipment.

For understanding and predicting the combustion behaviour
of coals and for modelling of reactions in flames coal
properties should be related to the individual steps of
coal combustion like devolatilization, ignition and burning
of volatiles, char formation, and burning of char. The
significant properties relevant for these steps are listed
in figure 41. Here the forefold and partly complex influence
of particle size is especially remarkable. Particle size
influences devolatilization, ignition and burning of volatiles, ignition of char at high temperatures and the coaction of all individual steps of coal combustion.

As to the devolatilization,the rate of the chemical pyrolysis reactions is determining for particles with sizes
below, let say 100 μ, for heating rates of combustion
in suspension firing systems. For greater sizes transport
processes become rate limiting with the consequence that
tar yield becomes lower and overall rate decreases. For
ignition of volatiles the amount of volatiles evolving
from a single particle is the dominant parameter. It decreases with decreasing particle diameter. Therefore the
concentration of volatiles around the particle is so small

for very small particles that volatiles can not ignite
(figure 17). For the burn-out rates of char particles
at very high temperatures particle size is the dominant
parameter (figure 32). The coaction between the single
steps of coal combustion is mainly governed by volatiles
and particle sizes. For high volatile coals and suitable
- not too small particle sizes - ignition of volatiles
takes place at low temperatures, so the particle is heated
up very fast by the burning of volatiles and can burn
itself with high rates. For low volatile coals, however,
not enough volatiles are evolved, therefore they can not
ignite, the particle itself is only ignited at high temperatures and burning times can only be shortened by small
particle sizes.

Some further properties significant for the environmental
impact of combustion are the content of pyritic and organic
sulfur on the point of view that removal of both kinds
of sulfur before combustion needs different measures.

For predicting NO formation during coal combustion the
total nitrogen content, the nitrogen distribution in gases,
tar and char during pyrolysis, the molecular weight distribution and the chemical structure of nitrogen containing
compounds in tar compounds are important. Regarding the
behaviour of nitrogen components during combustion it
can be assumed that the nitrogen in the volatiles is mostly
responsible for the first step of coal burning in air
staged combustion, while the nitrogen in the char governs
the second step of coal combustion. Here oxidation of
nitrogen mainly leads to NO and the kinetics of this reaction are strongly correlated to that of CO and CO_2 formation. Furtheron the reactivity of char regarding the
reduction of NO to nitrogen is of great interest.

References

1. Essenhigh R.H., "Fundamentals of Coal Combustion"
    in Chemistry of Coal Utilization second supplementary
    volume 1981 p 1153 - 1312

2. Ceely F.J., E.L. Daman "Combustion Process Technology"
    in Chemistry of Coal Utilization, second supplementary
    volume 1901 p 1313   1387

3. Jüntgen, H., Klein, J. Erdöl und Kohle 28 (1975)
    H. 2, S. 65-73

4. Hanbaba, P., H. Jüntgen, Advances in Organic Geochemistry
    1968, Pergamon Press Ltd., Oxford (1969) p. 459-471

5. van Krevelen D.W., Coal-Typology-Chemistry-Physics-
    Constitution, Elsevier-Publishing Comp. 1961

6. Teichmüller M., R. Teichmüller "Fundamentals of coal
    petrology" in Stach's Textbook of Coal Petrology,
    Gebrüder Borntraeger, Berlin, Stuttgart 1982

7. Neavel R.C., S.E. Smith, E.J. Hippo, R.N. Miller,
    Proceedings Intern. Conference on Coal Science Düssel-
    dorf 1981, p. A 1/1 - A 1/5

8. Jüntgen, H., Codata 1983, Data for Science and Tech-
    nology, Amsterdam, S. 241 - 251

9. Ruhrkohlen-Handbuch, 5. erweiterte Auflage, Verlag
    Glückauf, Essen 1969

10. Zelkowski, J., Kohleverbrennung - Brennstoff, Physik
    und Theorie-Hrsg VGB Technische Vereinigung der Groß-
    kraftwerksbetreiber, 1. Aufl. 1986

11. Tonczew I., Energoprojekt, Sofia Dezember 1977
    "Über die Bestimmung der mittleren Werte der qualita-
    tiven Grunddaten der Kohlen in Bulgarien"

12. Jüntgen, H., "Physikalisch-chemische Eigenschaften
    und Reaktionsverhalten der Steinkohle" in: Winnacker-
    Küchler Chemische Technologie 4. Aufl. Bd 5 Organi-
    sche Technologie I (1981) S. 285-291)

13. Melchior E., H. Luther, FUEL Vol. 61 (1982) p. 1071-1079

14. Agroskin A.A., Coke Chem. (USSR), (1980) Nr. 2, S.14-25

15. Mott R.A., C.E. Spooner, FUEL vol. 19 (1940) No 10
    p. 226-231, No. 11 p. 242-251

16. Gums W., Kurzes Handbuch der Brenstoff- und Feuerungs-
     technik, Springer-Verlag Berlin/Göttingen/Heidelberg,
     3. verbesserte Auflage 1962

17. Field M.A., Combustion and Flame 14 (1970) p. 237-248

18. Sun Yilu, Liu Wenzhen, Proceedings 1985 International
     Conference on Carbon, Sydney, . 379-382

19. Zhang Mingchuan, Xiang Daguang, Wang Chunchang,
     Proceedings 1985 International Conference on Carbon,
     Sydney p. 343-346

20. Lin Hao, Proceedings 1983 International Conference
     on Carbon, Pittsburgh, p. 561-564

21. Vleeskens, J.M., Proceedings 1983 International
     Conference on Carbon, Pittsburgh, p. 599-602

22. Smith S.E., R.C. Neavel, E.J. Hippo, R.N. Miller,
     FUEL Vol. 60 (1981) p. 458-462

23. P.H. Given Prog. Energy Combustion Sci 10 (1984),
     p. 149-158

24. Nandi, B.N., Brown, T.D. and Lee, G.K. FUEL 56 (1977)
     p. 125-130

25. Jüntgen, H., EPE Tijdschrift Primaire Energie, Vol.
     XII, Nr. 3-4, 1976-77, S. 11-24

26. Koch, V., H. Jüntgen, W. Peters, Brennstoff-Chemie 50
     (1969) Nr. 12, S. 369-373

27. Koch V., H. Jüntgen, W. Peters, VDI-Berichte Nr. 146
     (1970), S. 44-47

28. Koch V., K.H. van Heek, H. Jüntgen, Dynamic Mass
     Spectrometry Vol. 1, Heyden and Son Ltd. (1970) London,
     p. 15-21

29. Stahlherm D., K. H. van Heek, H. Jüntgen, VGB-Kraftwerks-
     technik 53 (1973) H. 1, S. 35-40

30. Stahlherm, H. Jüntgen, W. Peters, Erdöl und Kohle 27
     (1974) H. 2, S. 64-70

31. Wender I., L.A. Heredy, M.B. Neuworth, I.G.C. Dryden,
     Chemical reations and the constitution of coal" in
     Chemistry of Coal Utilization, second supplementary
     volume 1981, second supplementary volume 1981 p.425-521

32. Berkowitz, N.,The Chemistry of Coal, Amsterdam: Else-vier 1985, Coal Science and Technology Vol. 7

33. P.H. Given, A. Marcec, W.A.Barton, L.J. Lynch and B.C. Gerstein FUEL (86) 65, p. 155-163

34. D.D. Whitehurst, T.O. Mitchell, M. Farcasiu Coal Liquefaction, Academic Press New York, London, Toronto, Sydney, San Francisco 1980, 17

35. I.G.C. Dryden, FUEL 41 (1962), 301

36. Spiro c.L., FUEL Vol. 60 (1981) p. 1121 - 1126 Spiro C.L., P.G. Kesky, FUEL Vol. 61 (1982), p. 1080-1084

37. B.P. Tissot and D.H. Welte, Petroleum Formation and accurence Second Edition, Springer Verlag Berlin Heidelberg New York Tokyo 1984

38. R. Hyatsu, R.E. Winans, R.G. Scott, L.P. More and M.H. Studier, FUEL 58 (1978), p. 541

39. M. Vahrman, FUEL 49 (1970), p. 15

40. G. Ourisson, P. Albrecht and M. Rohmer Spektrum der Wissenschaft, Oktober 1984, p. 54-64

41. K.D. Bartle, T.G. Martin and D. Williams FUEL 54 (1975), p. 226 - 235

42. Liphard, K.G. et al, Erdöl und Kohle, Erdgas, Petro-chemie 34 (1981), p. 488 - 491

43. K 2 G - Seminar "Geochemie, Inkohlung und Gasbildung tiefliegener Steinkohlen" 27.-28.2.1986 Aachen

44. Kaiser, M., "Primär- und Sekundärreaktionen bei der Pyrolyse und Hydropyrolyse von Steinkohlen unter hohen Aufheizgeschwindigkeiten", Thesis, University Essen, 1986

45. Howard, J.B. "Fundamentals of Coal Pyrolysis and Hydropyrolysis in Chemistry of Coal Utilization, Second Supplementary Volume John Wiley, New York, 1981

46. Gavalas, G.R. "Coal Pyrolysis" Elsevier, Amsterdam 1982

47. Jüntgen, H., Erdöl Kohle 1964 17 (4), 105

48. van Heek, K.H., Jüntgen, H. and Peters, W. Bericht Bunsengesellsschaft f. Phys. Chemie 1967, 71, 113

26

49. van Heek, K.H., Jüntgen, H. and Peters, W. Brennst.
     Chem. 1967, 48, 163

50. Jüntgen, H. and van Heek, K.H. FUEL 1968, 47, 103

51. Hanbaba, P., Jüntgen, H. and Peters W., Brennst. Chemie
     1968, 49, 368

52. Jüntgen, H., van Heek K.H., Brennstoff-Chemie 1969
     50, 172

53. Jüntgen, H., van Heek, K.H. FUEL Process. Technol.
     1979, 2, 261

54. Jüntgen, H., FUEL 63 (1984), p. 731-737

55. Jüntgen, H., Erdöl und Kohle - Erdgas - Petrochemie
     Brennstoff-Chemie 38 (1985), H. 10, S. 448-455

56. Hanbaba, P., Jüntgen, H., Peters, W. Ber. Bunsenges.
     phys. Chemie 72 (1968) Nr. , S. 554 - 562

57. van Heek, Druckpyrolyse von Steinkohlen, VDI-For-
     schungsheft Nr. 612, 1982

58. P. Chiche, S. Durif and S. Pregermain FUEL 44 (1965)
     p. 5 - 28

59. G. Löwenthal, W. Wanzl and K.H. van Heek FUEL 65
     (1986) p.346-353

60. Jüntgen, H., K.H. van Heek "Kohlevergasung - Grund-
     lagen und technische Anwendung", Verlag Karl Thiemig
     München 1981

61. I.W. Smith, Gas Reactions (Australia) 1976, Nineteeth
     Symposium on Combustion / The Combustion Institute
     1982, p. 1045-1065

62. M.F.R. Mulcahy and I.W. Smith, Rev.Pure and Applied
     Chem. 19 (1969), p. 81-108

63. Beyer M., H.G. Ebner, J. Klein, Appl. Microbiol.
     Biotechnol. (1986) in print

64. Beyer M., H.G. Ebner, J. Klein
     Fundamental and Applied Biohoydrometallurgy" Process
     Metallurgy 4 Lawrence R.W., R.M.R.Branion, H.G. Ebner
     (Editors), Elsevier (1986), p. 151-164

65. Butler B.J., A.G. Kempton, R.D. Coleman, C.E. Capes,
     Hydrometallurgy, 15 (1986) p. 325-336

66. Ensley, B.D.Jr. Present affiliation: Amgen, Inc.
     Thousand Oaks, California, Editor: Microbial Degra-
     dation of Organic Compounds, Marcel Dekker, New York,
     (1984)

67. Isbister J.D., A.E. Desouza, E.A. Kobylinski, G.L.Ans-
     pach, J.F. Kitchens, Presented at Pittsburgh Coal
     Conference 1984

68. Finnerty W.R., 9th Energy Technology Conference (1982)
     p. 883 - 890

69. Isbister J.D., E.A. Kobylinski, Processing and utili-
     zation of high sulfur coals, Editor: Attia, Elsevier
     Amsterdam/Oxford (1985)

70. Wendt, J.O.L., D.W. Pershing, J.W. Lee, J.W. Glass,
     17th Symposium (International) on Combustion 1978

71. Glass, J.W., J.O.L. Wendt, Nineteenth Symp. (Intern.)
     on Combustion, The Combustion Institute (1982) p.
     1243-1251

72. Chen S.L., M.P. Heap, D. W. Pershing, G.B. Martin,
     FUEL (1982), vol. 61, p. 1218-1224

73. Kennedy L.A.,Proceedings Intern. Conference on Coal
     Science Düsseldorf 1981, p. B 16/284 - B 16/289

74. Fenimore C.P., Combustion and Flame 26 (1976), p.249-256

75. Miller J.A., M.C. Branch, W.J. Mc Lean, D.W. Chandler,
     M.B. Smooke, R.J. Kee, Sandia Combustion Research
     Program, Annual Report 1984, p. 3-15/3-19

76. Klein, J. "Untersuchungen zur Abspaltung von Wasser-
     dampf, CO, CO_2, N_2 bei der nicht-isothermen Steinkoh-
     lenpyrolyse unter inerter und oxidierender Atmosphäre,
     Thesis, University Aachen, 1971

77. Kobayashi, H. "Devolatilization of pulverized coal
     at high temperatures Ph. D. Thesis, Cambridge Mass.,1976

78. Solomon, P.R., M.B. Colket, FUEL (1978) Vol. 57,
     p. 749-755

79. Meier zu Köcker, H. Internal report: Applied funda-
     mental research BF/BMFT 1986

80. Radecke, R., Internal report: Applied fundamental
     research BF/BMFT 1986

81. Schuler J., Internal report fundamental research
     BF/BMFT 1986, Thesis University Essen 1986

28

82. van Heek K.H., H.J. Mühlen, Brennstoff-Wärme-Kraft 37 (1985) Nr. 1/2, S. 20-27

83. Schulz W., H. Kremer, Brennstoff-Wärme-Kraft 37 (1985/ Nr. 1-2, S. 29-35)

84. Kremer H., W. Schulz, Paper presented at the 21st Symposium (Int.) on Combustion, Munich, August 3-8, 1986.

"RANK" AS PARAMETER OF THE DEGREE OF COALIFICATION:

RANK	INCREASING
CLASS OF COAL (SIMPLIFIED)	LIGNITE
	SUBBITUMINOUS
	BITUMINOUS
	ANTRACITE

CLASSIFICATION OF "RANK":

COAL PROPERTY	RELATION TO RANK
MOISTURE OF COAL IN THE SEAM	DECREASING WITH RANK (SUITABLE FOR LOW RANK COALS ONLY)
CARBON-CONTENT	INCREASING WITH RANK
VOLATILE MATTER	DECREASING WITH RANK
REFLECTANCE	INCREASING WITH RANK

FIGURE 1: Classification of coal according to rank

Rank	Typical coal	Ultimate analysis (S,N-free)						H- and O- atoms per C-atom
		(% by weight)			Atomic ratios			
		C	H	O	C	H	O	
	Lignite	74.5	6.5	19.0	6.20	6.5	1.18	$C_1 H_{1.04} O_{0.19}$
	Coking	85.0	6.0	9.0	7.08	6.0	0.56	$C_1 H_{0.84} O_{0.08}$
	Hard coking	89.0	5.0	6.0	7.41	5.0	0.37	$C_1 H_{0.67} O_{0.05}$
increasing	Anthracite	94.0	4.0	2.0	7.83	4.0	0.12	$C_1 H_{0.51} O_{0.02}$

FIGURE 2: Typical ultimate analysis of coals of different rank

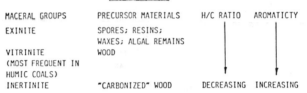

FIGURE 3: Precursor materials and chemical composition
of maceral groups

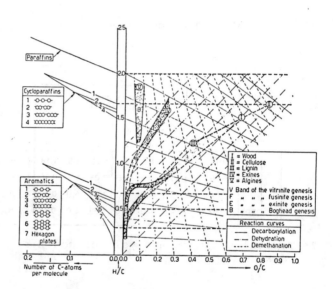

FIGURE 4: The H/C versus O/C diagram
according to D.W. van Krevelen 1950

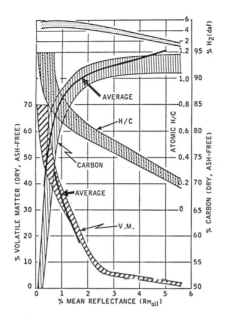

FIGURE 5: Relationship between Vitrinite reflectance and
different chemical rank parameters according
to M. and R. Teichmüller (1975) average lines
for volatiles matter and for carbon from D.W.
van Krevelen (1961)

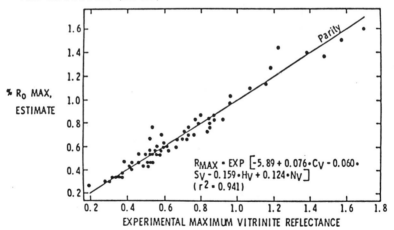

FIGURE 6: Vitrinite reflectance versus elemental composition
according to R.C. Neavel et al (1981)

PROPERTY	RELATION TO COAL RANK	OTHER CORRELATIONS
MOISTURE CONTENT	ONLY FOR LOW RANK COALS (AS RECEIVED FROM THE SEAM)	-
ASH CONTENT	NO	-
MINERAL MATTER	NO	RELATIONSHIP TO ASH CONTENT

ULTIMATE   C
ANALYSIS   H
            O
	} TO VOLATILES
	  TO MEAN REFLECTANCE	

S N	} NO	NO
TRUE AND APPARENT DENSITY	TO VOLATILES	
SPECIFIC HEAT OF COAL	-	TO $\dfrac{H_{AR}+1,3\ O}{C}$ AND TEMPERATURE
SPECIFIC HEAT OF CHAR OR COKE	-	TO TEMPERATURE
CALORIFIC VALUE	TO VOLATILE	TO C-,H-,S-,O- CONTENT

FIGURE 7: Properties needed for heat and mass balance

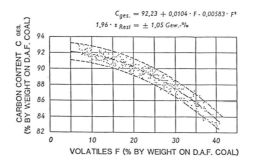

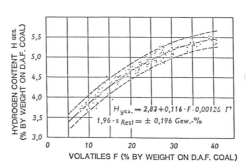

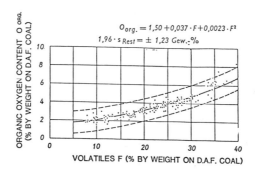

FIGURE 8: C-, H-, and O-Content as a function of volatiles for coals of Ruhr district according to Ruhr-kohle Handbuch (1969)

34

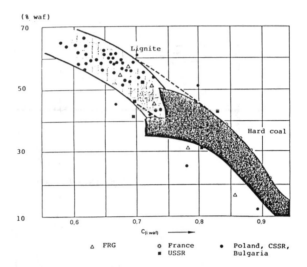

FIGURE 9: Volatiles as a function of C-content of coals from different areas according to J. Zelkowski (1986)

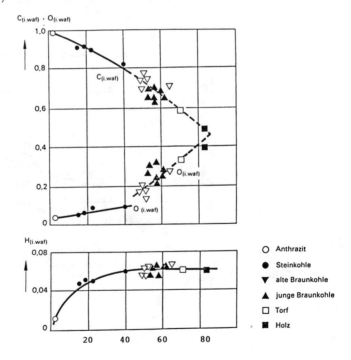

FIGURE 10: C-, H-, and O-content as a function of volatiles. Solid fuels from Bulgaria according to I. Tonczew (1977)

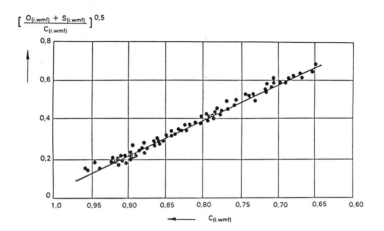

FIGURE 11: Correlation between the sum of O- and S-content
and the C-content according to I. Tonczew (1977)

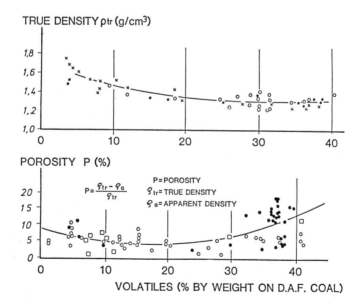

FIGURE 12: True density and porosity as a function of
volatiles according to Jüntgen (1981)

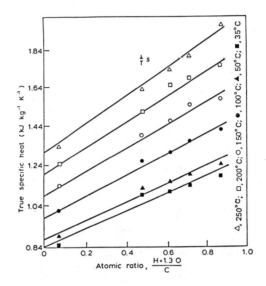

FIGURE 13: True specific heats of the bituminous coals
(waf) investigated in relation to the atomic
ratio according to Melchior et al. 1982

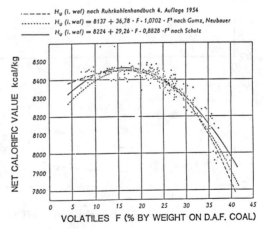

FIGURE 14: Net caloric value dependent on volatiles according
to Ruhrkohle-Handbuch (1969)

37

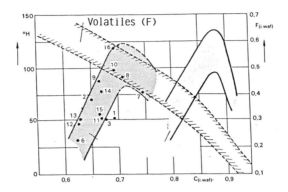

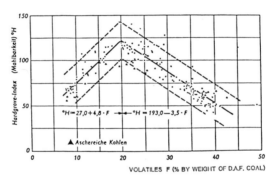

FIGURE 15: Grindability of lignites and hard coals: Hard-
grove-Index versus C-content for lignites and
hardcoals (above) and versus volatiles for
hard coals (below) according to Ruhrkohle-Handbuch (1969)

38

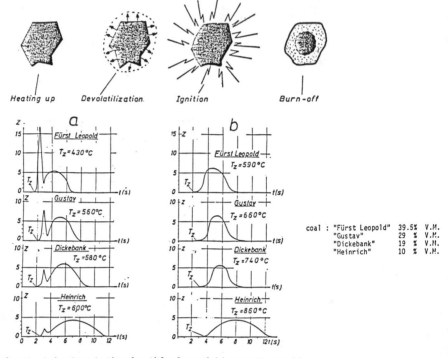

CO₂ formation during the combustion of particles from coal (a) and their cokes (b)

FIGURE 16: Mechanism of combustion of individual coal particles

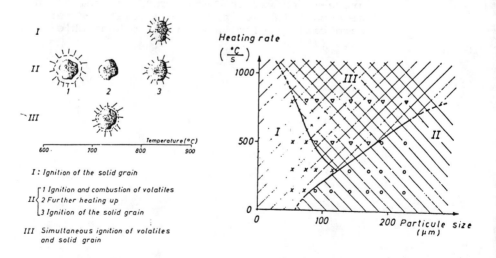

FIGURE 17: Mechanism of ignition as a function of grain
size and rate of heating (coal "Walsum"
39,5 % WM)

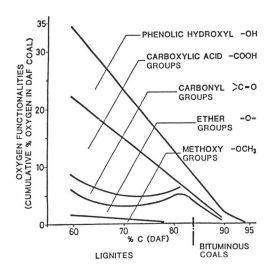

FIGURE 18: Mean content of different oxygen functionalities
in coals according to Withehurst (1980)

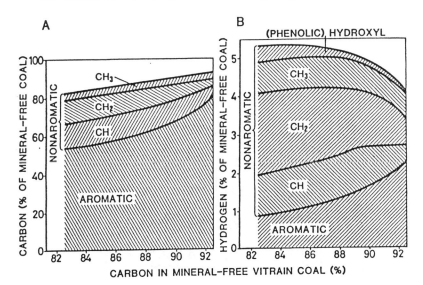

FIGURE 19: Distribution of carbon (A) and hydrogen (B)
in vitrain coals according to Dryden 1962)

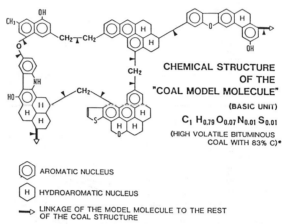

AROMATIC NUCLEUS

H HYDROAROMATIC NUCLEUS

→ LINKAGE OF THE MODEL MOLECULE TO THE REST
OF THE COAL STRUCTURE

► WEAK CHEMICAL BONDINGS

FIGURE 20: Interconnections of aromatic-hydroaromatic
structures by ether or aliphatic bridges according
to Heredy and Wender (1981)

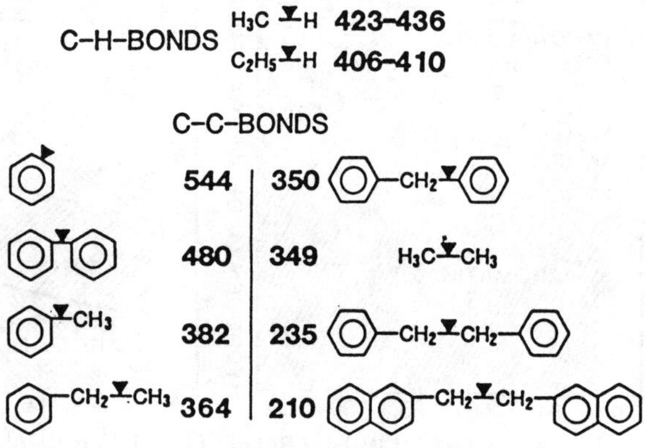

C-H-BONDS H₃C ⊻H 423-436
C₂H₅⊻H 406-410

C-C-BONDS

544 | 350

480 | 349

382 | 235

−CH₂−CH₃ 364 | 210

FIGURE 21: Values of Bond-energy of hydrocarbons (kJ/mol)

t = 10 s, m = 9000 K/s, Körnung: 0,063-0,1 mm

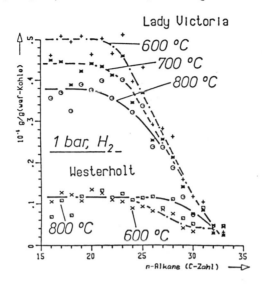

FIGURE 22: Yields on n-alkanes during shock heating of
Westerholt (12 % exinite) and Lady Victoria
(45 % exinite) - coal of a grain size 0.063 -
0.1 mm in a Curie-Point apparatus according
to Kaiser (1986)

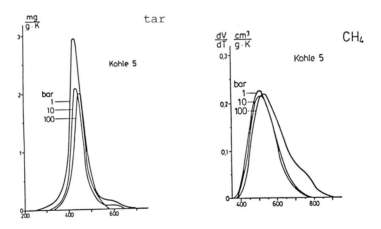

FIGURE 23: Comparison of tar and methane formation (coal
35.8 % volatiles daf, slow heating rate)
according to van Heek (1981)

42

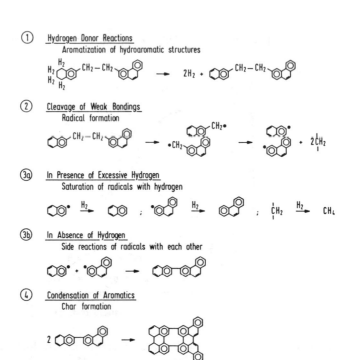

FIGURE 24: Degradation of macro-molecule - role of CH-distribution

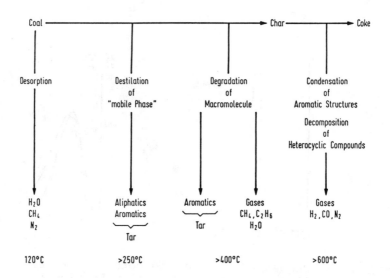

FIGURE 25: Main reactions during pyrolysis

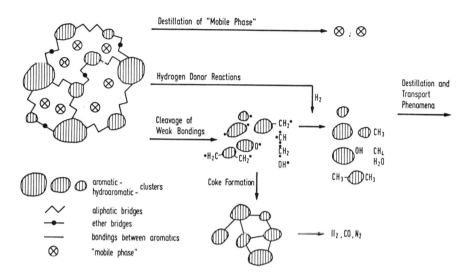

FIGURE 26: Mechanism of pyrolysis

CH-FUNCTIONALITIES			PYROLYSIS PRODUCTS
CH_3	)	SIDE CHAINS	
CH_2	)	AND	CH_4, C_2H_6, $CH_2 = CH_2$ ETC.
CH	)	BRIDGES	
CH_2 HYDROAROMATIC			HYDROGEN
CH AROMATIC			TAR OR COKE DEPENDING ON
			NUMBER OF AROMATIC RINGS H_2 AVAILABLE

FIGURE 27: CH-distribution and degradation behaviour of
macro-molecule

44

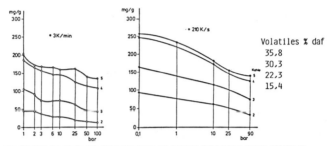

TAR YIELDS AT SLOW AND FAST HEATING UP DURING PYROLYSIS UNDER NITROGEN

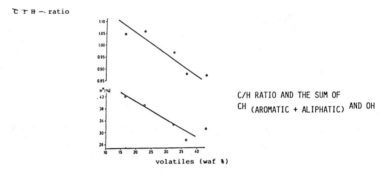

FIGURE 28: Amount and quality of tar during slow pyrolysis under nitrogen according to van Heek 1982

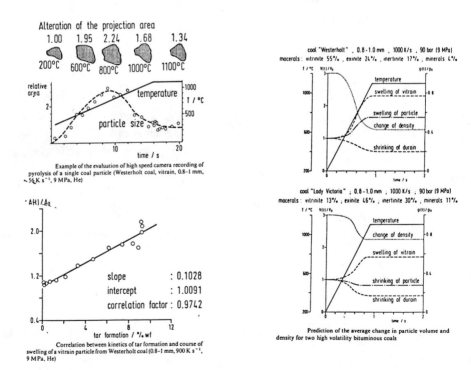

FIGURE 29: Change of particle size and density during pyrolysis according to G. Löwenthal (1986)

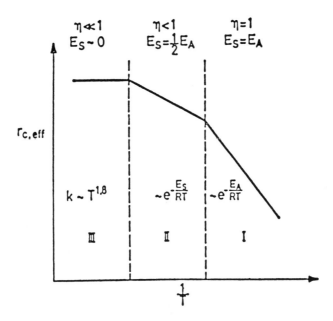

FIGURE 30: Temperature dependence of the effective burning
rate $r_{c,eff}$ in a wide temperature range

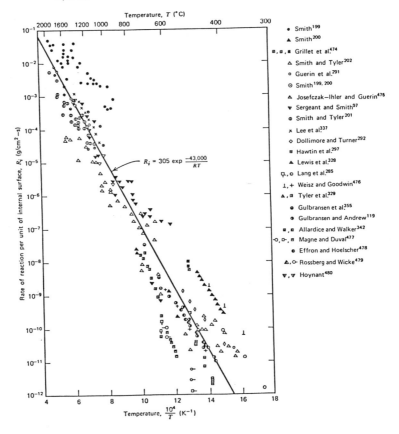

FIGURE 31: Intrinsic rate data of carbons in air per unit
area of internal surface as a function of temperature according to I.W. Smith (1982)

46

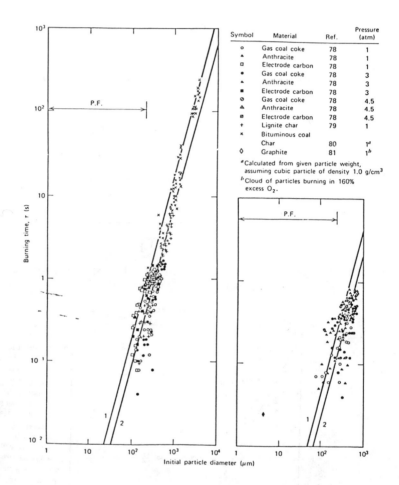

FIGURE 32: Comparison of experimental and theoretical
particle burning times (a) 21 % v/v O_2, (b)
98 % v/v O_2 for 2000 K, Curves 1: concentration
= 2,0 g/cm^3; curves 2: = 1,0 g/cm^3 according to M.F.R. Mulcahy (1969)

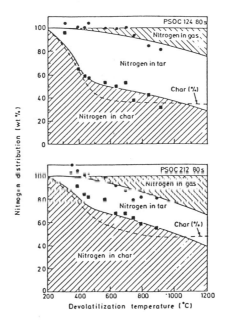

FIGURE 33: Nitrogen distribution in devolatilization: percentage
of parent-coal nitrogen retained in char and evolved
with gas and tar vs. devolatilization temperature, for
two bituminous coals. Symbols, experimental points:
solid curves, predictions from devolatilization model.
According to P.R. Solomon et al (1978)

48

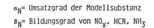

α_N= Umsatzgrad der Modellsubstanz

β_N= Bildungsgrad von NO_X, HCN, NH_3

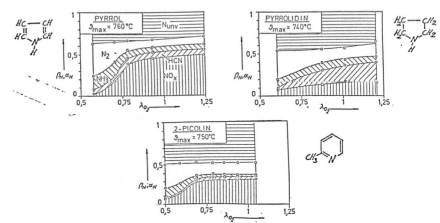

FIGURE 34: Typical results of formation of N containing species during combustion of model compounds according to H. Meier zu Köcker (1986)

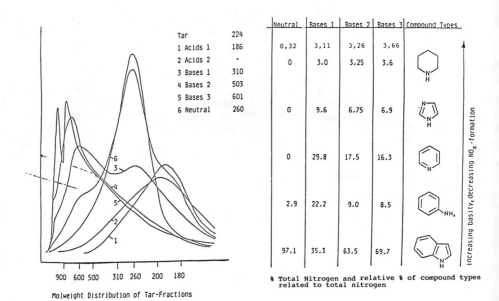

Molweight Distribution of Tar-Fractions

FIGURE 35: Characterization of N-compounds in tar (Fischer-carbonization) coal Westerholt according to R. Radecke (1986)

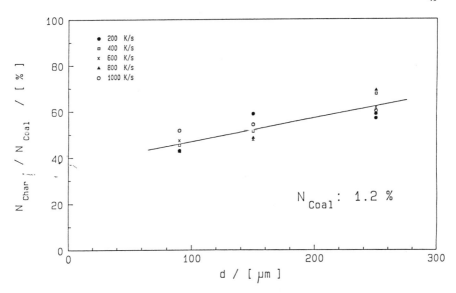

FIGURE 36: Remained N in char during pyrolysis dependent on grain size and heating rate according to J. Schuler (1986)

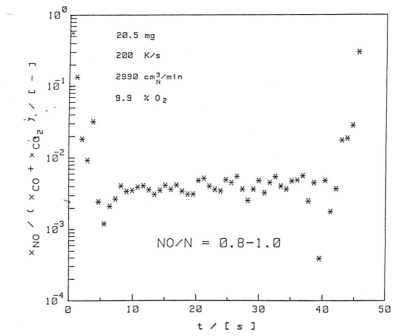

FIGURE 37: Formation of NO during combustion of char using a wire-net technique according to J. Schuler (1986)

50

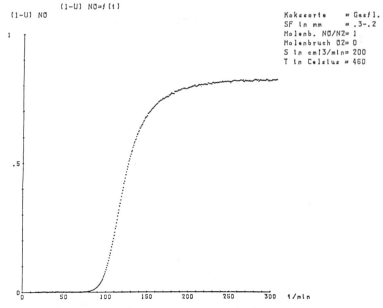

FIGURE 38: NO-reduction by pyrolysis char versus time
according to J. Schuler (1986)

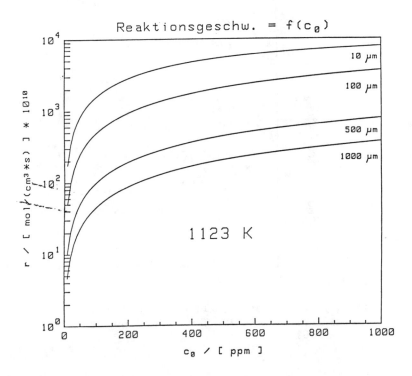

FIGURE 39: Rate of NO-reduction at pyrolysis char dependent
on concentration and particle size according
to J. Schuler (1986)

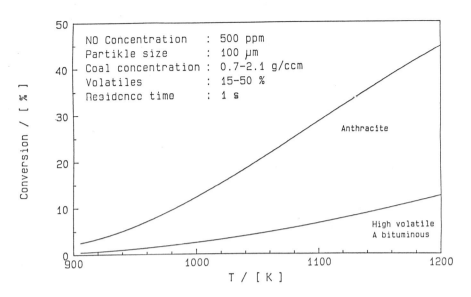

FIGURE 40: Model calculation of NO-reduction in fluidized
beds with different carbon concentration according to J. Schuler (1986)

DEVOLATILIZATION	MACERAL CONTENT
	CONTENT OF MOBILE PHASE
	CH DISTRIBUTION OF MACROMOLECULE
	PARTICLE SIZE
IGNITION AND BURNING OF VOLATILES	TAR FORMATION RATE DEPENDENT ON VOLATILES CONTENT AND PARTICLE SIZE TAR QUALITY
CHAR FORMATION (POROSITY, DENSITY, PARTICLE SIZE)	SWELLING BEHAVIOUR DEPENDENT ON LITOTYPES AND TAR FORMATION
BURNING OF CHAR LOW TEMPERATURE REGION	REACTIVITY DEPENDENT ON DEGREE OF GRAPHITIZATION = f (TEMPERATURE, RANK) CATALYTIC EFFICIENT COMPOUNDS (FROM ASH)
HIGH TEMPERATURE REGION	PARTICLE SIZE
COACTION OF DIFFERENT STEPS	VOLATILE CONTENT PARTICLE SIZE

FIGURE 41: Coal properties related to the individual steps
for coal combustion

DISCUSSION

T.F. Wall

From your understanding of coal chemistry what opportunities do you see for simplifying the feed material prior to
experiments (e.g. by chemical means or by mild pyrolysis) to
simplifying the interpretation of experimental results.

H. Jüntgen

Mild pyrolysis with low final temperatures between 400
and 520°C as prestep seems to alter coal in the sense of formation of higher rank coals with more simple devolatization
behaviour (P. Hanbaba, H. Jüntgen, W. Peters, Brennstoffchemie
48 (1968), S. 368-376). Also by extraction or mild hydrogenation the devolatilization behaviour can be changed (G. Löwenthal, W. Wanzl, H. Jüntgen, paper at seminar Organische
Kohlechemie, 11.9.85 TU Clausthal). These experiments may
contribute to a better understanding of the complex devolata-
lization behaviour of coal in relation to its chemical structure.

E. Suuberg

1) The presentation suggested a molecular hydrogen (H_2)
intermediate in formation of tars. Other workers have postulated direct abstraction of hydrogen by neighboring species,
involving no H_2 intermediate. Then, too, there is increasing
evidence as you presented for the role of "Guest" molecules in
formation/release of high molecular weight species during
devolatilization. Since the question of hydrogen transfer will
be key in future, more detailed models of tar formation, can
you clarify your position regarding the nature and importance
of H-transfer reaction. Do you have direct evidence of an H_2
intermediate during tar formation ?

2) It is well known that coals have high free spin densities as revealed by E.S.R., but these were not really suggested in the coal structure that you presented. Do you feel
that these radicals are important in the combustion, devolatilization processes of interest here ?

H. Jüntgen

1) The most important step of pyrolysis is the cleavage
of weak bonds resulting in the formation of radicalic species.
The consecutive reactions depend very strongly on the extend
of hydrogen available to saturate the radicals with hydrogen.
In the model suggested (figure 24) as main source of hydrogen
the aromatization of hydroaromatic rings in the coal molecule
is discussed. Here details of the mechanism of hydrogen transfer are omitted. I agree that it will be a topic of future
research to more clarify the nature of hydrogen transfer
reactions. In contrast to the tar formation for distillation
of "mobile phase" no hydrogen transfer is necessary. It can be

assumed that most small molecules of mobile phase have left
the particle before reactions of degradation take place, as
shown in figure 23.

2) I assume that free spin densities are related to
great aromatic clusters of coal macromolecule. These also can
in principle be saturated by hydrogen transfer during pyrolysis, if enough hydrogen species are available. On the other
hand big molecules have slow diffusities in the coal particle
with the consequence that they tend to react each with another, after cleavage of the bridges during pyrolysis. Therefore they preferently form char and not tar.

P.R. Solomon

1) On the basis of studies on model polymers in our laboratory, hydrogen in ethylene bridges is also donatable. Also
substitution on the rings, especially oxygen containing substituents is very effective in lowering bond energies.

2) To what extent is bond breaking involved in the release
of the "mobil phase" either to lower molecular weights to
enhance diffusion or to detach some bonds to the matrix ?

H. Jüntgen

1) I agree, that ethylene bridges can contribute to hydrogen transfer by disproportionation reactions, however hydroaromatic CH_2 - groups may play a preferential role as hydrogen
donors. - I further agree that the binding forces between
C-C-atoms in the coal macromolecule depend very strongly on
oxygen containing substituents in the aromatic rings of the
macromolecule. This question could not treated here in detail.

2) This question cannot be answered with last certainity. I
assume that in most cases no bond cleavage is involved. On the
one hand, the single molecules of "mobile" phase have a
relatively low molecular weight and are only trapped in the
matrix of the macromolecule, on the other hand, diffusities of
e.g. long chain paraffines in "bottle neck pores" are strongly
temperature dependent and can reach values high enough for
diffusion at 300 to 400°C. Maybe that a fast diffusion of
three to five ring systems (e.g. hopan derivatives) can only
be reached if first bridges of macromolecules are broken.

G. de Soete

With respect to your study on char-nitrogen transformation into NO.
What effect have :
1) Temperature
2) CO/CO_2 ratio
on the fraction of char-nitrogen converted into NO ?

H. Juntgen

We have investigated the effect of temperature in the range between 1000 and 1100°C. In this range we have not detected significant differences of NO-formation during oxidation of char-nitrogen. The relationship between N-conversion into NO and the CO/CO_2-ratio is not investigated in detail till now using the wire net technique. Therefore we cannot answer this question related to the behaviour of the single coal particle. However we have observed that the NO-formation during oxidation of char-nitrogen is dependent on the lenght of the char layer, if studied in a fixed bed reactor. Here NO-formation during N-oxidation and NO-reduction by the carbon of char plays together. Hence it can be expected, that in carbon layers e.g. in fixed bed or fluidized bed reactors relations between CO/CO_2 ratio and extend of NO-formation exist, however this relation may be also influenced by the oxygen available.

J. Lahaye

In commenting data published by I. SMITH in 1982 you mentionned that there is a correlation between intrinsic reactivity and degree of graphitisation. The degree of graphitisation is often difficult to determine. Walker, Laine and Vastola developped a method to determine the active surface areas (A.S.A.). Are these results correlating intrinsic reactivity and A.S.A. ?

H. Juntgen

In principle the parameters influencing intrinsic char reactivity are porosity and accessible internal surface (under reaction conditions), active sites for the oxidation reaction, structure of graphitic lattice in the char and catalytic influences of ash components. The most experimental studies (see also right hand picture of figure 16) show that char reactivity can be related to temperature of char formation and rank of the parent coal. It is not quite clear, in which way temperature of char formation and rank of parent coal influence the structure parameters mentioned above and which parameters are most important governing the intrinsic reactivity. It could be that the active sites for the oxidation reaction are identical with the lattice defects of the graphitic structural units in the char. We found that intrinsic char reactivity could not be correlated to porosimetry and internal surfaces of chars made from hard coals measured at room temperature. We have not used the ASA method developed by Vastola and Walker and can not report on correspondent results of correlations between char reactivity and ASA (accessible surface area).

J.B. Howard

You discussed changes in the particle size distribution due to swelling and break up of particles. From available information on kinetics of softening, swelling, and resolidi-

fication, and considering particle concentrations and levels
of turbulence of interest in pulverized coal flames, one can
predict that some particles may agglomerate during the pyrolysis stage of pulverization. Such predictions are of course
very difficult and uncertain. Would you comment on this
question of particle agglomeration.

H. Jüntgen

We have not investigated this question in detail. I
would assume that the agglomeration of particles in coal
flames depends on the probability, that particles can stick
together during softening in the pyrolysis zone of the coal
flame. Hence it follows that the density of the air-born dust
cloud, the velocity of the single particles, and the time
between softening and resolidification of the particle, which
latter depends on heating rate, are important parameters
governing the agglomeration of particles. Summing up it can be
concluded that particle agglomeration in coal flames can not
predicted from the behaviour of an individual particle alone.

A. Williams

In your lecture you did not place any emphasis on the
role of the inorganic species present in coal in relation to
the way in which it can influence the pyrolytic reactivity of
the coal macromolecule and labile species. The inorganic
species, in particular the sulphur groups, can bond with these
components and influence their rates of reaction, and there is
some eivdence from microwave pellating experiments with coal
to support this supposition. Please can you comment ?

H. Jüntgen

Certainly inorganic species can influence reaction behaviour of coals and coal related chars very much. Effects are
depending on the chemical kind and the distribution of
inorganic species in the coal particle. Examples are the
catalytic effect of pyritic sulphur on hydrogenation rates,
the catalytic effect of alkaline and earthalkaline species on
gasification with steam and/or on char oxidation. In general
alkaline species are movable in the particle under the conditions of gasification or oxidation and therefore always
effective. Ca species are not movable, therefore their effect
depends on distribution and binding in the coal macromolecule.
Ca ions in humic acids of lignites have a high catalytic
effect on gasification reactions, in contrast to Ca in coarse
ash particles is not very effective.

P.J. Jackson

1) Have you information available on the chemical behaviour of elements such as sodium and calcium, which are either
adsorbed on the internal surface of coal, or combined with the
organic structures ?

56

   2) (Not spoken). What is the evidence that inertinite is
derived from carbonized wood ?

H. Jüntgen

   1) The answer is given in the comments to the question
of Professor A. Williams.

   2) In contrast to the precursor materials of the macerals of the vitrinite and the exinite groups it seems to be
difficult to relate the macerals of inertinite groups to
distinct and well defined precursor materials. According to
Mackowsky (Microsc. Acta 77 (2), 114 (1975)) fusinite and
semifusinite can be related to "carbonized" woody tissues,
while micrinite can only be related to unspecified detrital
matter and sclerotinite to fungal spores and mycelia.

J. Beér

   You have mentioned two phenomena observed in the screen
experiments : heterogeneous ignition in small coal particles
and changes in particle size during pyrolysis. Drop tube
experiments of Timothy at MIT show that homogeneous ignition
is maintained in 38 μm bituminous coal particles even at 100 %
O_2 concentration for freely suspended particles. Also, drop
tube experiments show that devolatilizing particles are
subject to significant centrifugal acceleration (about 100 to
200 g) which can be instrumental to the explosive fragmentation of freely suspended particles. Do you agree that the
screen-heating technique might not provide full information on
particle ignition mechanism and the size changes during
pyrolysis as it pertains to pulverized coal combustion.

H. Jüntgen

   I assume that screen heat up techniques give comparative
results of ignition behaviour using coals of different volatiles and particle size and demonstrate in principle the role of
the heating rate. In this way the complex interplay of devolatilization, ignition of volatiles and ignition of char can be
understood in principle and the results can be transferred to
the more complex behaviour in drop tube experiments. According
to figure 17 (right hand picture) it can be expected, that for
38 μm particles (of 39.5 % VM) at heating rates of more than
1000 K/s simultaneous ignition of volatiles and solid
particle takes place which is not in contradiction to the drop
tube experiment of Timothy at MIT, mentioned in your comment.

   Additional effects of e.g. centrifugal acceleration may
then be studied in more detail in drop tube experiments. At
Bergbau-Forschung results in drop tube and screen-heat up
techniques are compared at time.

   As to the explosive fragmentation of particles it also
can be observed by using screen-heat up techniques. The most
important parameters governing particle size change including

explosive fragmentation of particles during pyrolysis are the tar evaluation rate and the maceral composition.

John H. POHL

Are the result from experiments in which the coal is held difficult to extrapolate to the conditions of pulverized coal combustion ? Five potential reasons may make extrapolation difficult : 1) heating rate, 2) temperature, 3) sample proximity, 4) surrounding atmosphere and 5) time resolution of the apparatus. Given the complexity of coal and coal reactions and our current knowledge and possible change in dominate mechanisms ; the results are empirical, the rates of reaction change with extent of reaction, and the devolatilizative reactions have time constants of 10's ms at the conditions of pulverized coal combustion, while the resolution of held samples is on the order of seconds. Does this result in derivation of a average rate constant which is too low and is difficult to apply to pulverized coal flames ?

H. Jüntgen

I agree that extrapolation of results of combustion of thin coal layers using a wire net technique to flame conditions is difficult. However these experiments give a first comparative picture of the behaviour of different coals at different conditions. Parameters as mentioned in your comment as heating rate, temperature, surrounding atmosphere can be measured and also systematically varied in wire technique experiments (which is not always possible under flame conditions). High time resolution can be reached by using a flight of time mass spectrometer or a high speed film camera. It could be showed that the reaction parameters of coal devolatilization are not changed in a range of heating rates between 10^{-2} and 10^{5} K/min using small particles. By the theory of non-isothermal reaction kinetics results can be extrapolated to higher heating rates. (H. Jüntgen and K.H. van Heek, Fuel 47 (1968), 103-117). Therefore I believe that experiments using a wire net technique are very useful to explain and to interpret the behaviour of different coals in flames. However an intensive discussion is necessary to compare results gained in wire net experiments and in flames. In this connection it is most important to investigate the same coals by the different methods. I refer to the paper of P.T. Roberts showing that model calculations combined with laboratory measurements can predict the coal combustion efficiency in boilers.

M. HERTZBERG

The term "ignition" can have many meanings or definitions in combustion. It can refer to the temperature at which flame propagation is initiated, or, it can refer to the temperature at which a pile of particles will self heat in a smouldering mode. In addition, one can observe ignition of single particles, or, alternatively, of large clouds of particles.

In your presentation, you indicated that for small particle the ignition is heterogeneous. Would you please define the "ignition process" involved in that observation in terms of the experimental details.

H. Jüntgen

In the experiments described using a wire net technique ignition is observed by the beginning of CO_2 formation, measured by a time of flight mass spectrometer. The wire net with a thin layer of coal particles was heated electrically and the temperature was measured by a thermocouple combined with the wire net. Ignition of volatiles takes place at the time (and temperature) of first CO_2-formation of the first sharp CO_2 peak (I_z in figure 16, left hand picture), ignition of char takes place at the time of first CO_2- formation of char combustion (I_z in right hand figure of figure 16). Additional experiments to determine ignition of volatiles and char are performed by observation of the coal layer using a high speed camera. Here comparable results could be established. In this way ignition conditions could be studied as a function of rate of heating, particle size and volatile content of coals. Typical results are summed up in the right hand picture of figure 17.

M. Hertzberg

I should like to make a comment in support of Prof. Beér's earlier comment regarding the gas-phase ignition of coal particles. Our data strongly support the viewpoint he has expressed. We measure the auto-ignition temperature of coals injected into a preheated oven at a high dust concentration. The auto-ignition temperature for Pittsburgh sub bituminous coal is insensitive to particle size for diameters less than 50 μm. Above that size, the auto-ignition temperature increases. For the higher rank coals, the auto-ignition temperature increases markedly. The totality of the data for particle size and rank dependance strongly supports the viewpoint that the ignition process is controlled by the combustible volatile content and that it involves the volatiles as they are penetrated by pyrolysis and premix with the air. That autoignition process is a more accurate simulation of what would occur in a coal burner than the ignition of a single particle or a small array of particles held on a screen.

H. Jüntgen

For an answer on your comment a more accurate comparison of the results of the autoignition process and the results of the wire net technique is necessary possibly with the same coals. I am convinced that a comparative study would give useful hints of a better understanding of ignition behaviour of different coals and of parameters influencing it as particle size, volatile content, and heating rate.

COAL CHARACTERIZATION BY A LASER MICROPROBE (LAMMA)

J.F. MULLER
Laboratoire de Spectrométrie de Masse et Chimie Laser
Université de Metz - 57012 METZ CEDEX 01 - FRANCE

1) INTRODUCTION

The characterization of coals, which are complex and hete-rogenous polymeric materials, requires the implementation of numerous physicochemical or spectroscopic techniques.

The different organic phases of coals, called macerals (from latin macerare - to macerate), have different structures because they don't have the same origine : - exinite, the re-productive organs of vegetables (e.i. sporinite), - vitrinite, humic substances that come from lignite and cellulose, - iner-tinite, lignite and cellulose that have previously undergone carbonization (forest fires) or oxydation on peat bogs. These three big maceral groups can be distinguished morphologically by microscopic optics (petrographic analysis). Moreover, the vitrinite reflectance (V.R.), that increases with the rank of the coal an consequently with the degree of coalification, gives indications on coal reactivity. If there are relations between the V.R. and the degree of aromaticity mesured by Nuclear Magnetic Resonance (N M R) of the solid or by Fourier Transformed Infrared (IR-TF) (1) (2), the structural patterns (that are more of less polymeric) of each of the coal compo-nents are more difficult to identify.

Mass spectrometry potentially offers an alternative for reaching this objective. Unluckily, there is a parameter that cannot be circumvented : the desorption of such structural patterns generally requires soft ionization techniques, but does not allow to break up the cohesion (or the compacity) of coal that is linked to the bond energies (interaction of the Van Der Waals type). They can only be overcome by very high energies that cause too much fragmentation themselves.

Most of the soft ionization techniques have been more or less successfully tried :
- field desorption
- desorption followed by chemical ionization
- fast atom bombardment.
On the whole, only a little significant information has been obtained.

On the other hand, mass spectrometry by pyrolysis at the curie point (3) has provided interesting and reproducible re-sults :

A given coal or maceral, finely grinded, deminerralized, and then placed on a tungtene filament, is pyrolized by ultra-rapid induction under vacuum (T=610°C heating time : 6 sec.). The gaseous pyrolysis residues are ionized by an electron beam. The ions obtained are accelerated and analyzed by a magnetic deflector or a quadrupole mass spectrometer.

The ionized fraqments of the mass spectra are very representative of the different basic patterns (for example : alkyls, naphtyls) or of the carboned structure. Factorial analyses of the fragments allow to classify the coals with repect to their grade and their origin (4). But this technique cannot in any way give punctual information directly from the solid. This is the reason why the LAMMA impact laser microprobe (punctual laser pyrolysis) seemed to be a promising technique. Actually, the first tests done in our laboratory and by another team (5), have convinced us that it allows the characterizations of organic matter as well as its mineral residues on a scale close to the micrometer.

However, it was necessary to know before hand the different laser ionization parameters, because, as with the other techniques previously mentionned, the soft desorption - strong intermolecular bond energy duality cannot be avoided.

II) LASER IONIZATION PARAMETERS

The ponctual laser ionization of solids was developed quite early in the United States, in particular by Vastola (6), then in France by J.F. Eloy (7) ans in Germany by F. Hillemkamp and R. Kaufmann (8).

The principle is quite simple : the ions generated by the impact of a laser beam focused on a solid sample are analyzed under vaccum by different type of mass spectrometers : first by magnetic deflection, then by flight time, or by a combination of the two (9).

The LAMMA (LAser Microprobe Mass Analysis), developed by Leybold Heraeus, allows the focusing, by optical microscope, of the pulsed laser beam and the creation of a microplasma. The analysis of the ions obtained this way is done by a time of flight mass spectrometer with a high transmission coefficient, ensuring a spectral resolution M/ M close to 800 for the 208 isotope of lead. Two configurations have been finalized, one for ion transmission analysis (LAMMA 500), the other for solid material reflection analysis (LAMMA 1 000). An English company, "Cambridge Instruments", also sells a reflection microprobe named LIMA (Laser Ionization Mass Analysis) that also uses a time of flight mass spectrometer, but uses different focalizing and visualizing optics.

The ions (negative or positive) that are obtained from the lase impact point (microplasma) gain through the acceleration lense (V=+3KV) a kinetic energy and therefore a speed that remains constant during their trip down the spectrometer tube,

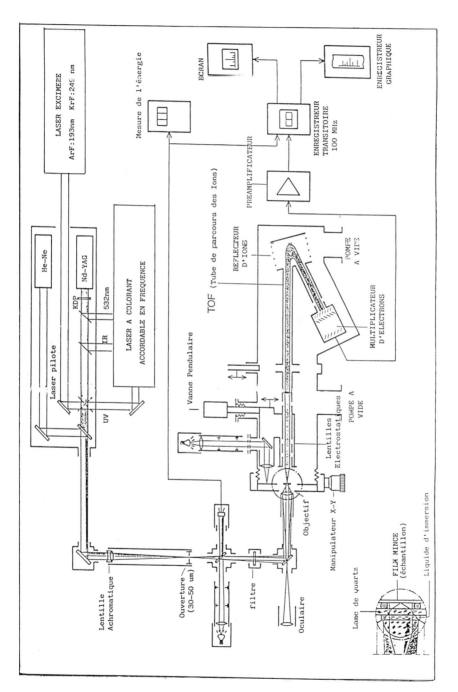

Figure 1 : Laser Microprobe LAMMA 500 (Leybold Heraeus) coupled with tunable dye laser and excimer laser.

inself brought up to a potential of 3 KV (Faraday trap).

$$q \ V = \frac{1}{2} \ mv^2 \qquad v = [2 \ (\frac{q}{m}) \ V]^{\frac{1}{2}} \qquad t = k \ (m/q)^{\frac{1}{2}}$$

The electrostatic reflector that is introduced half-way through allows to get rid of the initial energetic ion distribution because ions that have the same mass but slightly different energies are focalized on the 17 dynode electron multiplier made of a copper-beryllium alloy. These reflectors allow us for the moment to reach spectral resolutions close to 2 000.

The electric signals that correspond to each burst of ions of the same mass arriving upon the detector are amplified then stored in a fast transient recorder (100 or 200 MHz) that has a storing memory that is more or less large.

The larger it is (e.g. 32 Ko), the less it will limit the spectral resolution of the device. The benefits and the dynamics of the amplification are also important parameters. They must be high if we want to avoid signal saturation phenomena and approach the conditions for a real semi-quantitative analysis.

The specificity of the technique lies essentially in the laser-matter interaction phenomena, in which the type of laser is the dominating factor, but where the nature and the structure of the analyzed material in a solid state play a part that is just as important.

The main laser-matter interaction parameters can be divided into four types :

1) The irradiancy value of the laser beam expressed in W/cm^2

If the irradiancy value is less than 10^6 W/cm^2, proton adsorption by the material will respect the laws of Lambert and Bur. Our the contrary, if the diameter is superior to 10^6-10^7 W/cm^2, proton adsorption is not linear and is accompagnied by a series of procedures (Cf § II 4). At a very high irradiance 10^{10} W/cm^2, the plasma is considered as hydrodynamicand the ejection of particles is explosive.

2) The relative geometry of the laser beam and of the acceleration lenses with respect to sample surface

On the whole, the transmission analysis (LAMMA 500) is more sensitive, and leads to a better spatial resolution. The effects of the matrice are also minimized if the cuts are as fine as obtaining a spatial resolution of about 1 micron all the while keeping an acceptable sensitivity threshold.

In a reflection analysis configuration, it is difficult to

achieve an impact with a diameter of less than 2 μm, and the matrice effects are more sensitive. Also, ion ionization at 90° and extraction at 60° seems better than inversed geometry (ionization at 60° and extraction at 90°).

3) Interaction time

This is determined essentially by the duration of the laser pulse which, according to the type of laser, goes from one picosecond ot a few hundreds of microseconds (especially the CO_2 laser).

Interaction time also dependes on width, the depth and the vaporization time (which is linked to the physical parameters of the analyzed material: network energy, relative fusibility, conductivity, molar absorption coefficient) of the impact point, or on the phase changes during irradiation.

In most cases, since interaction time is less than 1 microsecond, there is a local thermodynamic equilibrium (LIE) at the impact point. In this system, the system that is characterized by a unique medium temperature is replaced by a serie of cub systems that each have a different temperature (atomization temperature, ionization temperature,...). Each subsystem locally checks the Boltzmann, Maxwell and Saha laws.

This L T E conception will work only if the temperature is high enough, and this requires a high electronical density of the microplasma and therefore an irradiancy superior to 10^6 W/cm^2. Under theses conditions, the degree of ionization can be determined by the simplified SAHA-EGGERT relation, in which the coulombian depression potential (or extraction potential) of a given element is not substracted from the ionization potential of the same element (10).

$$\frac{n_o n_i}{n_u} = 2,4 \times 10^{15} \ T^{3/2} 10^{-L(T)}$$

in which $L(T) = \dfrac{5.10^3 P_I}{T}$ P_I is the ionization potential

T is the Kelvin temperature.

Knowing the degree of ionization, it is possible to evaluate the microplasma temperature, which, according to the irradiancy and the type of material, can vary between 6 000° and 15 000°K.

4) Laser beam wavelenght

The laser ionization at medium irradiance (10^8 W/cm^2 < Ø < 10^{10} W/cm^2) is a series of complex phenomena whose respective effects are difficults to dinstinguish: no linear multiphonic absorption accomapgnied by photoelectron emission, creation of hole-electron couples, then surface loads followed by atom or molecule ionization and secondary electron emission. During the first nanosecond, important energy transfers inside the matter lead to the forming of a dense plasma which rapidly expands under vacuum (about 2 to 5 nsec.). During the expansion, many ion-molecule reactions occur, forming clusters (CF § III 2) or cationized molecular peaks (CF § III 3).

Multiphonic absorption by the material being the first step and microplasma expansion being the last, it is perfectly understandable that the influence of the wavelength, especially during absorption, is very often concealed by the characteristics of the microplasma. This explains why, up to the appearance of excimer lasers and the works of Egorov (11), many authors pretended that the type of laser had little importance for solid ionization.

In our L.S.M.C.L. Laboratory, we have to prove this influence and to benefit from it on the analytical level. The actividted Neodym-Yag (Q switchched, τ = 12-15 nsec), the wavelength emission (1 060 nm) of which is doubled (λ = 532 nm) by the intercalation of a thermostated K D P crystal, pumps a coloring laser, which, with respect to its nature, emits another monochromatic laser beam into the visible field. The adjunction of a frequency doubling and frequency mixing crystal (with a residual infrared) allows us to obtain a U V beam with two wavelengths that can be distinguished by a prism and be injected alternatively into the optical path of the LAMMA microprobe.

This system makes it possible to vary the irradiance of all wavelengths and to install four analytical protocols adapted either to the information that is wanted at the type of material:

i) High irradiancy resonant analysis, aimed on selectively decreasing the detection level of a given element. As for coals, these techniques can be applied for following an element that appears only as a trace in different organic or mineral matrices.

ii) High irradiancy non resonant analysis, that makes the most of ionized cluster detection and identification detection in the case of coals, whether they are hydrocarbonized ($C_nH_m^+$) or a metallic origine ($X_pO_q^-$)($X_pO_q^-$) (X = metallic element).

iii) Two step analysis (low, then high irradiancy) reserved for organic and then mineral analysis of microparticles ("fly ash" or individual airborn particles).

iiii) Finally, the low irradiancy desorption laser which is being developed and which we will not discuss here (Characterization of organic molecules).

III) APPLICATIONS EXAMPLES

1) Use of resonant ionization (12)

The sensitivity that is gained using resonant ionization is illustrated here in the analysis of a standard steel containing 0.538 % of Molybdène (I.R.S.I.D. standard n° 108/1). Under usual conditions ($\lambda = 266$ nm; $E \simeq 2 \mu J$) the major elements are detected and the relative added intensities of the main elements, corrected with respect to their respective ionization potentials (SAHA Law), allow to find the composition of the alloy within 3%. Molybdene can be detected by increasing the energy which causes a saturation of the other signals. On the other hand, if we vary the wavelength from 310 nm to 320 nm the intensities of the peaks that correspond to the different Molybdene isotopes are considerably increased to 313.26 nm and to 311 nm. These wavelength correspond to two Molybdene emission rays that are listed in the charts (13). Their respective intensities are inversely proportional to the Einstein coeffcient, i.e. the lifetime of the Molybdene intermediary metastable excited states. This is totally in accordance with the resonant ionization theory (R I S) (14). Indeed, in a $M(\omega, \omega, e)M^+$ type double photon mechanisme, the longer the lifetime of the intermediary state, the more it is likely that the second photon be absorbed.

Therefore, if the coloring laser wavelength is adjusted to 313.26 nm, the sensitivity threshold of Molybdene is divided by factor 50. Parallel to this, the other elements are less ionized and the corresponding peaks are less intensive.

This procedure has been successfully put into application with Cadmium, Erbium and Copper (15) (16) (17).

2) Analysis of coals or of hydrocarboned particles (18) (19) (20)

The analysis of a series of 19 coals (6 of which were French - CERCHAR Minibank) and a various kerogenes (I F P samples) was done according to the following protocol:

1 - The absorbance - density curves obtained after differential gradient centrifugation (D G C) of finely grinded and demineralized French coals show that the latter have relatively low differences of density between the different maceral groups and especially between vitrinites and inertinites... This allowed us to use on one hand perfectly demineralized coal samples and on the other very rich maceral fractions (especially for exinites and vitrinites). Several LAMMA microprobe analyses were implemented with these samples: demineralization control, Hydrocarboned cluster analysis, fly ash analysis.

a) Demineralization control

The sensitivity (1 to 10 ppm for most elements) of the

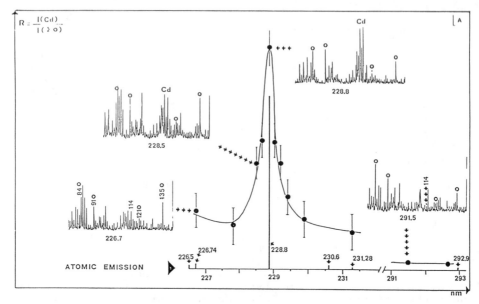

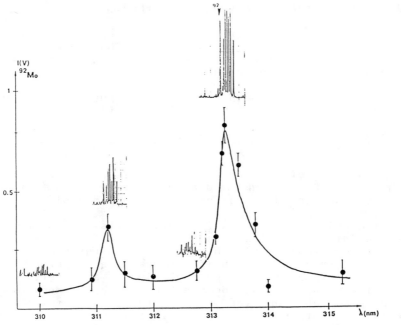

Figure 2 : Resonance Ionization (Lamma 500 coupled with tunable dye laser) of a polymer doped with cadmium
complex (CD+ 200 ppm) and a standart steel (IRSID
108-1) with 0.5% of Molybdenium.

LAMMA microprobe has proved that it is an excellent tool for studying the demineralization of finely grinded (2 to 5 μm) coals. Indeed, the minimum grinding time that ensures good HF/HCl demineralization must be found for a given coal, without degrading the organic matrice too much (depolymerization).

Moreover, before demineralization, it is possible to obtain more information from particles that are very rich in elements concerning the nature of the existing elements (pyrite, clay, gypse, etc...) and by joint analysis of the positive and negative spectra (CF § III 2).

b) <u>High irradiancy coal analysis</u> ($\Phi > 10^8$ W/cm²)

The average positive spectrum (after 30 laser impacts) of pyrite enrichened kerogene has several elements of information:

i) The nature of the present metal ions: sodium, magnesium, aluminium, potassium and vanadium have concentrations that can be evaluated between 50 and 200 ppm. On the other hand sulphur and iron have a higer concentration.

ii) This high concentration is seen in the detection of ionized clusters of the $(SC_{2p+1} H)^+$ (m/z=45, 69, 93, 117) and FeC_pH^+ (M:z=81, 105, 129, p=2,4,6,...). These ions only appear when an elements, submerged in a hydrocarboned matrice, goes above a concentration threshold of about 0.4 to 0.5%. With high irradiancy ($\emptyset > 10^8$ W/cm²), the S^+ or the Fe^+ ions formed in the microplasma are "solvated" by successive collisions with hydrocarbonated fragments (neutral or radicalar) that come from the matrice. Therefore, ionized clusters of the C_nH^+ type can also be detected in petroleum residues that are rich in vanadium ($\simeq 4\ 000$ ppm).

iii) The hydrocarboned matrice itself leads to the forming of clusters of the $C_nH_m^+$(or $C_pH_q^-$) type. The more the hydrogen level goes up, the higher the m (in the case of paraffins and of lignites), whereas to the contrary, m goes towards zero when it is very pure graphite.

iiii) This last particularity has been used to evaluate the C/H ratio of the hydrocarboned particles of mine dust coals or of combustion residues. In the case of most coals and kerogenes, this ratio is between 0.2 and 1.8 (cf. the Van Krevelen diagram). In anthracites, the C^+ and C_nH^+ ions are dominant, whereas in lignites or kerogenes ($C_nH_2^+$, $C_nH_3^+$or $C_nH_4^+$) are much more intensive while C_n^+ are almost unexisting.

After several tries, the even order clusters (even n) were eliminated, because of the risks of spatial interference with elements such as magnesium, aluminium, silicium or titanium.

Taking into account only the relative intensities of the odd order clusters and m=1 and 2, a linear relation was found between the H/C ratio determined by centesimal analysis and

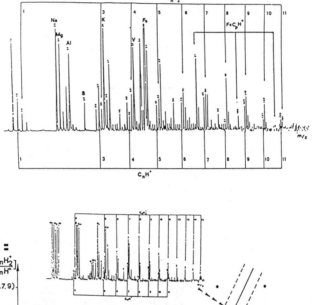

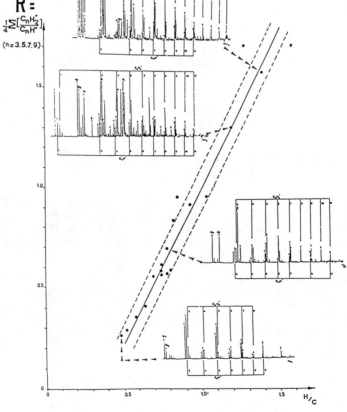

Figure 3 : a) Typical LAMMA spectrum of a coal particle
b) Relationships between relative cluster intensities and H/C ratio.

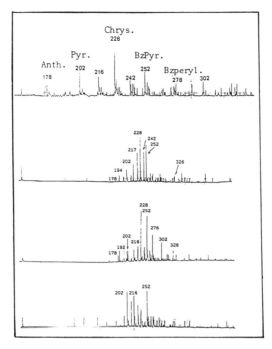

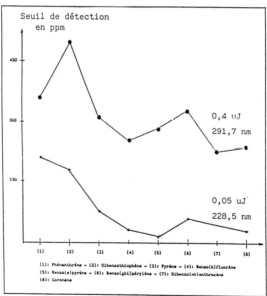

Figure 4 : a) Individual analysis of industrial airborne particles at very low energy and low wavelength
(228.5 nm)
b) Detection threshold of polycyclic aromatic hydrocarbons adsorbed on microparticles of hematite at two wavelengths (228.5 nm and 291.7 nm).

te intensity ratio R.

$$R = 1.45 \ (H/C) - 0.45 \quad (r = 0.97)$$

$$R = \frac{1}{4} \ \Sigma \ \frac{I(C_nH_2^+)}{I(C_nH^+)} \qquad \text{with } n = 3, \ 5, \ 7, \ 9.$$

This methodology should be very useful for evaluating on site the H/C ratio of the carboned materials included in the heterogenous systems such as rocks.

3) <u>Two time particle analysis</u> (22) (23)

When the irradiancy is too high, it is impossible to record a positive spectrum, then a negative spectrum from <u>one</u> particle with a diameter of about 1 to 3 µm. It is ejected after the first impact (a transfer of thermal energy into kinetical energy).

Very low wavelength irradiation ($\lambda < 240$ nm) allows to compensate for inconvenience. Indeed, desorption tests on hydrocarbon aromatic polycyclics (HAP) show that under these conditions ($\lambda = 228.5$ nm).

- The energy threshold at which the molecular peak M^{o+} appears was decreased of a factor close to 10 with respect to $\lambda = 286$ nm, their network energy was not too high.

- The ionization yield is higher than at 266 nm or 286 nm.

- Certain ions appear (M^{o+} or others) without any apparent damage to the surface of the sample. <u>Laser photoablation</u> can be considered.

In fact, studies made at 228.5 nm and 291.7 nm with absorbed H A P mixtures on amorphous carbon particles (2-3 µm) or on hematite have lead to several interesting results (23):

- at 228.5 nm, the ionization energy threshold goes down a factor 10, without causing any particle movement.

- with the same wavelength, the HAP detection threshold is divided by a factor going from 7 to 3 according to the type of PAH.

- these two decreases are not as distinct if the support particles are made of asbestos alumine or of pure silicium, and this shows the influence of the interaction between the absorbed organic molecules and the nature of the surface.

These experiments lead to the establishment of a two step low wavelength particle analysis proptocol. <u>The first consist in characterizing the eventual organic molecules absorbed at very low energy ($E \leqslant 0.05$ µJ). During the second with the same</u>

particle, a higher energy laser impulse (E> 2 ZJ, $\emptyset$ > 1(3 W/cm²)
is triggered, so as to detect the consecutive elements and the
nature of the majority mineral support by analyzing the clusters.

IV) CONCLUSION

The LAMMA laser impact microprobe is particularly adapted
to the analysis of either mineral or hydrocarbuned microparticles (21).

The characterization of individual particles is the field
in which this technique offers the most advantages: in addition to establishing histograms of preponderant elements or of
elements to appear only as a trace, it allows to do a two step
analysis with low energy and weak wavelength molecule desorption followed by an identification of high energy mineral
compounds. Moreover, it is possible to look for elements that
appear. as a trace (tracing elements that characterize a given
origin) by adjusting the laser beam wavelength to a ray that
belongs to one of them. Finally, the on site evalutation of
the H/C ratio opens interesting perspectives in the on site
analysis at bituminous rocks, or of various coals and
kerogenes.

All these elements put together allow to find the origin of
a particle, to follow its evolution, to evaluate its toxicity
level. In the more specific cases of coals, the LAMMA microprobe combines morphological recognition (optical microscopy)
and   on site analysis of organic matters with mineral
distribution.

V) BIBLIOGRAPHY

(1) L.B. ALAMANY, D. GRANT, R.J. PUGMIRE, L.M. STOCK,
    Fuel, 63, (1984), 513.

(2) D.W. KUEUN, R.W. SNYDER, A.D. PAINTER, P.C. PAINTER,
    Fuel, 61, (1982), 682.

(3) H.L.C. MEUZELAAR, A.M. HARPER, G.R. HILL, P.H. GIVEN,
    Fuel, 63, (1984), 640.

(4) a) H.L.C. MEUZELAAR, J. HAVERKAMP, F.D. HILEMAN
    Pyrolysis mass Spectrometry of recent and fossl.
    biomaterials, Vol. 3, Elsevier. Sci. Compagny - 1982.
    b) H.C.L. MEUZELAAR, A.M. HARPER, J. PUGMIRE, J. KARAS
    International J. of Coal Geology, 4, (1984) 143.

(5) T.K. DITTA, Y. TALMI,
    Fuel, 61, (1982), 1241.

(6) a) F.J. VASTOLA et A.J. PIRONE
    Advances in Mass Spectrometry, Vol 4, (1967), 107-111.
    b) B.E. KNOX et F.J. VASTOLA,
    Laser Focus, 3, (1967), 15.

(7) J.F. ELOY
Int. J. Mass Spectrom. Ion Physics, $\underline{6}$, (1971), 101.

(8) a) R. KAUFMANN, F. HILLENKAMP et E. REMY
Microsc. Acta, $\underline{73}$, (1972), 1.
b) R. KAUFMANN, F. HILLENKAMP, R. NITSCHE, M. SCHUERMANN
et R. WESCHUNG
Microsc. Acta, Suppl. 2, (1978), 297.

(9) J.F. ELOY, M. LELEU, E. UNSOLD,
Int. J. Mass. Spectrom Ion Process., $\underline{41}$, (1983), 39.

(10) a) N. FURSTENAU et F. HILLENKAMP,
Int. J. Mass. Spectrom. Ion. Physics., $\underline{37}$, (1981), 135.
b) H.W. DRAWIN et P.F. FELENBOK,
Data for plasma in local thermodynamic equilibrum,
Ed. Gauthier-Villars, (1966), Paris.

(11) S.E. EGOROV, V.S. LETOKHOV, A.N. SHIBANOV,
Chem. Phys., $\underline{85}$, (1985), 349.

(12) J.F. MULLER, F. VERDUN, G. KRIER, M. LAMBOULLE,
S. GONDOUIN, J.L. TOURMANN, D. MULLER et S. LOREK
C.R. Acad. Sci. Série 2, $\underline{299}$, (1984), 1113.

(13) W.F. MEGGERS, C.H. CORLISS, B.F. SCRIBNER
Tables of Spectral lines intensities-NBS Monograph, $\underline{32}$,
(I), (1961).

(14) G.S. HURST,
Applied, Atom. Coll. Phys., $\underline{5}$ (1982), 201.

(15) G. KRIER, F. VERDUN, J.F. MULLER,
Fresenius. Z. Anal. Chem., $\underline{322}$, (1985), 379.

(16) F. VERDUN, J.F. MULLER, G. KRIER
Laser Chem., $\underline{5}$, (1985), 297.

(17) F. VERDUN, G. KRIER, J.F. MULLER
Anal. Chem., sous presse.

(18) E. TOTINO, J.F. MULLER
C.R. Acad. Sciences, 301 II, (1986), 295-300.

(19) E. TOTINO, G. KRIER et J.F. MULLER
C.R. Acad. Sciences, $\underline{303}$ (II), (1986), 821.

(20) E. TOTINO
Thèse université de Metz (1986).

(21) R. KAUFMANN
"Physical and Chemical Characterization of Individual
Airborn Particles" (Ellis Horwood Ltd. Chischester)
p $\underline{227}$, (1986).

(22) J.F. MULLER, G. KRIER, F. VERDUN, M. LAMBOULLE, D. MULLER
     Int. J. Mass. Spectrom. Ion Process., 64, (1985), 27.

(23) F.VERDUN
     Thèse université de Metz (1985).

DISCUSSION

H. Jüntgen

You have reported two results : on the one hand the correlation between $C_nH_2^+/C_nH^+$ seems to show that C_nH^+ correlates to aromaticity and $C_nH_2^+/C_n^+$ the aliphatic groups of coal. On the other hand, the energy of laser treatment is so low that only desorption of molecules takes place. Are the results related to destruction of macromolecules or to identification of small molecules from the mobile phase ?

J.F. Muller

In this experiment (LAMMA), the laser energy is relatively high and is enough to distroy macromolecules and small molecules of the mobile phase. At the laser impact, we can observe phase. We can observe a thermal spike followed by a micro-plasma with formation of elemental ions (C^+ for example). These react with neutral species (or radicals) by ion-molecule processes and give cluster ions ($C_nH_m^+$). In this experiment, the energy is too high for the desorption of organic molecules from the solid phase. On the other hand, if the laser energy is reduced, no ion is detected : the lattice energy (or bond energy) of the solid coal particle is too high for desorption.

J.B. Howard

Please comment on the extent to which the distribution of species measured in the mass spectrometer is representative of the distribution of species formed in the pyrolysis.

J.F. Muller

In the LAMMA spectrometer, the distance between the laser impact point and the accelarating lens of T.O.F. masspectrometer is approximately equal to 3 mm. In this space, the micro plasma is in extension with collisions in the gas phase (formations of clusters ions or quasimolecular ions). The quasimolecular ions ($(M + H)^+$, $(M + Na)^+$, $(M - H)^-$ ... etc of organic molecules are produced by ions molecules reactions : the neutral organic molecules are desorbed from the peripheric surface of the thermal spike and react whith elemental ions (H^+, Na^+) produced in the center of the laser impact. We observed a delay between the zero time of the laser pulse and the quasimolecular ion formation (delay $\simeq$ 1 µs or more). Now we are testing and comparing the ion distribution of laser ionization with two kinds of spectrometers : first is the T.O.F. spectrometer and second is Fourier Transform Mass Spectrometer (FTMS 2000 from Nicolet). We are also testing the double ionization : one laser for thermalization and vaporization, second laser for ionization of gas phase in RIMS conditions (Resonant Ionization Mass Spectrometry).

DEVOLATILIZATION

KINETICS OF DEVOLATILIZATION

J. B. HOWARD, W. S. FONG, AND W. A. PETERS

Department of Chemical Engineering and Energy Laboratory

MASSACHUSETTS INSTITUTE OF TECHNOLOGY

1. INTRODUCTION
   Quantitative description of coal devolatilization is desired as
input for overall descriptions of coal combustion mechanisms.
Devolatilization is very complex, involving multiple chemical reactions coupled with transport processes, and a purely fundamental
description is not yet feasible. In practice we resort to the use of
simplified global devolatilization models, which can be an effective
approach if properly used. However, failure to observe the inherent
constraints of simplified models can lead to inaccurate conclusions.
These topics are addressed in the next section.
   More complete descriptions of devolatilization are being formula
ted in terms of global chemistry of primary decomposition of the coal
structure to form a high molecular weight intermediate, metaplast,
which participates in subsequent (secondary) reactions leading to char
or coke as well as lighter products including gases, or is transported
out of the coal. The transport process occurs in competition with the
secondary reactions and can significantly affect the distribution of
products formed. In the case of softening coals, the kinetics of the
development and decay of the softened or plastic condition as well as
the time-dependent viscosity of the melt can be described in terms of
the kinetics of generation and loss of metaplast. This information in
turn permits description of the coupling between transport and secondary reactions, and predictions of particle swelling and agglomeration
under combustion conditions.
   The last section of the paper presents data useful for the above
type of description, which were recently obtained for a bituminous
coal using an electrically heated wire mesh sample holder and a fast
response high temperature plastometer. The observed kinetics of
generation and destruction of pyridine extractables within the coal
under rapid pyrolysis conditions is closely related to the observed
kinetics of plasticity. The data demonstrate a strong contribution of
coking/repolymerization reactions in depleting the intra-particle
liquid-phase material. These reactions are found to occur on a time
scale comparable to that of the primary coal decomposition thereby
confirming their importance in devolatilization kinetics.

2. GLOBAL DEVOLATILIZATION KINETICS
   The global or overall first-order single reaction rate coefficient k for devolatilization may be expressed as
$$k = (dV/dt)/(V^*-V) \qquad (1)$$
where V is the amount of volatiles released up to time t, V^* is the
limit of V for very large t, and T is absolute temperature.
   An Arrhenius plot of k is presented in Fig. 1. The two sets of
points are for the same coal, but they appear to represent quite
different results. The rate coefficients indicated by the two

straight lines differ by more than a factor of 10^3 at 1100 K. However, as is explained below, all the points obey perfectly the same kinetics. The two sets differ simply because two different heating rates were used, and the single-reaction description of Eq. 1 does not adequately reflect the true behavior, which in this case involves multiple parallel first-order reactions.

Thus the apparent difference in Fig. 1 does not represent inaccurate data, such as poor temperature measurements as some authors have concluded in other cases from such Arrhenius plots. It is a natural consequence of applying a model too simplified to deal with two very different temperature-time histories with only one set of parameter values. This behavior points up both the need for care in the use of the single reaction model, which is further discussed later, and the opportunity for more effective global descriptions.

One of the most successful global descriptions of coal devolatilization, which is reviewed elsewhere[1], is based on the picture that the coal decomposes in a large number of roughly independent parallel first-order reactions. If subscript i denotes one particular reaction from which the amount of volatiles formed up to time t is V_i, then

$$dV_i /dt = k_i (V_i* - V_i)$$ (2)

and

$$k_i = k_{oi} \exp(-E_i /RT)$$ (3)

where V_i* is the limit of V_i as t becomes very large, and k_i is the rate coefficient with parameters k_{oi} and E_i.

If temperature is constant, solution of Eq. 2 gives

$$(V_i*-V_i)/V_i* = \exp(-k_i t)$$ (4)

When temperature varies with time, which is always the case in practice, the solution of Eqs. 2 and 3 can be written

$$(V_i*-V_i)/V_i* = \exp(-k_i t_{e,i})$$ (5)

where $t_{e,i}$ is the effective isothermal time. If the temperature-time history is approximated as a three-step sequence of heating at constant rate m to temperature T, holding at T for time τ, and cooling at constant rate m_c (i.e., $dT/dt = -m_c$ during cooling), and if the condition $E_i \gg RT$ is obeyed, then $t_{e,i}$ in Eq. 5 is found to be

$$t_{e,i} = \frac{RT^2}{mE_i} + \tau + \frac{RT^2}{m_c E_i} = \frac{RT^2}{m_e E_i} + \tau$$ (6)

where $m_e = mm_c /(m+m_c)$.

The total volatiles are the sum of the contributions of all reactions. If there are n such reactions then

$$V = \sum_1^n V_i$$ (7)

Substitution of Eq. 7 into Eq. 1, and use of Eq. 2 gives

$$k = (\frac{V_1*-V_1}{V*-V}) k_1 + (\frac{V_2*-V_2}{V*-V}) k_2 + \ldots\ldots (\frac{V_n*-V_n}{V*-V}) k_n$$ (8)

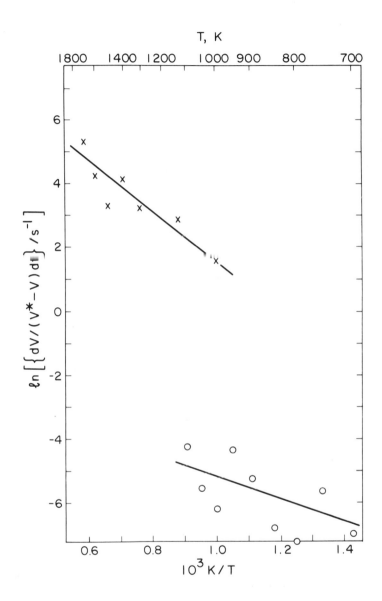

FIGURE 1. Arrhenius Plot of Global First-Order Single-Reaction Rate
Coefficient for Devolatilization of a Given Coal. Heating rate:
(x) 10^4 K/s; (o) 1 K/s.

From Eq. 5 the factors in Eq. 8 can be written

$$\frac{V_i{}^*-V_i}{V^*-V} = \frac{V_i{}^*\exp(-k_i t_{e,i})}{\sum\limits_{1}^{n} V_1{}^*\exp(-k_i t_{e,i})}$$ (9)

Therefore, Eq. 8 shows that the overall rate coefficient is a weighted average of the rate coefficients for the individual reactions, and from Eq. 9 the weighting factors are seen to be functions of the temperature-time history as represented by $t_{e,i}$ with Eq. 6. During a heating process each of the weighting factors decreases from its initial value, which is the fraction of the total ultimate volatiles yield that will be contributed by the reaction in question, to its value at a given time which is the fraction of the volatiles yet to be formed that will be contributed by the reaction in question. Also the individual rate coefficients vary with temperature in accordance with Eq. 3. Thus the temperature dependence of the overall rate coefficient is itself dependent upon temperature-time history.

To illustrate the above behavior, consider a hypothetical coal whose devolatilization is perfectly described in terms of three independent first-order parallel reactions, each having a pre-exponential factor k_0 of 10^{10} s^{-1}. The other parameter values are given below:

Reaction	$V_i{}^*/V^*$	E_i, kcal/mole
1	0.25	40
2	0.50	50
3	0.25	60

Figure 2 gives the devolatilization behavior, calculated from the above equations and parameter values, when this coal is heated at the indicated rate to the temperature given on the abscissa and then instantaneously quenched. Thus m is 1 K/s or 10^4 K/s, $\tau = 0$, and $m_c = \infty$ in Eq. 6.

The points in Fig. 2 are values of ln k which would be observed at temperature T, and which can be written as shown on the ordinate and computed from Eq. 8. The dashed lines give the values of the individual rate coefficients as computed from Eq. 3. The numbers on the points are the percentages of completion of devolatilization for which the calculation has been performed. Only selected points instead of a continuous curve have been computed so as to simulate the usual form of devolatilization data.

In the early stages of devolatilization the values of the overall rate coefficient are seen to follow in parallel with k_1, the individual rate coefficient of lowest activation energy, but to be lower than k_1 by a fraction which is initially $V_i{}^*/V^* = 0.25$. Thus the first term in Eq. 8 is initially dominant. As the fraction of the coal decomposing by Reaction 1 becomes depleted, i.e., as V/V^* approaches $V_i{}^*/V^* = 0.25$, Reaction 2 takes over and the trend of k then parallels that of k_2. At still higher conversions Reaction 3 dominates the overall behavior and the final values of k are identical to k_3 as would be expected from the limiting value of the last term in Eq. 8.

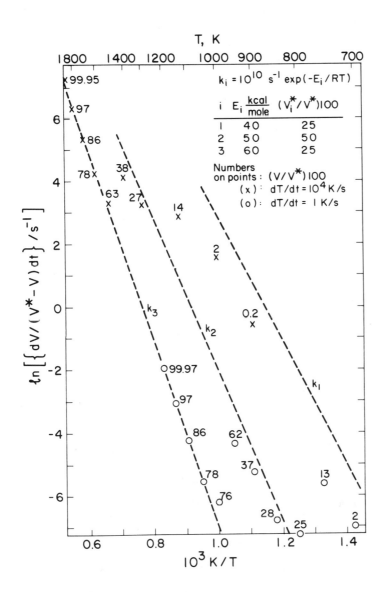

FIGURE 2. Arrhenius Plot of Global First-Order Single-Reaction Rate Coefficient for Devolatilization of Hypothetical Coal Described by Three Separate First-Order Reactions.

If a straight line were drawn in Fig. 2 so as to approximate the points for a given heating rate, as is sometimes done in fitting data points to a simple single-reaction model, the slope would clearly be less than that of either of the dashed lines for the individual reactions. Such lines are drawn in Fig. 1, where the points are identical to those in Fig. 2 for completions of devolatilization between 2% and 86%. Thus we see the well known tendency for overall single-reaction activation energies to be much less than those of the individual reactions of the set. Also it is clear, as was mentioned above, that greatly different values of the overall single-reaction rate coefficien for the same coal can simply be the consequence of different temperature-time histories being used.

Extension of the above analysis to the case of a large number of independent parallel first-order reactions, all having the same pre-exponential factor k_0, and activation energies in a Gaussian distribution with mean E_0 and standard deviation σ,[1] gives

$$k = k_0 \left[\frac{\int_0^\infty \exp\left\{ -\frac{k_0}{m} \int_{T_0}^T e^{-E/RT} dT - \frac{(E-E_0)^2}{2\sigma^2} - \frac{E}{RT} \right\} dE}{\int_0^\infty \exp\left\{ -\frac{k_0}{m} \int_{T_0}^T e^{-E/RT} dT - \frac{(E-E_0)^2}{2\sigma^2} \right\} dE} \right] \qquad (10)$$

Equation 10 gives the overall single-reaction rate coefficient as defined in Eq. 1 that would be observed at temperature T, to which the coal is assumed to be heated at constant rate m from initial temperature T_0.

Values of k computed from Eq. 10 for two different heating rates using experimental values of k_0, E_0 and σ for Montana lignite[2] are plotted as the solid lines in Fig. 3. Each line extends from 1% to 99% completion of devolatilization. Within this range the values of k do not deviate significantly from the straight line on the Arrhenius plot. Thus unlike the above case of only three reactions, the present large number of reactions smooth out the transitions occurring as different reactions become active. Arrhenius parameters for the single-reaction fit for each heating rate are given in Fig. 3. The pre-exponentials and activation energies are much smaller than the multiple reaction parameters k_0 and E_0 used in the calculation. Also, at a given temperature the single-reaction rate constant for the higher heating rate is more than a factor of 10^3 larger than the one for the lower heating rate. This behavior, which is discussed elsewhere[1], is equivalent to that seen above for the three-reaction case. The main point here is that the large difference between the single-reaction Arrhenius lines in Fig. 3 does not arise from dis-agreements in data or analyses, but is a well understood consequence of the different temperature-time histories, in this case heating rates, and the inability of the single-reaction model to describe

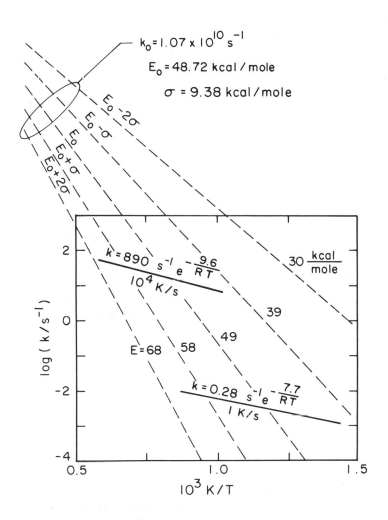

FIGURE 3. Arrhenius Plot of Global First-Order Single-Reaction Rate
Coefficient (k) for Devolatilization of Montana Lignite Described by
Multiple Parallel First-Order Reactions with Gaussian Distribution of
Activation Energies. Pre-exponential (k_o), mean (E_o) and standard
deviation (σ) of activation energy distribution from Anthony et al.[2]

multiple-reaction behavior for the different conditions with only one set of parameter values.

The overall single-reaction model was previously fitted[2] to the same data from which were also derived the multiple parallel reaction model parameters shown in Fig. 3. Three different sets of single-reaction parameters were obtained from three different sets of heating and cooling rates represented in the data, as is illustrated in Fig. 4. Also included in the figure are the two computed single-reaction cases from Fig. 3. With regard to the effect of temperature-time history, here represented by the effective heating rate m_e (Eq. 6), on the single-reaction parameters, the different sets of parameter values obtained previously using the different parts of the data separately (cases B, C and D) are consistent with those computed here (cases A and E) from the multiple parallel reaction parameters, which were peviously evaluated using all the data simultaneously. Thus the single-reaction parameters for the different cases, though differing substantially, are not in mutual conflict and can be very useful if each set is applied within its own range of conditions.

3. KINETICS OF PLASTICITY ACCOMPANYING DEVOLATILIZATION
3.1 Introduction to Plasticity Phenomena

Bituminous coals upon heating undergo melting and pyrolytic decomposition with significant parts of the coal forming an unstable liquid that can escape from the coal by evaporation to give tar and light oil or crack to form light gases, leaving behind a solid residue. Polymerication or cross-linking of the liquid molecules to form solids also occurs. All the liquid is eventually depleted, leaving a porous coke residue. The kinetics of these processes are strongly temperature dependent; therefore their relative contributions depend on temperature-time history.

The transient liquid within the pyrolyzing coal causes softening or plastic behavior that can strongly influence the chemistry and physics of the process. Bubbles of volatiles can swell the softened coal mass[3,4], in turn affecting the combustion behavior of the coal particles. Undesirable agglomeration of coal particles during combustion can occur when softened particles collide. Plasticity also affects the yields and quality of products since the transport of volatiles through the viscous coal melt affords opportunities for intra-particle secondary reactions. Figure 5 is a schematic of changes in coal physical structure and liquid volume fraction during the onset, continuation and destruction of plasticity.
3.2 Apparent Viscosity Measurement

A fast response, rapid-heating high-temperature plastometer has been developed in our laboratory[5,6]. The instrument determines the torque required for constant rotation of a thin disk embedded within a thin layer of coal heated between two parallel metal plates. Heating rates, final temperatures and sample residence times at final temperatures can be separately selected and controlled over the ranges 40-800 K/s, 600-1250 K, and 0-40 s. The coal layer thickness between the disk and each bounding plate is small (0.13 mm, or about two particle diameters for the coal particles tested) in order to achieve near isothermality across the sample layer under rapid heating conditions. By calibrating the instrument with viscous liquid standards, apparent viscosities can be obtained from the experimental torque curves.

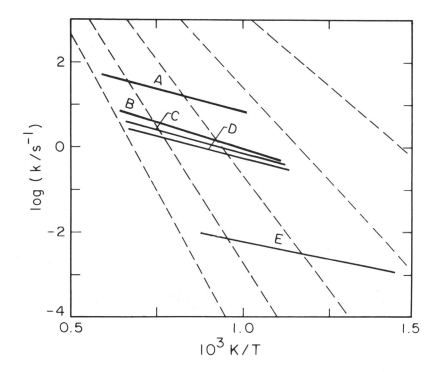

FIGURE 4. Single–Reaction Rate Coefficients for Montana Lignite Devolatilization under Different Conditions. Data, Atnhony et al.[2] Lines A and E computed from multiple parallel reaction model fitted to the data. Lines B, C and D from fitting single–reaction model directly to the data. Heating and cooling rates and Arrhenius parameters as follows:

	Heating and Cooling Rates (Eq. 6)			$E,$ $\dfrac{kcal}{mole}$	$k_0,$
Case	$m, K/s$	$m_c, K/s$	$m_e, K/s$		s^{-1}
A	10,000	∞	10,000	9.6	890
B	10,000	~ 200	196	11.1	283
C	3,000	~ 200	188	10.0	128
D	650	~ 200	153	9.0	60.7
E	1	∞	1	7.7	0.28

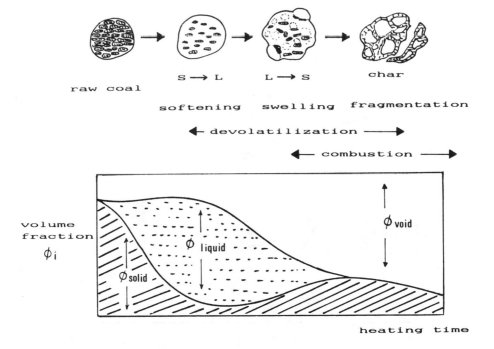

FIGURE 5. Physical Changes in Bituminous Coal during Devolatilization.

The coal studied was a high volatile Pittsburgh No. 8 Seam bituminous coal (Table 1). The coal was freshly ground under nitrogen, washed, vacuum dried at 383 K for four hours and sieved. Particles in the size range 63-75 μm were used. For the measurements presented here the coal was heated under 1 atm He.

Table 1. Composition of Pittsburgh No. 8 Seam Coal Studied (wt. %, dry basis, unless otherwise indicated)

Proximate Analysis: Ash, 11.5; VM, 39.4; F. C., 49.1; Moisture, 1.4 wt. % as-received.
Ultimate Analysis: C, 68.8; H, 4.9; O, 8.2; N, 1.3; S, 5.4; Ash, 11.5

Figure 6 shows typical viscosity curves for this coal as a function of heating time for different constant heating rates and holding temperatures. For runs at different heating rates the initial softening temperatures T_s are rather insensitive to the heating rate, indicative of physical melting. The temperature (T) at which the torque has dropped to a certain low value increases more significantly with heating rate, indicative of chemical decomposition. Apparent viscosities measured were in the range of a few thousand to one million poise. The shear rates were about 1.3 s^{-1}.

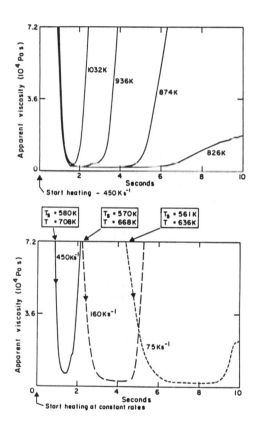

FIGURE 6. Apparent Viscosity of Coal during Heating. (Above) Coal heated at constant rate and then held at temperatures shown on curves. (Below) Coal heated for extended period at constant rate shown on curves. T_s, temperature at initial drop of torque. T, temperature when viscosity drops to 7.2×10^4 Pa s.

The apparent viscosities reflect the resistance to shear of a complex, reacting suspension of partially reacted coal, mineral matter and coke in a polymerizing viscous liquid. Description of the rheological properties of this system in fundamental terms is difficult. However, one can relate the relative viscosity to the solids volume fraction, and hence to the kinetics of liquids formation and depletion in the coal. The existence of bubbles in the melt has to be taken into account when such a correlation is made. Therefore, the interpretation of these plasticity curves is more difficult than for those of less complex liquid-solids suspensions.

Since plasticity arises from the formation of liquids within the coal continuum, valuable complementary information can be obtained by determining the amount of liquid in a rapidly pyrolyzing coal particle. A quantitative study of intra-particle liquids formation and depletion reactions was therefore performed as described below.

3.3. Liquid Formation and Depletion Kinetics

Coal samples were pyrolyzed using an electrical screen heater reactor and the liquid left in the solid residue after cooling was removed by solvent extraction. The reactor[7] was an enlarged and modified version of the one described by Anthony et al.[8] About 20 mg of 75-90 μm diameter particles of the same coal used in the plastometer were spread evenly in the central region of two layers of 10.5 cm x 5.3 cm, 400 mesh stainless steel screen mounted between two electrodes. Heating was achieved by sequentially passing two constant pulses of current of preselected duration and magnitude through the electrodes. The volatile products readily escaped from the screen and were quenched and diluted by the ambient helium gas at 0.85 atm. At the end of the preset heating interval, pressurized liquid nitrogen was sprayed onto the screen by automatically opening a cooling valve, resulting in a quenching rate of 1100-1500 K/s. Other runs featured cooling rates above 6000 K/s (using a different valve), or below 250 K/s (when the screen was allowed to cool by natural convection).

The weight loss of the sample was determined by weighing the loaded screen before and after the experiment. This corresponds to the volatiles (tar, light oil and gas) yield. The extract yield was determined by Soxhlet extraction of the char left in the screen using pyridine at its boiling point, drying the extracted char and screen in a vacuum oven at 383 K for four hours and reweighing. This quantity is taken as the intra-particle inventory of liquids at the time of quenching.

Figure 7 shows the evolution of the volatiles, pyridine extract, and the pyridine insoluble char as a function of the heating time, for two sets of experiments with rapid quenching. The corresponding temperature-time histories for each set are shown in the lower parts of the figure. For example, the vertical dotted lines connect two data points from a single experiment, with the temperature-time history indicated by the bold line. The solid curves near the data points are from a kinetic model described below.

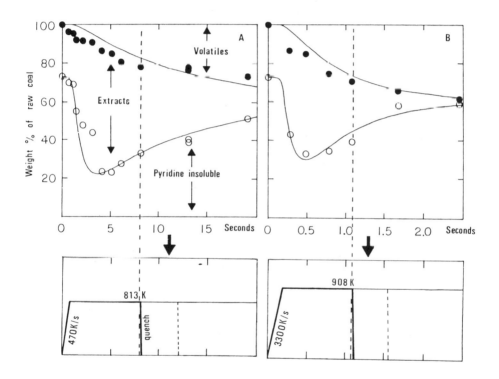

FIGURE 7. Effects of Temperature-Time History on Yields of Volatiles, Extract, and Pyridine Insolubles.

An extract yield of about 25% was obtained from the raw coal.   As
the coal was heated, the amount of extract increased to a maximum and
then decreased, the rate of this process being strongly dependent on
temperature, as indicated in Fig. 8.   The time interval over which the
extract concentration was high (> 30 wt % of coal) decreased rapidly
with the final holding temperatures of the experiments.   The maximum
amount of fluid material (extract plus volatiles) was as high as 80%
by weight of the raw coal.   However, this quantity decreased with
increasing holding temperature, or with increasing holding time at a
given holding temperature, primarily due to conversion of the extractables to coke for part of the reaction cycle, the weight average
molecular weight of these extracts has been found to increase in
parallel with these yield decreases[6] (see also Fig. 10).   This implies
that polymerization is an important pathway in extract depletion and
coke formation.   Evaporation and escape from the coal also contributes
to extractables depletion, as noted above, thereby presumably increasing the extractables molecular weight.

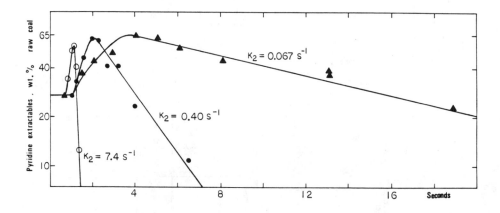

FIGURE 8.   Extract Generation and Depletion.   Heating rate and holding
temperature: triangles, 470 K/s and 813 K; solid circles, 446 K/s and
858 K; open circles, 514 K/s and 992 K.   Cooling rate: all cases, 1100
K/s.   First-order rate constants for extract depletion, indicated on
curves.

As shown in Fig. 9, the respective times for initial generation and near final depletion of extractables within the coal, corresponds rather well with the times for initiation and termination of the coal's plastic phase. This supports our assumption that the pyridine extract of the rapidly quenched coal is a reasonably accurate measure of the amount of liquid product within the pyrolyzing coal at the time of quenching.

However, the fact that even without heating, 25 wt % of the raw coal can be extracted, means that "extract" and "liquids" are not identical materials. The dried coal is a solid at room temperature and contains no organic liquid. It is quite likely that the initial 25% extract corresponds to some loosely bonded, low melting point material. Upon heating to a sufficiently high temperature, this material undergoes physical melting. This would be consistent with the fact that initial softening temperatures recorded in our plastometer show little effect of heating rate. Other workers found the Gieseler plasticity during the early softening stage to be reversible as the temperature is raised or lowered[9]. At higher temperatures, pyrolytic bond breaking generates significant quantities of liquid which cannot be reversibly converted back to resolidified coal.

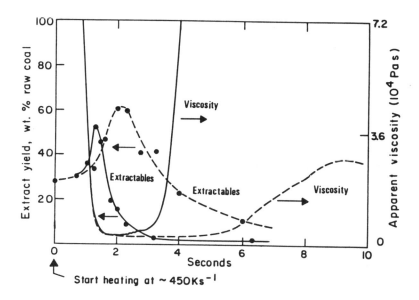

FIGURE 9. Comparison of Viscosity and Extract Curves. Heating rate: 450 K/s. Holding temperature: solid curves, 910 K; dashed curves, 858 K.

92

It is thus clear that better understanding of the reactions
occurring within the coal can shed light on the underlying mechanisms
of coal plasticity. An important measurement is the molecular weight
distribution of the extract as a function of the extent of pyrolysis.
A two-column gel permeation chromatography (GPC) system calibrated
with different size fractions of coal tars was employed. Figure 10
shows the weight average molecular weight of the pyridine extract at
different heating times for the bituminous coal and a heat-treated
coal tar.

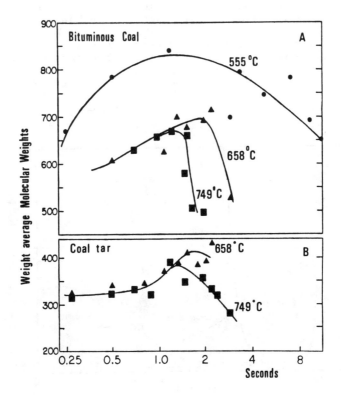

FIGURE 10. Molecular Weight of Extracts for Different Temperature-
Time Histories. Holding temperatures indicated on curves.
Heatingrate: circles, 736 K/s; triangles, 465 K/s; squares, 455 K/s.
Cooling rate: circles, ~200 K/s (natural cooling); triangles and
squares, 5000 K/s.

The trends are strong functions of temperature-time history. In general, with increasing time at a given holding temperature, the molecular weight increases to a maximum, and then decreases. The increase is due to evaporation of the lighter components, and polymerization reactions. The decrease apparently arises from secondary cracking of the liquids since further generation of lower molecular weight liquids by primary decomposition of the parent coal is unlikely at these severities (see Figure 7). At higher holding temperatures, more severe cracking drives the molecular weight to even lower values. The maxima of these curves are thus determined by competition between evaporation and cracking reactions. If the cracking reactions have higher activation energies, then increasing the holding temperature would enhance their relative importance and thus decrease the value of the maxima in the curves, which is consistent with our data.

The molecular weight of the coal extract is much higher than that for the coal tar extract. This probably reflects two differences expected in their thermal behavior because of differences in their structure and initial state. Cracking of the high molecular weight coal extract, especially in the presence of surfaces of mineral matter, unsoftened macerals and already available coke, can give more residual coke. A larger liquid molecule also has a lower vapor pressure, and thus a slower evaporation rate. Both are consistent with the lower pyridine insoluble yield and the faster evaporation rate observed in the coal tar pyrolysis. More detailed kinetic interpretation of these GPC data will be undertaken in the future.

3.4. Mathematical Modeling of Plasticity Kinetics

The kinetics data on liquids generation and depletion are fitted with a model in which softening coal is assumed to undergo both physical melting and pyrolytic bond breaking to form a liquid intermediate, metaplast. This liquid is assumed to crack to form a volatile product which escapes from the coal and a heavier component that quickly repolymerizes into coke. The weight ratio of coke to volatiles thus formed is assumed to be constant (=a). The material that can and does undergo physical melting is assumed to be identical with the pyridine extract from the unheated coal. This material is assumed to melt over a narrow range of temperatures. In the present calculations, a Gaussian distribution of melting points with a mean at 623 K and standard deviation of 30 K is assumed. This temperature range is where initial softening of coal is observed. The rate of generation of liquids by melting would therefore be the product of the heating rate and the distribution function. The ash (weight fraction f_a) is assumed to be inert. The reaction scheme and rate expressions are,

$$\text{coal} \xrightarrow{k_1} \text{metaplast} \xrightarrow{k_2} \text{volatiles} + \text{coke}$$

$$\frac{dC}{dt} = -k_1 (C - f_a) - r_m \tag{11}$$

$$\frac{dM}{dt} = k_1 (C - f_a) + r_m - k_2 M \tag{12}$$

$$\frac{dV}{dt} = k_2 M (1/1 + a) \tag{13}$$

$$\frac{dE}{dt} = k_2 M \ (a/1 + a) \tag{14}$$

where C, M, V, E are the weight fractions of the unreacted coal, metaplast, volatiles, and coke, respectively, and r_m is the rate of physical melting. At a late stage of pyrolysis when most of the coal has reacted and all the material that can physically melt has done so, the rate of liquid depletion is

$$dM/dt - k_2 M \tag{15}$$

The rate coefficient k_2 can thus be found from a plot of $\ln(M)$ vs time for constant temperatures. Typical values were obtained from Fig. 8, and from an Arrhenius plot k_2 is found to be

$$k_2 = (1.9 \times 10^{10} \ s^{-1}) \ \exp(-21200 \ K/T) \tag{16}$$

The rate coefficient k_1, obtained from a best fit to the experimental data using Eqns. 11-14 and 16, is

$$k_1 = (6.6 \times 10^7 \ s^{-1}) \ \exp(-14500 \ K/T) \tag{17}$$

Values of C and M observed in the screen heater runs, and corresponding predictions from Eqns. 11-14, using Eqns. 16 and 17 and an "a" value of 1.25 (estimated from long holding time data) are compared for two sets of reaction conditions in Fig. 7.

3.5. Physical Processes during the Plastic Stage

Reliable treatment of intra-particle liquids formation and depletion kinetics is essential for mathematical modeling of process phenomena occurring during the plastic stage. To illustrate one application, a preliminary model of the agglomeration of pyrolyzing plastic coal particles entrained in an isotropic turbulent flow field is presented. Such flows might arise in dense phase fluidized beds, or in certain types of coal feeding operations.

The agglomeration of a system of particles with a monodisperse particle size distribution is modeled as a binary collision mechanism:

$$dn/dt = - K_c \xi n^2 /2 \tag{18}$$

$$n(o) = n_0 \tag{19}$$

where K_c is the collision frequency, ξ is the probability of forming an agglomerate, n is the number of particles per unit volume at any time t, and $n_0 = n$ at $t = 0$. Permanent sticking of two particles to form an agglomerate is assumed to occur as follows:
(i) Particles collide with a frequency[10]

$$K_c = 1.3 \ d^3 (\varepsilon_0 /\nu)^{1/2} \tag{20}$$

where ε_0 is the turbulent energy dissipation rate, ν the kinematic viscosity of the gas, and d the particle diameter.

(ii) The particles are assumed to be irregularly shaped with a mean d. Two particles may approach at many different orientations resulting in a large number of possible contact areas. For simplicity, the agglomerate is considered as two distorted spheres with a neck of diameter x at the plane of contact. The quantity $c_i = x/d$ is assumed to be normally distributed with mean $\bar{c_i}$ and standard deviation δ.

(iii) The strength of the neck region is assumed to be

$$\sigma_n = \sigma_0 M \qquad (21)$$

where σ_0 is a constant (i.e. only the liquid portion of the contacting surface is assumed to adhere).

(iv) The madimum stress imposed by the flow field on a pair of particles is the bending stress[11]

$$\sigma_b = 10.8\mu_{gas} (\varepsilon_0/\nu)^{1/2}/c_i^3 \qquad (22)$$

(v) If σ_b is larger than σ_n, then no agglomeration results ($\xi = 0$); otherwise $\xi = 1$. Thus the critical c_i for agglomeration is

$$c_i \geq 2.21 (\mu_{gas}/\sigma_0 M)^{1/3} (\varepsilon_0/\nu)^{1/6} \qquad (23)$$

Given a temperature-time history for the coal particles and the kinetic parameters for liquid formation and depletion, the instantaneous liquid content of the agglomerating particles can be calculated. Summing over all agglomerates which satisfy the condition in step (v), an overall efficiency of agglomeration can be obtained. The overall efficiency is a function of the liquid content, the intensity of turbulence, and the distribution of c_i. Once the energy dissipation rate, ε_0, and the initial number density, n_0, are specified, the binary agglomeration equation can be solved.

To account for particle swelling, a single volatile bubble is assumed to grow in a viscous, spherical droplet. The viscosity of the coal melt is related to the solids fraction by a concentrated suspension model. The volatiles in the bubble are assumed to be in equilibrium with, arbitrarily, 0.8 mole fraction of the liquid.

The liquid formation and depletion rate parameters presented above are used. An initial particle number density of $1000/cm^3$, and particle radii of 100 μm are assumed. The particles are heated at 10^4 K/s to a holding temperature of 1000 K. Figure 11 shows the liquid content, swelling ratio, and number of unagglomerated particles for two different levels of turbulence intensity, all as functions of time, with heating started at t=0. The predictions are qualitatively consistent with data of Tyler[12] and McCarthy[13].

Extension of the model to multi-particle agglomeration is possible. Calculations with 3 or 4-particle agglomerates show the same trends as in Fig. 11. If the particles are entrained in a laminar flow, similar expressions can be derived. Additional development of the model will be presented in the future.

96

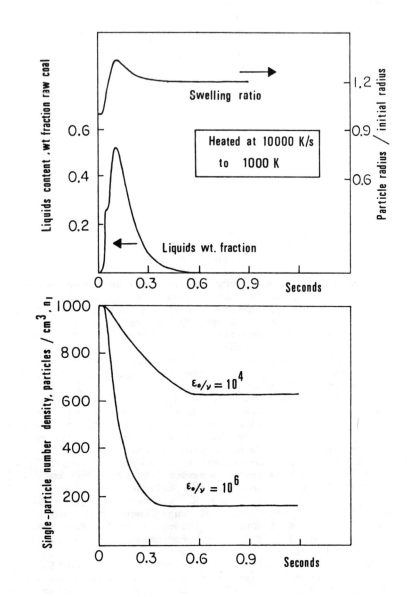

FIGURE 11. Predicted Swelling and Agglomeration Behavior of Softened
Coal Particles Entrained in Turbulent Flow Field. ε_0/ν = turbulent
energy dissipation rate/kinematic viscosity, in s^{-2}.

ACKNOWLEDEMENTS
     Financial support of this work by the United States Department of
Energy under Contracts No. DE-AC21-82MC-19107 and DE-AC22-84PC-70768
is gratefully acknowledged. We also wish to acknowledge the contributions of Dr. Jean-Louis Saint-Romain to studies on heat treated coal
tar, and of Mr. Jaanpyng Hsu to mathematical modeling of agglomeration
kinetics.

REFERENCES

1. Howard JB: Fundamentals of Coal Pyrolysis and Hydropyrolysis.
   Chapter in Chemistry of Coal Utilization, Second Supplementary Volume, Elliott MA (ed), pp. 665-784.   New York: John
   Wiley and Sons, 1981.
2.   Anthony DB, Howard JB, Hottel HC, and Meissner HP: 15th
     Symp. (Int'l) on Combustion, p. 1303,   The Combustion
     Institute, 1975.
3.   Oh MS: "Softening Coal Pyrolysis". Sc.D. Thesis, Department
     of Chemical Engineering, MIT, Cambridge, Massachusetts,
     1985.
4.   Oh MS, Howard JB, and Peters WA: "Modeling Volatiles
     Transport in Softening Coal Pyrolysis", AIChE National
     Meeting, San Francisco, November, 1984.
5.   Fong WS, Peters WA, and Howard JB: Rev. Sci. Instrum. 56,
     586 (1985).
6.   Fong WS, Khalil YF, Peters WA, and Howard JB: Fuel 65, 195
     (1986).
7.   Fong WS, Peters WA, and Howard JB: Fuel 65 251 (1986).
8.   Anthony   DB, Howard JB, Meissner HP, and Hottel HC: Rev.
     Sci. Instrum. 45, 992 (1974).
9.   Waters PL: Fuel 41, 3 (1962).
10. Saffman PG, and Turner JS: J. Fluid Mech. 1, 16 (1956).
11. Yoshida T, Kousaka Y, and Okuyama K: Aerosol Science for
     Engineers. Tokyo: Power Co. Ltd, 1979.
12. Tyler RJ: Fuel 58, 680 (1979).
13. McCarthy DJ: Fuel 59, 563 (1980).

DISCUSSION

E. Suuberg

1. The decrease in molecular weight of extractables with increasing temperature has been attributed to cracking, but it should be pointed out that it may well be a consequence of polycondensation processes as well. Polycondensation theory predicts decreasing molecular weight of extractable with increasing degree of condensation.

2. The ineffectiveness of pyridine in extracting the lignite is surprising. Has a substantially better solvent been identified ?

3. Is the implication of the simple kinetic modelling you showed at the outset that no one set of data at one heating rate should generally be used for deriving kinetic constants? In terms of comparing different sets of data obtained under wildely varying temperature/time conditions, is one not in trouble comparing entrained flow reactor results with heated grid results, because the former involves considerable secondary reactions not present in the latter? Thus would it be fair to suggest that the heated grid type systems, which have the ability to be used over a wide range of temperature, time conditions, might be the best choice for further kinetic work? The main disadvantage of these systems might be the inability to access high temperatures (> 1500°C) at high heating rates (> 10^5 c/s), conditions under which it is suggested that other mechanisms might come into play.

J.B. Howard

1. One would indeed expect both polymerization and cracking to contribute to the change in molecular weight of the extractable material. In addition, evaporation and diffusion contribute by the preferential removal of lower molecular weight material. In order to monitor and to rationalize the overall change in molecular weight distribution during the heating of a coal, it is important to study both the material leaving the coal as well as the material remaining within the particles, and to consider all three of the processes mentioned above.

2. It is well known from the literature that the effectiveness of pyridine decreases as coal rank decreases. For lignites, pyridine is indeed ineffective, and the following order of effectiveness is reported (D.W. van Krevelen, "Coal", Elsevier, New York, 1961) : diethylenetriamine > monoethanolamine > ethylenediamine > pyridine.

3. One set of good data at one heating rate can be useful for making predictions at other heating rates if a suitable model, such as the multiple parallel reaction model, is used. The single reaction model, or a two-reaction model, would not be suitable for this purpose. If possible, it is preferable to have data for a range of heating rates. In answer to your question about flow reactor data, I agree with you that much

care is needed, including the taking into account of secondary
reactions, in comparing such data with screen-heater results.
The screen heater technique does have some strong advantages,
as you point out. Not only can temperature-time history,
including heating rate, be varied over considerable ranges as
is desirable in kinetics studies, but temperature can be measured more accurately than is generally possible with presently available flow reactor techniques.

H. Kremer

Is there an influence of the ambient atmosphere on the pyrolysis reactions if one thinks, for example, of large coal
particles in fluidized bed combustors?

J.B. Howard

Probably not in the case of coal particle in a fluidized
bed combustor. However, such influences do occur when pulverized coal is heated in the presence high-pressure hydrogen,
which is known to penetrate into the coal and to redirect the
pyrolysis product pathways (see Ref. 1 of the paper). If
pulverized coal is preheated in the presence of air at 1 atm.
to a temperature of 150 to 200°C for a few minutes, the yields
of products when the coal is subsequently pyrolyzed at higher
temperature are significantly affected. Based on this fact one
might expect that the pyrolysis occuring after pulverized coal
in a flame is heated to of order 1000°C in 0.1 s might be
affected by the oxygen of the combustion air in contact with
the coal. Whether such an effect actually occurr has not been
determined. In the absence of better information it is reasonable to neglect such an effect in pulverized coal combustion.

K.H. Van Heek

Uur experiments on the swelling of single coal particles
(Fuel, 65, 1986, p. 346) show that whether a particle shape
goes through a maximum depends on the possibility for the
volatiles to escape before the outer shell of the particle
solidifies. That means, all conditions inhibiting transport
(pressure, particle diameter and heating rate) lead to the
fact that the particle stays on its maximum values when these
parameters are increased. Especially for heating rates as
expected under fluidized bed or entrained phase conditions the
maximum swelling size has to be taken into account for the
char grains. These results have been found for vitrain particles, whereas durain particles show a shrinking only.
My question is, to what extent reactor conditions and maceral composition of the coal can be included in the model presented in your lecture.

J.B. Howard

Exploratory calculation of the effects of pressure, heating
rate and particle size with our model give qualitatively
reasonable trends. The effect of maceral composition could

presumably be addressed by proper choice of coal properties, but we have not attempted such calculations.

P.J. Jackson

In the presentation (Fig. 11), reference was made to the significance of turbulence levels, denoted 10^6 - 10^4 , etc. What are the units used here?

J.B. Howard

The units are s^{-2} , referring to the ratio of turbulent energy dissipation rate to kinematic viscosity.

R. Cyprès

You showed curves of the decrease of average molecular weight with increasing temperature. Could you tell us something about the distribution of the individual components of the tar, taking into consideration the known fact that more heavy compounds are formed with increasing temperature, by association reaction. There must be a compromise between cracking and association.

J.B. Howard

We have not measured individual compounds in the tar, but we have performed NMR analyser of the size fractions recovered from GPC, or size exclusion chromatography. The data show evidence of formation of aromatic C-H bonds from the products of cracking of aliphatic structures. In experiments such as ours where tar, defined as liquid products transported from the coal, and extract, defined as material left in the sample and subsequently removed by a solvent, are separated during the experiement, the molecular weight change of both tar and extract must be studied in order to assess the overall change. The molecular weight decrease that you refered to is the overall change that pertains to the tar and extract together.

F. Kapteijn

With regard to heat effects associated with devolitilization : At the Institute of Chemical Technology (University of Amsterdam, The Netherlands) recently the heat of pyrolysis of coals of different rank have been measured by DSC (up to 800°C) combined with TG and gas analysis.
Related to your presentation of two remarks are made:
1) The heat associate with pyrolysis is endothermic and small compared to the heats of combustion.
2) The observed endothermic effects are mainly associated with the formation of small molecules (H_2, CH_4, CO, Cl_2) and much less with the release of volatile matter ("tars").
(Presented at the Rolduc Coal Science Symposium, April 1986 and to appear in Fuel Processing Technology).

J.B. Howard

Thank you for these comments. To these remarks I would add
that pyrolysis can be exothermic, but still with a small heat
change compared with the heat of combustion of the coal, if
secondary reactions occur to a sufficient extent. The association reactions that accompany cracking are exothermic, and the
heat thus generated can more than offset the heat consumed in
the bond-breaking cracking reactions. Thus a bituminous coal
whose pyrolysis at atmospheric pressure under conditions of
minimal secondary reactions may be slightly endothermic, can
under high pressure with extensive secondary char-forming
reactions exhibit exothermic pyrolysis.

E. Saatdjian

What values do you suggest for the heat of pyrolysis? How
does it depend on coal rank? and type?

J.B. Howard

The heat of pyrolysis is generally small, only a few hundred calories per gram of coal, and endothermic. However, as
mentioned in the response to the comments of Kapteijn, the
overall pyrolysis (referring to primary pyrolysis and secondary reactions) can be slightly exothermic if the conditions
favor substantial secondary reactions. The dependence on coal
rank or type is not well known but the trend is toward somewhat higher endothermicity for lower rank coals at moderate
temperatures.

J. Lahaye

In work on the behavior of liquid tars, tars are characterized by their viscosity i.e. by a rheological property. The
behavior of liquid tar can also refer to a property at equilibrium : surface tension. Don't you believe that it would be
useful to introduce surface tension parameter in models
describing liquid tars?

J.B. Howard

I agree that surface tension should be considered along
viscous forces in the analysis of molten coal particle behavior. Our mathematical modelling tar in fact included surface
tension, but according to our calculations the effect of
surface tension on bubble escape rate is small compared to
that of the viscous forces. However, better property information and/or a focus on other aspects of molten coal behavior
could result in a larger importance being seen for surface
tension.

H. Jüntgen

I can really confirm all your considerations on the significance of taking into account secondary reactions in kinetics

of coal pyrolysis. You have reported on changes of overall yields of tar on heating rate. Have you also investigated the chemical composition of tar dependent on heating rate? The second question is regarding the metaplast formation: What is the definition of metaplast in a chemical sence? Is this reaction step necessary for modelling? What is the role of mobile phase in metaplast formation?

J.B. Howard

We have not investigated the chemical composition of tar as a function of heating rate. Regarding the definition of metaplast, we have simply adopted this term as it is commonly used in the coal literature. Our need for a quantitative relationship between viscosity of the molten coal and time and temperature led us to describe the molten coal as a solid (undecomposed coal) in a liquid which is formed and consumed by reactions whose kinetics we can model. This liquid, assumed to be approximated by metaplast is formed partly by physical melting early in the heating of some coals, and by chemical breaking of crosslinks in the coal structure.

H. Jüntgen

Can we define metaplast very simply by species in coal grain formed by chemical degradation minus species of this kind transported away at this time of observation?

J.B. Howard

Yes. This definition is the one we have been using at MIT.

J.H. Pohl

Please comment on the status of qualitatively predicting the distributions of gases, soot, tar (if any) and solids at the conditions of pulverized coal flames. Also, to what extent can the rates of devolatilization and product distribution be quantitatively estimated at pulverized coal flame conditions.

J.B. Howard

The behavior of coal for which the product distribution have been measured int the laboratory at temperatures and heating rates approaching those of pulverized coal flames can probably be predicted to within perhaps 20 %. In our own work the accuracy with which product yields are predicted within the ranges of experimental conditions covered in the laboratory is 10 to 15%. Product yields for pulverized coal flame conditions are not available, so one can of course, only speculate on the accuracy with which a model that is good to within 10 to 15% in the range of conditions studies can be extended to the flame conditions. One encouraging observation is that the tar yields predicted for flame heating rates and temperatures are close to the soot yields actual-

ly observed for there conditions in the absence of oxygen,
which they should be.

T.F. Wall

One advantage of multiple reaction schemas in combustion
models is that each pyrolysis product may be associated with a
rate equation. The progressive total heating value of the volatile matter may then be calculated by summation. For more
simple kinetic schemas how would you recommend that the estimation of this heating value be made ?

J.B. Howard

The use of a highly simplified kinetic schema does not justify a detailed description of volatiles heating value. A
simple approach that would probably be within the error limits
of the kinetic model would be to assume that the heat of
pyrolysis is zero and that the original heat of combustion of
the coal is distributed between the volatiles and the char in
proportion to the d.a.f. masses of these two global products.
Thus, at any stage of devolatilization the rate with which
fuel heating value is carried from the coal in the volatiles
could be calculated from the simple kinetic model being used.

J.M. Beér

You have stressed the importance of viscous forces in the
formation of the cenospheric char particle. The plastic coal
material's viscosity will be affected by the shear rate as it
is treated as a non Newtonian slurry. The shear rate could be
calculated from the kinetics of cenosphere formation. Do you
think that measurements of the effective viscosity might be
made at the high shear rates so estimated and be used in the
model?

J.B. Howard

We have done some exploratory measurements of effective
viscosity at different shear rates, but not enough to draw
quantitative conclusions. More such measurements would indeed
be valuable, and the results could be used in the model.

J. Lahaye

With respect to the questions of Professors Juntgen, Suuberg and Cypres on molecular structure of tars, I hope you do
no mind if I mention that in the frame of a joint programme
C.R.P.C.S.S. CNRS/MIT on pitch used as binder, we are comparing results obtained by Gel Permeation chromatography, by
G.C.M.S. (Gas chromatography coupled to Mass Spectrometry) and
Differential scanning calorimetry. These methods and comparisons might be applied to tars produced during devolatilization.

A MICROSCOPIC AND KINETIC STUDY OF COAL PARTICLE DEVOLATILIZATION IN A
LASER BEAM

MARTIN HERTZBERG And DANIEL L. NG

U. S. Department of the Interior
Bureau of Mines, Pittsburgh Research Center
Pittsburgh, Pennsylvania 15236

1. ABSTRACT

  Data are presented for the devolatilization weight loss of particles
of Pittsburgh seam bituminous coal (36 pct volatility by the ASTM
method) exposed to a laser beam in a nitrogen environment. The
devolatilization rate was measured as a function of input laser flux,
which was varied between 75 and 400 W/cm^2., and as a function of
particle diameter, which was varied between 51 and 310 μm. The measured
"half-life" for the devolatilization process was directly proportional
to the particle diameter and inversely proportional to the absorbed
flux. This new experimental method, of direct weight loss measurements
for single particles or small arrays of particles, permitted the
devolatilization data to be correlated directly with scanning electron
microscopic observations of the morphological changes resulting from
each exposure to the laser heating flux. For coal particles exposed to
the laser flux in an air environment, essentially the same
devolatilization process is observed initially; however, when
devolatilization is complete, it is followed by a heterogeneous char
oxidation process which occurs on a much slower time scale. From the
totality of the data on morphological records, particle size, and flux
dependence, it appears that the reaction mechanism for devolatilization
involves the simple, inward progression of a laminar
pyrolysis-devolatilization wave from the exposed surface. Simple
thermodynamic transport constraints require that the pyrolysis wave
velocity be determined by the sum of the absorbing heat flux necessary
to raise each element of the solid to its devolatilization temperature
(500° to 600° C), plus the flux required to supply the heat of
devolatilization. The reaction rate for a particle is then simply
controlled by the time required for that devolatilization wave to
traverse through it, and that time is predicted to be directly
proportional to the particle diameter and inversely proportional to the
absorbed flux, as is observed.

2. INTRODUCTION

  An accurate knowledge of the mechanism and rate of coal particle
devolatilization during thermal pyrolysis is important in all stages of
coal technology, from mining the coal to its end use in power generation
and industrial production. The coal mining industry has experienced
both explosions and coal mine fires; such accidents are tragic wastes of
human and material resources [1-3]. Such explosions and fires are
propagated by an overall reaction mechanism whose first step involves
the heating and devolatilization of coal either as a fine dust particle
in the case of explosions [4,5] or as a larger macroscopic surface in
the case of fires [6].

The combustible volatile content of a coal is the major factor
controlling the explosivity of its fine dust [4], and it is the
devolatilization process that limits or controls the flame propagation
rate at high dust concentrations and through coarse coal particles [5].
Most of the coal produced in the world is eventually pulverized and
burned in boilers, furnaces, and kilns, and the stability of the burner
flames that power those systems depends on the particle devolatilization
rate and its combustible volatile yield. The same pyrolysis process
also plays a major role in the production of coke, where the objective
is to maximize the char yield and porosity of the carbon product [7].
The rate of char burnout in boilers depends on microscopic porosity, and
that micropore structure of the char is generated during the
devolatilization process in the burner flame that powers the boiler [8].
Several new technologies for the efficient gasification and liquefaction
of coal also depend on the devolatilization process as a critical first
step.

Industrial operations such as those just mentioned involve such
large-scale systems and such varied dynamic interactions that it is
usually impractical to isolate or accurately control any single,
fundamental variable in order to study its intrinsic effect on system
performance. Such variables are, however, readily controlled in
laboratory systems, and the data have shown that the basic parameters
involved in the devolatilization of coal under thermal pyrolysis are
convective and/or radiant heat flux, exposure temperature, flow
structure, particle size, rank of coal, oxidizing or reducing
atmosphere, pressure, exposure time, and exposure path. Several reviews
of the fundamentals of thermal decomposition, pyrolysis, and char
oxidation have summarized the methods used by previous investigators and
their results [7-9]. The laboratory techniques that have yielded the
most data are the laminar flow furnace [10], the flat flame reactor
[11], and wire mesh heating [15].

One motivation for conducting this research was the desire to obtain a
better understanding of dust explosion phenomena. The mechanism of
flame propagation in heterogeneous dust-air explosions is substantially
more complex than that for homogeneous mixtures. For homogeneous
methane air, only the normal sequence of gas phase, hydrocarbon flame
oxidation reactions is involved; however, for coal dust flames those
reactions are preceded or paralleled by several additional processes.
Those processes are particle heating, particle devolatilization, and
mixing of volatiles in the air space between particles. Those
additional processes occur on two microscopic scales that are irrelevant
for homogeneous systems. The first microscopic scale involves the
particle itself and the necessity of resolving the chemical reactions
and the thermal, mass, and momentum transport processes within its own
diameter. The second relevant microscopic scale involves the distance
between particles, a scale that is necessary to resolve the cooperative
interactions of those processes in the space between adjacent particles.
This research will focus on that first microscopic scale: The particle
diameter and its internal structure during devolatilization.

Recent optical studies of coal particle devolatilization in flat flame
reactors [12,13] have given a realistic picture of the particle
pyrolysis, devolatilization, and combustion process. The structure data
obtained here with a Scanning Electron Microscope (SEM) provide

additional data with which to obtain further insights into the dynamics of the devolatilization process and its microscopic realities. Current concepts used to describe the devolatilization kinetics are generally based on the assumption that the process is chemically rate-controlled [9,10,20]; hence, they tend to ignore such microscopic structure factors. As will be shown, the data reported here do not support the assumption of chemical rate control; instead they suggest that the pyrolysis-devolatilization process is controlled mainly by the constraints of thermodynamic transport. An approximate steady-state model is presented which is consistent with the rate data and their particle size and flux dependencies.

3. EXPERIMENTAL PROCEDURES

The laser pyrolysis technique has evolved as a means of studying heterogeneous combustion processes [16-18]. Since coal particle devolatilization is the first step in the flame propagation mechanism of dust explosions, it was logical to use the laser pyrolysis technique to try to focus on that process by itself. In a propagating flame front, the enthalpy flux density that activates the unburned mixture ahead of the flame is given by $S_u c \rho (T_b - T_u)$, where S_u is the burning velocity, c is the heat capacity, ρ is the density of reactant mixture, and $T_b - Tu$ is the temperature difference across the flame front. For coal dust particles entering a coal dust-air flame at its maximum burning velocity, the corresponding heating flux is in the range of 150 W/cm^2. The time available for devolatilization is α/S_u^2, where α is the effective diffusivity of the flame zone. For a limit burning velocity of S_u = 3 cm/sec, the maximum devolatilization time is 60 msec [5]. It is possible to simulate the flame fluxes and exposure times for the coal pyrolysis and devolatilization by using a laser heating flux whose magnitude is comparable to the flame heating flux. Operating in a pulse mode controls the exposure time to match the residence time within the flame front. The experimental arrangement is shown in fig. 1. Details of the positioning of particles, substrate, and laser beam are shown in the fig. 1 insert. The laser used to supply the heating flux was a model 42 CO_2 laser manufactured by Coherent Radiation and operating at its normal wavelength of 10.6 μm. The initial mass of coal particles and their corresponding weight losses during pyrolysis were determined with an electronic balance, Perkin-Elmer, Model AD-2. The microscopy of the coal and char particles was investigated with a JEOL, U3 scanning electron microscope. The laser was operated in the TEM_{00} mode with an unfocused beam of 8-mm, nominal diameter. The laser power was measured with a Coherent Radiation Model 201 power meter. The power meter was modified by replacing its entrance cone with a water-cooled, circular slab of copper having a central aperture of 5-mm as shown in the insert in fig. 1. The aperture allowed only the central portion of the gaussian beam with its more uniform power density to reach the power meter. The sample particles were located well within the central aperture. The average input power flux density at the sample was obtained by dividing the power meter reading by the area of the aperture. The modified power meter-sample holder was then mounted on a movable stage for easy alignment in the center of the laser beam.

Sample coal particles to be studied were placed on a substrate that fit over the 5-mm aperture. Most of the data were collected with the sample immersed in a flow of nitrogen which prevented oxidation of the

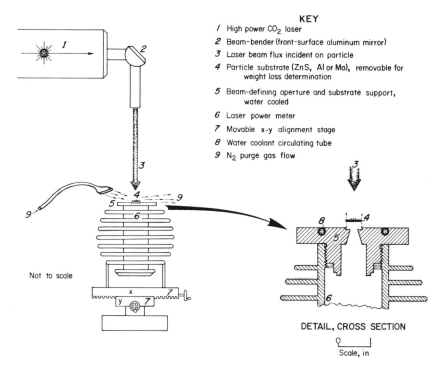

KEY

1 High power CO_2 laser

2 Beam-bender (front-surface aluminum mirror)

3 Laser beam flux incident on particle

4 Particle substrate (ZnS, Al or Mo), removable for weight loss determination

5 Beam-defining aperture and substrate support, water cooled

6 Laser power meter

7 Movable x-y alignment stage

8 Water coolant circulating tube

9 N_2 purge gas flow

Not to scale

DETAIL, CROSS SECTION

Scale, in

FIGURE 1. - The experimental configuration for studying coal particle devolatilization in a laser beam.

char residue or the emitted volatiles. Some data were also obtained in air as the char residue was allowed to oxidize almost to completion during its exposure to the laser beam flux.

The substrate material had to withstand the spatial thermal stresses and temporal thermal shocks produced by the laser flux. It had also to be very light in weight so that the initial sample mass of coal particle(s) could be accurately determined by difference, and the devolatilizing weight had to be measurable within the noise, drift level, and reweighing accuracy of the electronic microbalance. Thin Pyrex glass substrates were not suitable because they strongly absorbed the 10.6 μm radiation and shattered too easily. Three types of substrate were found to be suitable: Zinc sulphide, aluminum (Al) foil, and molybdenum foil. The ZnS is suitable because it is highly transmissive at 10.6 μm, whereas the metal foils were suitable because they were almost totally reflecting. Either property produced the desired result of little energy being absorbed in the substrate to produce thermal stresses or shocks that resulted in its destruction. Some limited breakage still occurred with ZnS, especially at higher fluxes and longer exposure times. With the Al substrate, damage occurred only if the substrate temperature reached the foil's melting point. Molybdenum was used for pyrolysis experiments with inhibitors whose decomposition products were highly corrosive. Results of those studies are reported elsewhere [21].

Three coal particle sizes were studied. The coarsest diameter coal particles were 45 x 50 mesh with a mean diameter of 310 μm. This coarse particle was easily manipulated and transferred on or off its substrate by means of an electrostatically charged synthetic fiber hair. The smaller sizes studied had surfaced-area-weighted mean diameters of 105 μm and 51 μm respectively. The detailed size distributions for the finer dusts were measured with a Coulter Counter, and their approximate distributions have already been published [5]. The coal used was the same Pittsburgh seam bituminous (36 pct volatility) whose properties were described earlier [4]. The particles were air-dried at room temperature, but otherwise no particular effort was made to remove the intrinsic moisture content (1 pct).

For some of the experiments with 310 μm particles, single particles were used. More often, a small array of dispersed particles was used and located centrally in the laser beam. They were dispersed with a fiber and individually separated initially. Typically, the mass of the particle sample was between 200 and 310 μg for a microbalance whose sensitivity was 0.1 μg. The experimental weight-loss method had a reproducibility within 2 μg.

4. DEVOLATILIZATION RATE DATA

The data for the weight loss of 105 μm coal particles as a function of heating time at various laser power densities are shown in fig. 2. All

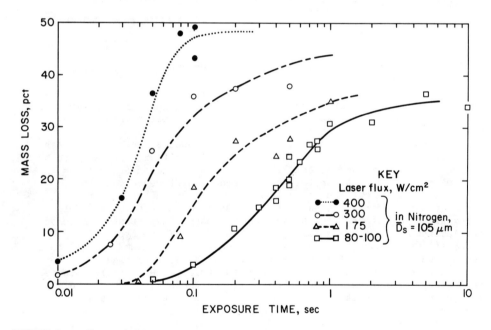

FIGURE 2. - The weight loss as a function of exposure time for 105 μm diameter Pittsburgh seam bituminous coal particles exposed to a laser beam in a nitrogen environment. The devolatilization weight loss is shown at several laser flux levels that are representative of normal flame heating fluxes.

exposures were in a nitrogen atmosphere; the measured weight losses
represent only the evolution of volatiles since there is no char
oxidation in such an atmosphere. The curves have characteristic
S-shapes. In the early stages there appears to be an induction period,
which is followed by a linear portion of significant devolatilization,
which eventually levels off to some final volatility yield of $V(\infty)$ as
$t \rightarrow \infty$. For the lower flux level of 80 to 100 W/cm the final volatility
was 35 pct, which is in good agreement with the value measured by the
standard ASTM proximate analysis. High flux levels, however, gave final
volatility yields that were in excess of that "standard" value, as has
been observed by others using different experimental methods [9,10,14].
Those higher $V(\infty)$ values are also apparent from the data in fig. 3,
where the mass loss results are plotted as a function of exposure time
for various particle diameters at a constant laser flux level of
300 W/cm^2. The coarsest particle ($\bar{D}_S$ = 310 μm) gave a final volatile
yield of 45 pct, and the finest particle ($\bar{D}_S$ = 51 μm) gave a $V(\infty)$-value
of 50 pct. The highest yield measured in these experiments was 55 pct
for the 51 μm diameter particles in nitrogen at a flux level of 400
W/cm^2.

    After the induction period, the shapes of the weight loss curves in
figs. 2 and 3 are reasonably well described by the equation

$$V(t) = V(\infty)[1 - e^{-t/\tau}] \tag{1},$$

where $V(t)$ is the volatile yield for any exposure time t, and $V(\infty)$ is
the maximum volatile yield as $t \rightarrow \infty$. In a nitrogen atmosphere, the mass

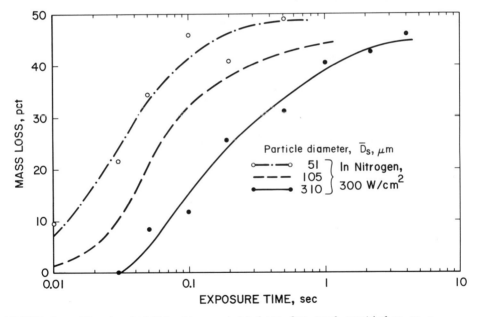

FIGURE 3. - The devolatilization weight loss for coal particles as a
function of exposure time for three particle sizes at a constant laser
flux of 300 W/cm^2.

loss ordinate is identical with the volatile yield, $V(t)$, and is given by $100[m_0 - m(t)]/m_0$, where m_0 is the initial mass of the sample at $t = 0$ and $m(t)$ is the sample mass after an exposure time of t. The constant τ in equation (1) is the characteristic time constant for the evolution of volatiles. It is proportional to the "half life" for volatiles evolution, such that $\tau = t_{1/2}/\ln 2$.

The argument is usually made that such data fit the standard first order rate equation

$$\frac{dV(t)}{dt} = k[V(\infty) - V(t)] \tag{2},$$

and that, accordingly, the devolatilization rate is chemically controlled by an isothermal, unimolecular rate process. While it is true that the integral of equation (2) gives equation (1) (if the unimolecular rate constant is replaced by its reciprocal, $1/\tau$), the data reported here do not support the argument that the process is chemically controlled under isothermal conditions. As will be shown, many other rate-limiting factors or processes can approximate the time-dependent behavior observed in figs. 2 and 3.

All the data obtained in these experiments for the laser flux and particle diameter dependence are summarized in fig. 4. The data points are the devolatilization half-lives, $t_{1/2}$, plotted as a function of laser flux for the three particle sizes studied. The curves representing theory will be discussed later. The data points show clearly that the devolatilization time is inversely proportional to some net flux level, which, as will be shown, is the difference between the input laser flux and some loss flux. The data also show that the devolatilization time is directly proportional to the particle diameter. The particle size dependence measured here is more pronounced than those previously reported [9] and is directly attributable to a superior experimental method, which allows for better intrinsic control of that variable than is available with other methods.

Those other methods involve furnace exposure [10] or wire-mesh heating [15]. An insight into the problems involved in isolating the particle size variable is provided by the following observation. It was not possible, even in these laser pyrolysis experiments, to obtain reliable data for particles with diameters less than 51 µm because of several experimental problems that are, as yet, unresolved. The accuracy of the experimental equipment and method used did not allow single particle weight loss measurements for such small masses. Thus, for the finer sizes, particle arrays were required for each exposure time. With such arrays, the finer particles tended to agglomerate and it was difficult to space the arrays evenly so that all particles were uniformly exposed to the laser heating flux. For those arrays of finer particles there was an increasing tendency for them to cover or "shadow" one another. Similar problems may be involved in furnace exposure experiments. This problem of the initial shielding of particles from their heating source is most severe for the wire-mesh heating technique. In the case of laser heating, there is also the problem of diffraction losses. As particle sizes decrease and approach the laser wavelength of 10.6 µm, diffraction losses can become more significant. Such diffraction losses can be counteracted by increasing the number density of the particle array, but that merely enhances the probability that particles will shadow one another.

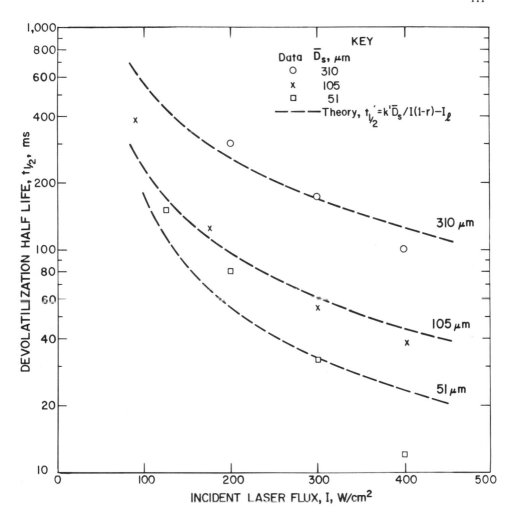

FIGURE 4. - Summary of all the laser pyrolysis data. The devolatilization half-life in nitrogen is plotted as a function of laser flux for the three particle sizes studied. The data points are compared with the theoretical predictions of equation (4).

Recent experiments with single particles injected in a flat flame reactor were reported in which the rate of evolution of internal surface area was measured for various exposure times [11]. The surface area development of a coal-char residue is directly related to the evolution rate of volatiles. For given flame conditions, the rate of evolution of surface area for 40 μm coal particles was observed to be four times that for 80 μm particles [11]. As will be shown later, that result is consistent with the particle size dependence in the data reported here.

The maximum flux levels used in these studies were limited by the power level available from the unfocused laser beam and also by the need to avoid extremes. Very high flux levels can be obtained by rather moderate focusing; however, extreme flux levels would give data that are not relevant to the normal devolatilization process. At extreme fluxes the residual char surface may reach temperatures so high that direct vaporization of solid carbon can become significant. Such extremes were avoided, and these data were limited to flux levels of 400 W/cm^2 and below. That value is characteristic of the maximum heating flux in a typical pulverized coal flame, and the maximum temperature reached by the char surface at those fluxes is well below the point of significant carbon loss by direct evaporation.

5. MICROSCOPIC STRUCTURE DATA

The correlation of the weight loss data with the microscopic observations for devolatilization in nitrogen is shown in fig. 5. The volatility yields for 310 μm particles are plotted for three separate flux levels, and the letters A through F for those data points indicate the exposure times and fluxes for which photomicrographs of the exposed particles were obtained with the SEM. The photomicrographs themselves are shown in fig. 6. Some preliminary SEM data for 84 μm particles exposed to a laser flux of 80 W/cm^2 were shown previously [5] in order

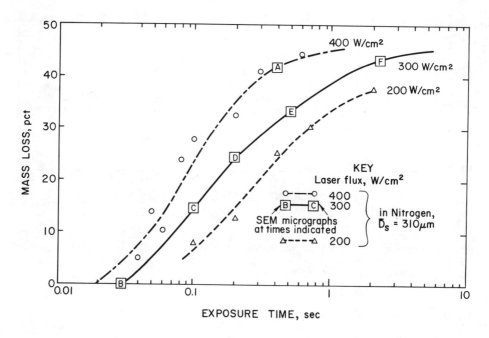

FIGURE 5. - The devolatilization weight loss for coal particles as a function of exposure time for 310 μm coal particles exposed in nitrogen to the laser fluxes shown. The weight loss data are to be correlated with the morphological changes to be shown in fig. 6 according to the exposure time and fluxes indicated by the points |A| through |F|.

113

to compare them with the char residues from dust explosions. The new
data in fig. 6 are more extensive and revealing. Photomicrograph A,
corresponding to the data point A in fig. 5, is the residue of a coal
particle that was exposed to 400 W/cm² for 400 msec. That 310 μm
particle lost 41 pct of its initial mass and was devolatilized to almost
its maximum extent. The residual rounded char structure is pockmarked
with blow holes of various sizes and shapes.

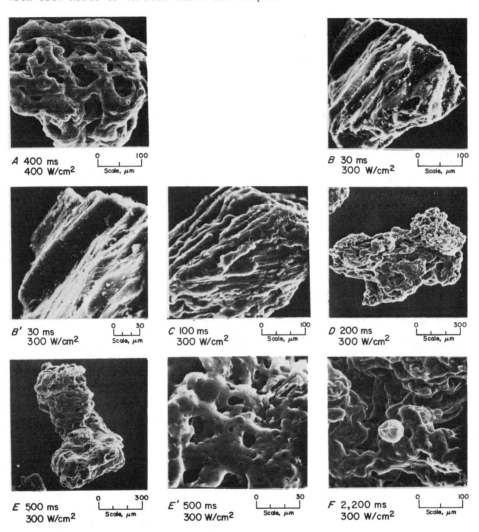

FIGURE 6. - SEM photographs of 310 μm diameter Pittsburgh seam coal dust
particles exposed to nitrogen to a laser flux for the indicated exposure
times. The scale for each SEM photograph is indicated, and the
corresponding weight loss for each exposure can be obtained from points
[A] through [F] in fig. 5.

When bituminous coal particles are subjected to such heating fluxes, they pyrolyze and soften or melt into a plastic phase. The heavier, higher molecular weight pyrolysis products are liquid, and surface tension forces cause the liquid mass to try to assume a spherical shape. However, the lighter molecular weight products that are being formed at the same time are volatilizing. The internal pressures resulting from the vaporization of those lighter components or tars generate bubbles within the liquid, plastic mass of the heavier components. But that liquid mass is simultaneously generating both a solid char matrix and even lighter molecular weight gases. At the same time that the lighter pyrolysis products of tars and gases are being emitted through bursting bubbles, the competing condensation reactions are generating a solid char matrix from the puffed liquid in the bubble wall. The final result seen in fig. 6A is typical [23,24,25]: A deformed char residue pockmarked with "frozen" blow holes of various sizes and shapes.

Photomicrograph B in fig. 6 is essentially the unexposed or "unburned" particle since its short exposure time of 30 msec at 300 W/cm² has resulted in no measurable weight loss or change in structure. The sharply cleaved edges or ledges of the particle that were formed by mechanical pulverization are clearly seen, and they presumably reveal the bedding planes of the coal seam. That anisotropic structure of the original seam is more evident under higher magnification in fig. 6B'. For particle C, the exposure is 100 msec at 300 W/cm², and the weight loss was 14 pct. There are now significant changes caused by laser heating. Liquid bitumen was formed and recondensed at the cleaved edges, rounding them out as the particle cooled. Small blow holes are the visible trails through which has passed the 14 pct of the particle's mass that was vaporized during the exposure. Note, however, that the original outlines of the cleaved edges are still preserved despite the significant extent of devolatilization, and that those outlines still reveal the anisotropic matrix of bedding planes. It appears that much of the liquid bitumen has "oozed" out from between the bedding planes. An even earlier stage of this process will be seen in the micrograph to be shown in figs. 8H and H'. Note also in fig. 6C that while some parts of the particle including its exposed ledges have been softened and rounded, other parts of the particle that were less exposed seem almost unaffected. The particles shown in figs. 6D and 6E under lower magnification are about half to two-thirds devolatilized. The particles appear to have been completely molten, and adjacent particles have touched and fused at their points of contact. Blow holes now cover the entire structures, and there are only some vague outlines of the external shapes of the original particles. In fig. 6E', the details of the porous char particle structures are seen under a high enough magnification to reveal the size distribution of blow holes. The completely devolatilized structure after a 2.2 sec exposure is shown under high magnification in fig. 6F. Its weight loss, 43 pct, was comparable to that shown in 6A for a shorter exposure time at a higher laser flux. The final char structures in the two cases appear to be similar.

For particles exposed in an air environment, the correlation of the weight loss data with the microscopic observations is shown in fig. 7. The mass loss data are for 310 μm particles, and the letters G through N indicate the exposure times for which photomicrographs were obtained. The weight loss curves in air show the two sequential processes: Devolatilization followed by char oxidation. The lower flux curve at 75 to 100 W/cm^2 shows a brief hiatus near the standard volatility of 36 pct. The devolatilization half-life for that portion of the curve is 330 msec. In air, the curve does not level off to its $V(\infty)$-value but shows a continuing weight loss attributable to the slower oxidation of the char, which requires an additional 10 sec before the char is essentially consumed and only an ash remains.

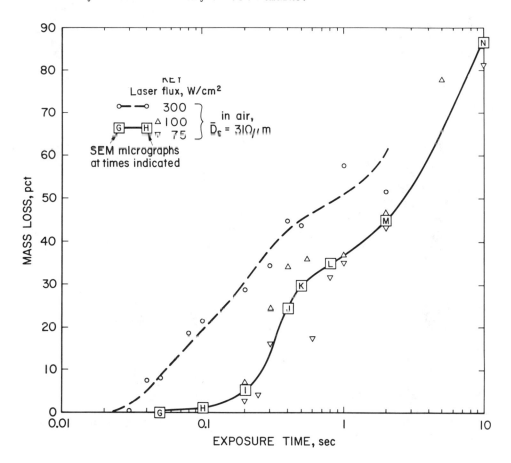

FIGURE 7. - The devolatilization and char oxidation weight loss for coal particles as a function of exposure time for 310 μm coal particles in air. The weight loss data are to be correlated with the morphological changes to be shown in fig. 8 according to the exposure times indicated by the points G through N.

116

Figure 8G is a photomicrograph of a 310 μm particle seen under high magnification after a short, 50 msec exposure. Essentially nothing has happened to the particle in such a short time at the low flux. The orthogonal fracture cracks and striations are normal features of the mechanically pulverized particle in its original state. A "suspicious" region was isolated which shows some "bumps" or "pimples" that may or may not represent some surface softening with subsurface

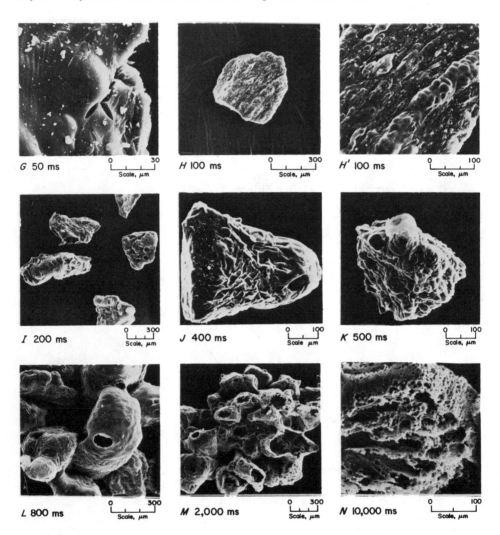

FIGURE 8. - SEM photographs of 310 μm diameter Pittsburgh seam coal dust particles exposed in air to a laser flux of 75 to 100 W/cm² for the indicated times. The scale for each SEM photograph is indicated, and the corresponding weight loss for each exposure can be obtained from the points G through N in fig. 7.

devolatilization. Photomicrograph 8H is for a 100 msec exposure at 75 to 100 W/cm^2, and even this low magnification view now shows significant effects of laser pyrolysis. Even though the weight loss is only 1 pct, there is clear evidence that liquid bitumin was formed, that it was oozing out from between the bedding planes, and that it solidified as the particle cooled after laser exposure. Photomicrograph 8H' is that same particle under higher magnification, which clearly reveals a large number of unbroken bubbles and several small blow holes in the ridge of resolidified bitumen. The bubbling, liquid mass of "metaplasts" had been oozing out from between the bedding planes, and its resolidification left a pattern of such ridges oriented along the bedding planes. The structure in fig. 8H' is an earlier stage of the same process previously seen in fig. 6C. A later stage of devolatilization is shown in fig. 8I, where several particles are seen under low magnification after an exposure of 200 msec and an average weight loss of 5 pct. Only upper portions or the sharpest corners of each particle have been devolatilized; other parts in the lower regions seem unaffected by the laser flux exposure. A particle that is about two-thirds devolatilized is shown in fig. 8J. Only the right side of the particle, which we assume was the upper portion, seems to have been devolatilized; the left side or lower portion is unaffected with a structure similar to that of the original coal surface. The upper portion was devolatilized into a dome or bubble, which, after it was vented through a blow hole, seems to have started to collapse under its own weight just before it solidified, leaving a wrinkled skin residue. The interpretation given here is that in the 400 msec exposure, the devolatilization wave was "frozen" after it had traversed half way through the particle. An alternative explanation [22] is that two different macerals comprise that one particle, and that the unreacted part of the particle consists of less reactive maceral. The particle in fig. 8K is almost completely devolatilized after a 500 msec exposure. The entire particle appears to have been liquid, and many blow holes are visible. The particle array in fig. 8L is for an exposure time of 800 msec, which is at the hiatus of the weight loss curve. At 800 msec, devolatilization is essentially complete and char oxidation is about to begin. In fact, the bright ring around the rim of the blow hole crater in the central particle suggests that char oxidation has already begun at that rim. For the particle array in fig. 8M, the char residue has been exposed to the laser flux in an air environment for some 1.2 sec after devolatilization was complete. The char oxidation process is clearly evident; it seems to start at the crater rims of the blow holes and to enlarge them. The enlargement clearly reveals the hollow structure of the char residues left by the devolatilization process. The char oxidation process continues to enlarge the blow holes until after about 10 sec only a flakey, porous structure is left. Figure 8N is a higher magnification view of that porous structure after some 86 pct of the original mass of the coal particles has been lost. Of the spongy structure remaining in fig. 8N, about half is ash and half is unburned char.

The microstructure data shown in fig. 8, together with the correlative weight loss data in fig. 7, clearly show that the normal process for bituminous coal combustion in air is a two-step process: Rapid devolatilization followed by a much slower oxidation of the char.

Although the experimental method used to obtain these time-resolved
microstructure data is entirely new, the final structural features
observed for the completely devolatilized coal particles are quite
similar to those of earlier investigators. Lightman and Street [23],
Street, Weight, and Lightman [24], and McCartney [25] have reported
similar structures for the char residues obtained for pulverized coals
heated in a variety of ways: drop tube furnaces, shock tubes, flames,
and electrically-heated mesh screens. Similar structures were reported
for the char residues from coal dust explosions at constant volume [26].
Some researchers have even attempted to correlate the observed
structural changes with maceral types within the coal structure [23-25].
No attempt was made in these studies to control or isolate the various
macerals either within or among the particles studied.

6. ANALYSIS AND DISCUSSION

While it is beyond the scope of this work to make a detailed
comparison of these data with those of previous investigators, it should
be pointed out that the absolute devolatilization rates reported here
for the 51 μm coal particles compare favorably with the rates reported
by Kobayashi, Howard, and Sarofim [10] for 40 μm particles exposed in a
laminar flow drop furnace. Their data at different furnace temperatures
can be transformed into equivalent heating fluxes by adding their
radiant furnace flux to the conductive-convective flux of their inert
carrier gas. A geometric correction factor is also required in order to
transform their omnidirectional, furnace heating flux, which is
uniformly incident on the entire surface area of the particle ($4\pi r_0^2$),
into an equivalent, directed laser flux that is incident only in the
projected area (πr_0^2) of the particle.

For this discussion, it suffices to say that the comparison thus made
between these data and those reported by Kobayashi [10] shows reasonable
agreement between the two families of curves despite the marked
differences in both the exposure systems and the experimental methods
used to obtain the weight losses. The theoretical analysis of the data
to be presented here is, however, quite different from the chemical rate
control models of the previous investigators [9,10,14,20]. Here, one
considers a planar, solid surface volatilizing or subliming in an
incident laser flux, I. For the steady-state regression of that surface
at the laminar rate $\dot{x}_0$, the first law of thermodynamics requires that

$$I(1 - r) - I_\ell = \dot{x}_0\rho[\int_{T_0}^{T_s}C(T)dT + \Delta H_v], \qquad (3),$$

where r is the average particle reflectance, and I_ℓ is the flux lost to
the surroundings by conduction-convection, reradiation, and transmission
through the particle. The flux absorbed by the solid is $I(1 -r) - I_\ell$,
and it supplies the power necessary to bring the solid to its
sublimation temperature T_s and to vaporize the solid at the linear rate
$\dot{x}_0$. The density of the solid is ρ, and its heat of vaporization is ΔH_v.
The heat capacity of the solid is $C(T)$, and the integral is taken from
the ambient temperature T_0 to the surface sublimation temperature, T_s.
The ambient temperature T_0 is assumed to be maintained in steady state
at points in the solid that are far from the devolatilizing surface.

For a cubic particle with sides of width a_o in a radiant flux normal to one of its faces, the devolatilization half life is simply the time required for the devolatilization or regression wave to travel half way through the particle. Thus,

$$t_{1/2} = \frac{a_o}{2\dot{x}_o} = \frac{a_o \rho [\int_{T_o}^{T_s} C(T)dT + \Delta H_v]}{2[I(1 - r) - I_\ell]} = \frac{k'\bar{D}_s}{I(1 - r) - T_\ell}. \quad (4)$$

Comparison of equation (4) with the data in fig. 4 suggests that this simple analysis correctly predicts the measured linear particle size and inverse flux dependences. This same steady-state analysis also correctly predicts the quadratic particle size dependence for $t_{1/2}$ measured by others in flame reactors [11]. For flame reactors, the heating flux I is almost entirely by conduction from the high-temperature flame gases. In that case, $I = k''\Delta T/\bar{D}_s$, where the constant is proportional to the Nusselt member. Substituting into 4, gives $t_{1/2} \sim \bar{D}_s^2$, which is the dependence reported by Samuelson, Seeker, Heap, and Kramlich [11].

This steady-state analysis is necessarily oversimplified. First, the steady state assumption can be accurate only if δ, the devolatilization wave front thickness, is much smaller than the particle's width, a_o. Prior to exposure, the entire particle is at temperature T_o; hence the incident flux must first bring the exposed surface to T_s before any steady-state devolatilization can begin. A finite induction time t_o, is required to do that. It is only when $\delta << 1/2a_o$ that t_o is negligible compared to $t_{1/2}$.

One can attempt to use the data reported here to compare measured t_o values with measured $t_{1/2}$ values. For example, for the 51 μm diameter particle at a flux of 125 W/cm^2, the laser pyrolysis data gives a t_o of about 60 to 80 msec, compared to a $t_{1/2}$ value of 160 msec. For the 105 μm particles in fig. 2 at a comparable flux, t_o is again 60 to 80 msec; however, $t_{1/2}$ is now about 400 msec. For the 310 μm particles at 300 W/cm² (shown in fig. 3) t_o is about 30 msec whereas $t_{1/2}$ is 180 msec. Clearly, the data suggest that the assumption that $\delta << 1/2a_o$ is more accurate at larger diameters and higher fluxes. In any case, the morphological structures revealed in figs. 6C, 8H, 8I, and 8J would not be possible unless δ were significantly smaller than a_o.

One must be cautious, however, in interpreting the threshold time for the first appearance of volatiles as the induction time for bringing the particle surface to T_s. The measured curves have appearance "tails" which may simply reflect their nonspherical shapes and may thus have no relation to t_o. A particle with a sharp corner exposed to the laser beam may devolatilize at that corner well before the bulk of the particle is heated to the devolatilization temperature. Nonspherical particles are representable in terms of a "Fourier spectrum" of particle sizes. The early appearance of volatiles and the presence of a "tail" can obscure the induction time for the bulk of the particle. The "tail" simply reflects the distribution of sizes or radii of curvature in the Fourier spectrum needed to describe the complex shape of any given particle.

There is a second oversimplification involved in the steady-state approximation requiring that δ should be much less than $1/2a_0$. The approximation assumes that the absorbed energy does not accumulate in time. Heat accumulation would decrease the heat capacity integral and accelerate the devolatilization rate. Thirdly, the particles are not cubes, nor is the incident flux oriented perpendicular to any particular face, so the real geometry involved is more complicated than that assumed. Fourthly, for coal particles, there is a char residue which can shield the pyrolysis wave from the incident flux as devolatilization proceeds. The presence of a char probably increases the loss flux as the wave proceeds through the particle.

These complications do, however, tend to counteract one another so that the a_0- and $(I - I_\ell)$-dependence predicted by equation (4) may still be maintained even as $\delta \to 1/2a_0$. In the case of coal it should also be realized that the magnitude of ΔH_v is uncertain and that it most likely displays a path-dependent hysteresis effect. The simple theory of equation (4) gives a ramp-type weight loss curve with a fixed slope during pyrolysis. The above complexities, if corrected for, should transform the ramp into the observed S-shaped curve.

To proceed further with this steady-state analysis, it is assumed that these complications tend to counteract one another, or that any noncancelling effects can nevertheless be accounted for by a proper choice for the constants k' and I_ℓ in equation (4). The loss flux I_ℓ is taken as some average steady-state value; i.e., the sum of the convective loss flux to the cold air surroundings, given by $Nu\lambda(T_s - T_0)/\bar{D}_s$, plus the reradiative loss flux of σT_s^4, where Nu is the Nusselt number, λ is the thermal conductivity of the gas, and σ is the Stefan-Boltzmann constant. For these small particles, $Nu = 2$. Taking the devolatilization temperature as $T_s = 550$ C (823 K) gives $I_\ell = 26$ W/cm^2 for the 105 μm particle. The coal reflectance is taken as 7 pct. The data in fig. 4 for $\bar{D}_s = 105$ μm then give a reasonable fit to equation (4) for a k'-value of 1.46 kjoule/cm^3. That good fit is shown as the 105 μm theory curve in fig. 4. The other two theory curves for the particle diameters 51 μm and 310 μm are obtained using the same k'-value. Their respective loss fluxes are calculated the same way, and they are 51 and 10 W/cm^2, respectively. The reasonable agreement in fig. 4 between the data points and the theory curves predicted by equation (4) tends to confirm the reasonableness of its simple derivation from the energy conservation equation.

A more careful examination of the differences between the data points and the theory curves reveals that I_ℓ-values are overestimated for the higher fluxes and underestimated at the lower fluxes. If the real, time-dependent loss flux values, $I_\ell(t)$, were used rather than their average steady-state values, there would be better agreement with the theory. For high fluxes and short devolatilization times, $\delta < < a_0$, only part of the particle loses heat to the surroundings and the steady-state I_ℓ-values used here overestimate the losses. For low fluxes and longer devolatilization times, δ approaches a_0, and conductive losses to the solid substrate may need to be added to I_ℓ.

The data and this analysis suggest that the coal devolatilization rate process may be no more a "chemically controlled phenomenon" than is the evaporation rate of any solid near its sublimation temperature. Similar data have been previously reported for macroscopic particles of ammonium perchlorate under laser-induced combustion and pyrolysis by two

independent investigators [16-18]. The absolute macroscopic regression
rate of NH_4ClO_4 and its flux dependence are similarly predicted by
equation (3), and those data suggested that the dissociative sublimation
process for NH_4ClO_4 was similarly controlled by the constraints of
thermodynamic heat transport [19]. Recently, data were reported for
coal devolatilization rates with particle size and heating rate
dependences of volatility yields that were not properly explained by
present kinetic models [20]. Those data were obtained by a mesh heating
technique with rapid quenching. The laser pyrolysis data reported here
suggest that it is not a matter of simply replacing one chemical kinetic
rate control model by another. Rather, it is a matter of realizing that
at high devolatilization rates, the rate process is more likely
controlled by the thermodynamic transport constraints.

7. REFERENCES

1. Humphrey, H. B., Bureau of Mines Bulletin 586, 1960.
2. Richmond, J. K., Price, G. C., Sapko, M. J., and Kawenski, E. M.,
   Bureau of Mines Information Circular 8909, 1983.
3. U.S. Congress, Federal Mine Health and Safety Act of 1977. Public
   Law 95-164, Nov. 9, 1977.
4. Hertzberg, M., Cashdollar, K. L., and Lazzara, C. P., Eighteenth
   Symposium (International) on Combustion, p. 717, The Combustion
   Institute, 1981.
5. Hertzberg, M., Cashdollar, K. L., Ng, D. L., and Conti, R. S.,
   Nineteenth Symposium (International) on Combustion, p. 1169, The
   Combustion Institute, 1982.
6. Lee, C. K., Singer, J. M., and Chaiken, R. F., Combust. Sci.
   Technol., 16, 205 (1977).
7. Merrick, D., Fuel, 62, 534 (1983).
8. Smith, I. W., Nineteenth Symposium (International) on Combustion,
   p. 1045, The Combustion Institute, 1983.
9. Howard, J. B., Fundamentals of Coal Pyrolysis and Hydropyrolysis,
   Chapter 12, p. 665, Chemistry of Coal Utilization, John Wiley and
   Sons, 1981.
10. Kobayashi, H., Howard, J. B., and Sarofim, A. F., Sixteenth
    Symposium (International) on Combustion, p. 411, The Combustion
    Institute, 1976.
11. Samuelsen, G. S., Seeker, W. R., Heap, M. P., and Kramlich, J. C.,
    The High Temperature Decomposition and Combustion of Pulverized Coal,
    in A. F. Sarofim (ed.) EPA Project Decade Monograph, Ch. 2,
    Government Printing Office, 1983.
12. Seeker, W. R., Samuelsen, G. S., Heap, M., and Trolinger, J. D.,
    Eighteenth Symposium (International) on Combustion, p. 1213, The
    Combustion Institute, 1981.
13. McLean, M. J., Hardesty, D. R., and Pohl, J. H., Eighteenth Symposium
    (International) on Combustion, p. 1239, The Combustion Institute, 1981.
14. Badzioch, S., and Hawksley P. G. S., Ind. Eng. Chem. Process Design
    and Dev. 9, No. 4, 521 (1970).
15. Suuberg, E. M., Peters, W. A., and Howard, J.B., Seventeenth
    Symposium (International) on Combustion, p. 117, The Combustion
    Institute, 1979.
16. Hertzberg, M., Combust. Sci. Technol., 1, 449 (1970).
17. Hertzberg, M., Oxidation Combust. Rev., 5, 1 (1971).

122

18. Pellett, G. L., Fifteenth Symposium (International) on Combustion, p. 1317, The Combustion Institute (1973).
19. Hertzberg, M., Comment to reference 18 above.
20. Niksa, S., Russel, W. B., and Saville, D. A., Nineteenth Symposium (International) on Combustion, p. 1151, The Combustion Institute, 1983.
21. Hertzberg, M., Cashdollar K. L., Zlochower, 1., and Ng, D. L., Twentieth Symposium (International) on Combustion, p 1691, The Combustion Institute, 1984.
22. Street, P. J., private communication, 1984.
23. Lightman, P., and Street P. J., Fuel, 47, 7 (1968).
24. Street, P. J., Weight R. P., and Lightman P., Fuel, 48, 343 (1969).
25. McCartney T. J., Fuel, 50, 457 (1971).
26. Ng, D., Cashdollar, K. L., Hertzberg, M. and Lazzara, C. P., Bureau of Mines Information Circular 8936, 1983.

DISCUSSION

M. Morgan

In order for your model to be accurate, you need good data on C_p and ΔH_{R+V} for the pyrolyzing coal particle. Could you comment on the derivation and accuracy of your data ?

M. Hertzberg

The theoretical predictions shown in Fig. 4 are based on a constant k'-value of 1.46 Kjoule/cm² for all three particle sizes. That k'-value is consistent with a heat capacity integral from ambient temperature to T of 550°C that is comparable to the values given in reference 7 ; and to a ΔH_v that is endothermic, and a factor of two or three larger than heat capacity integral.

J.H. Pohl

Char particles from drop tube experiments performed by Sarofim, Kobayashi, and myself were cross sectioned. Cross section of lignite and bituminous coals devolatilized at 1500 K did not show an unreacted core. How does the particle sitting on a heat sink affect your conclusion that heat transfer controls devolatilization. In addition, please quantitatively compare your weight losses with those of Kobayashi and state the temperature (estimated) for Kobahashi's results.

M. Hertzberg

Our detailed estimates of heat losses from the particle to the surroundings during laser pyrolysis are given in the paper. For the most part they are estimated to be caused mainly by conduction convection to the surrounding gas. The heat losses are most probably time - dependent so that as the wave approaches the cold metal susbtrate conduction loss to the substrate may become important. Those refinements are discussed briefly in the paper.

Since the particle probably never reaches the furnace temperature during devolatization we prefer to transform Kobayashi's furnace exposure temperatures to equivalent heating fluxes. As indicated in the paper when that transformation is made, there is quite good agreement between our rates and those reported by Kobayashi et al. for comparable particle sizes. But we have measured the particle size dependence which led us to a different interpretation of the data.

P.R. Solomon

Do your data on the dependence of pyrolysis rate flux and particle size require a nonisothermal dependence ? The same dependences would be expected for isothermal particles under heat transfer control. Your SEM data show zones of 40μm isothermal thickness so perhaps 40 μm diameter particles are reasonably isothermal.

M. Hertzberg

The devolatilization rate data are most simply explained by thermodynamic heat transport constraints to the particle from the incident laser heating flux. The microscopic structure data for 200 and 300 μm particles at laser fluxes as low as 100 w/cm² show clearly that those particles are not isothermal during pyrolysis since one part of the particles is seen to be devolatilized while the other part of the particle is unaffected. The devolatilization wave thickness that we see at 100 W/cm² is about 40 μm, but we are uncertain whether that thickness is the true wave thickness or whether it represents the depth of penetration of the absorbed laser radiation. A wave thickness of about 50 m for a heating flux at 100 W/cm² is consistent with the wave thickness of 1000 μm at a heating flux of 8 W/cm² reported by Lee, Singer and Chaiken in reference 6. In neither case is the particle isothermal since no heat could be transfered through the surface of the particle or through the devolatilization wave if the particle were isothermal.

E.M. Suuberg

Is the resolution of the apparent controversy perhaps ascribable to the enormous surface heating rate of the present experiment compared to the usual pyrolysis experiments ? Even using a lumped analysis (clearly inapplicable here) the heating rate would be more than 10^4 K/s, very possibly much higher in near-surface layers of the coal. Under these conditions, a heat transfer limitation is probable, but to the results of other experiements, is inappropriate and unnecessary. The important question, it would seem, is to ask if given the kinetics derived at lower heating rates, will devolatilization be controlled by heat transfer under most pulverized coal combustion conditions ?

M. Hertzberg

I would prefer to use the incident heating flux rather than the "heating rate" as the important variable. The "heating rate" is not known unless one actually measures the temperature of the particle, and since there is usually a temperature gradient within the particle, the "heating rate" is not uniquely definable for the particle as a whole. Furthermore, there are other problems with expressing heating rates in terms of K/s. What, for example, is the heating rate of a water droplet while it is vaporizing in an incident heating flux ? Clearly, it is zero. The heating rate in K/s is zero in that case even though an important dynamic process is occuring as the droplet vaporizes. Clearly, it is the incident heating flux (in watts/cm²) that is driving that process and that variable is a more meaningful one than the "heating rate" in K/s.

Our absorbed laser fluxes are comparable to the sum of

radiative and conductive heating fluxes in the drop furnace experiements of references (10) and (14). Those laser fluxes were chosen because they are comparable to those in pulverized coal burners. For maximum burning velocities of coal air flames that flux is about 150 W/cm^2. Our absorbed laser fluxes thus cover the range of fluxes of interest in pulverized coal burners and are also comparable to those experienced by particles in drop furnaces at temperatures near 1500°C.

EVALUATION OF COAL PYROLYSIS KINETICS

Peter R. Solomon and Michael A. Serio

Advanced Fuel Research, Inc., 87 Church Street, East Hartford, CT 06108, USA

INTRODUCTION

To develop accurate predictive models for coal combustion or gasification, it is necessary to know the rate and amount of volatiles release as a function of the particle temperature. The volatiles, which can account for up to 70% of the coal's weight loss, control the ignition, the temperature and the stability of the flame, which can, in turn, affect the subsequent reactivity and burnout of the char. Unfortunately, there is still controversy concerning the rate of coal pyrolysis. For example, at particle temperatures estimated to be 800°C, rates reported in the literature (derived using a single first-order process to define weight loss or tar evolution) vary from less than 1 s^{-1} [1-4] to more than 100 s^{-1} [5-11] with values in between.[12-14]

Accurate combustion models cannot be developed with this wide range of values. The problem was evident in the papers on coal pyrolysis and combustion modeling presented at the 20th Symposium (International) on Combustion. Such models calculate the particle temperatures and predict the weight loss using one of the reported rates. Several papers[15-17] reported reasonably accurate modeling of results using very different pyrolysis kinetics, yet the individual results were sensitive to which rates were assumed. For example, Lockwood et al.[15] found the rates of Badzioch and Hawksley[5] and of Anthony et al.[1] (distributed rate) acceptable, while the lower rates of Kobayashi et al.[2] and of Anthony et al.[1] (single rate) were not. Truelove[16] successfully used the high rates of Badzioch and Hawksley, while Jost et al.[17] employed the lower rates determined by Witte and Gat.[18] Also reported at the Symposium was a rate by Niksa et al.[3] which was close to that of Anthony (single rate model) and one reported by Maloney and Jenkins[14] which was close to that of Badzioch and Hawksley.

An important objective of coal pyrolysis research is to identify the source of the variations in these reported rates and provide an accurate separation of the chemical kinetic rates, heat transfer rates and mass transfer rates which combine to produce the observed results. This paper reviews the experimental data which is available and considers the conclusions which can be drawn from these data.

BACKGROUND

A summary of pyrolysis rates for a number of experiments is presented in Fig. 1 and Table I. In Fig. 1 the rates in sec^{-1} which describe, by various models, the weight loss or tar loss are plotted as a function of reciprocal particle temperature. The activation energies, E_o, and frequency factors, k_o, which describe the rates by Arrhenius expressions are summarized in Table I. In some cases, a Gaussian distribution of activation energies[1] has been used to describe multiple parallel processes. This mode requires the additional parameter, σ, to describe the width of

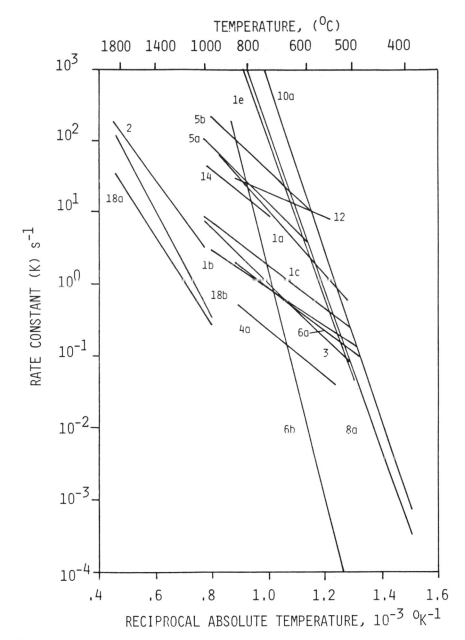

Figure 1.   Comparison of Kinetic Rates for Weight Loss (or tar loss) from
Several Investigators.   The Numbers Next to each Line Indicate the
Reference and Case in Table I.

TABLE I

KINETIC RATES AND MODELS FOR COAL PYROLYSIS

Author	Reference (case)	Model Rate For
Anthony et al. (1975)	1(a)	Weight Loss from Lignite at 10,000 K/sec (1 atm, excluding cooling)
	1(b)	Weight Loss from Lignite at 10,000 K/sec (1 atm, including cooling)
	1(c)	Weight Loss from Bituminous Coal at 650 K/sec (1 atm)
	1(d)	Weight Loss from Lignite (all pressures)
	1(e)	Weight Loss from Bituminous Coal (69 atm)
Kobayashi et al. (1977)	2	Weight Loss from Lignite and Bituminous Coal
Niksa et al. (1984)	3	Weight Loss from Bituminous Coal
Solomon and Colket (1978)	4(a)	Tar Evolution
	4(b)	Aliphatic Gas Evolution
Badzioch and Hawksely (1970)	5(a)	Weight Loss for Coal B
	5(b)	Weight Loss for Coal F
Suuberg et al. (1979)	6(a)	Tar Evolution from Bituminous Coal
	6(b)	Tar Evolution from Bituminous Coal
Solomon and Hamblen (1985)	8(a)	Tar Evolution from Lignite, S. Bit & Bit. Coal
	8(b)	Aliphatic Gas from Lig. S. Bit. & Bit. Coal
Solomon et al. (1985)	10(a)	Tar from Lignite and Subbituminous
	10(b)	Aliphatic Gases from Lignite and S Bit. Coal
	10(c)	Weight Loss from Lignite and S. Bit. Coal
Freihaut (1980)	12	Weight Loss Rate for 50% Reaction Completion, S. Bit. Coal
Solomon et al. (1982)	13(a)	Tar from Lignite, S. Bit. & Bit. Coal
	13(b)	Aliphatic Gas from Lignite, S. Bit. & Bit. Coal
Maloney and Jenkins (1984)	14	Average for Initial Weight Loss
Witte and Gat (1983)	18(a)	Weight Loss for Subbitiminous Coal
	18(b)	Weight Loss for Subbituminous Coal
Serio et al. (1986)	23,24(a)	Tar from Lignite, S. Bit. & Bit. Coal
Solomon et al. (1985)	23,24(b)	Aliphatic Gases from Lig., S. Bit. & Bit. Coal
	23,24(c)	Weight Loss from Lig., S. Bit. & Bit. Coal
Solomon and King (1984)	30,31(a)	Tar from Ethylene-Bridged Anthracene Polymer
Squire et al. (1986)	30,31(b)	Tar from Ethylene-Bridged Naphthalene Polymer
	30,31(c)	Tar from Ethylene-Bridged Benzene Polymer
Stein (1981)	42	Bibenzyl Decomposition in Tetralin

TABLE I (continued)

KINETIC RATES AND MODELS FOR COAL PYROLYSIS

Experiment	Frequency Factor k_o (sec^{-1})	Activation Energy, E_o K cal/mole	Width of Activation Energy Distribution σ (K cal/mole)	Model
Grid	2.9×10^5	20.0	0.0	S
Grid	283	11.1	0.0	S
Grid	1800	13.3	0.0	S
Grid	1.67×10^{13}	56.3	10.9	G
Grid	1.67×10^{13}	50.7	7.0	G
EFR	6.6×10^4	25.0	0.0	S
Grid	70.5×10^2	17.2	0.0	S
Grid	7.5×10^2	15.8	0.0	S
Grid	4.2×10^3	17.7	0.0	S
EFR	1.14×10^5	17.5	0.0	S
EFR	3.12×10^5	17.5	0.0	S
Grid	0.7×10^7	13.2	0.0	S
Grid	2.3×10^{15}	68.9	11.4	G
Grid, EFR & EGA	4.5×10^{13}	52.0	3.0	G
Grid, EFR & EGA	1.7×10^{14}	59.1	3.0	G
HTR	8.57×10^{14}	54.6	3.0	G
HTR	8.35×10^{14}	59.1	3.0	G
HTR	4.28×10^{14}	54.6	0.0	G
Drop tube	1.0×10^3	7.6	0.0	S
Grid & EFR	4.5×10^{12}	52.0	3.0	G
Grid & EFR	1.7×10^{14}	59.1	3.0	G
EFR	1.9×10^4	15.0	0.0	S
Laser	2.25×10^4	28.0	0.0	2P
Laser	2.95×10^5	22.4	0.0	2P
TGA/EGA, HTR & EFR	8.6×10^{14}	54.6	3.0	G
TGA/EGA, HTR & EFR	8.4×10^{14}	59.1	3.0	G
TGA/EGA, HTR & EFR	4.3×10^{14}	54.6	0.0	S
TGA/EGA	1.0×10^{15}	49.5	0.0	S
TGA/EGA	1.0×10^{15}	56.2	0.0	S
TGA/EGA	1.0×10^{15}	61.0	0.0	S
Tubing Bomb	7.9×10^{15}	65.0	0.0	S

EFR – Entrained Flow Reactor
HTR – Heated Tube Reactor
Grid – Heated Grid Reactor
Laser – Laser Heating Experiment
EGA – Evolved Gas Analysis at 0.5 K/sec
TGA/EGA – Thermogravimetric and Evolved Gas Analysis at 0.5 K/sec

S – Single Rate Model
G – Gaussian Distribution of Activation Energies Model
2P – Two Parallel Rate Model

the distribution. In this case, the rate shown in Fig. 1 is the mean of the distribution. Data have been included from five different types of experiments.

Several studies have been done in heated grid reactors[1,3,4,6] where a thin layer of particles is contained in an electrically heated wire mesh. The time-temperature history of the coal is assumed to be the same as for a thermocouple bead which is either attached to the screen or placed within its folds. By proper adjustment of the current to the screen, the heating rate and holding time can be independently controlled.

A second type of experiment is the entrained flow reactor, where coal particles are injected along the axis of a hot furnace tube and entrained in preheated gas.[2,5,7-9,13,14] The residence time of the particles is determined by adjustment of the separation of the water-cooled injector and collector. Both moveable collector and moveable injector experiments have been done. For these experiments, the particle time-temperature history is usually calculated from heat transfer/fluid mechanics models.

Either heated grid or entrained flow reactors have been used in nearly all of the attempts to obtain kinetic rate information for coal pyrolysis under rapid heating ($>10^3$K/s), high temperature ($> 600°C$) conditions. However, nearly all of these attempts were experiments in which coal particle temperature was calculated or inferred from thermocouple measurements, rather than directly measured. This lack of firm knowledge of particle temperature is one of the most important reasons for the wide variation in reported rates from entrained flow reactor experiments. It is of concern for heated grid experiments, as well, especially at low pressures (< 1 atm) and high heating rates (> 1000 K/s).

Recognition of this problem has led to recent studies in which particle temperatures were measured.[18] Three-color pyrometry has been used in a laser heating experiment.[18] A new technique for measuring the temperature and emissivity of small coal particles using Fourier Transform Infrared (FT-IR) emission and transmission (E/T) spectroscopy[8,10,19-23] has also been applied to the determination of pyrolysis kinetics.[10,11,23,24]

The latter technique was first applied to determine particle temperatures[8] in an entrained flow reactor (EFR) in which an extensive amount of pyrolysis data had been obtained.[7-9,13] However, the complex geometry and hydrodynamics of such systems made validation of the technique difficult. To solve this problem, a new reactor system was designed to provide a geometry which simplified the prediction and measurement of particle temperatures.[10,11] The reactor consists of an electrically heated metal tube into which a premixed stream of coal and gas is injected. The residence time is varied by moving the electrode positions. The stream exiting the hot tube is either ejected from the tube end for FT-IR temperature measurements, or is quenched (in a time, short compared to the hot residence time) in a water-cooled section of tube for mass balance measurements. FT-IR measurements are made close enough to the tube so that the particles do not cool substantially. Particle velocity determinations were made using FT-IR transmission spectroscopy and also by measuring, with phototransistors, the transit time of a pulse of coal particles through the tube. The particles (-200 +325 mesh) are found to move at about 70% of the average gas velocity due to wall collisions.

The reactor was operated in two different modes. In the first mode, the residence time is chosen so that the tube wall and gas/coal stream reach a constant temperature, which indicates that the particles, gas, and wall are equilibriated. From the known particle temperature, FT-IR E/T spectroscopy can be used to determine the spectral emittance of the particles. Once the spectral emittance is known, the reactor can be

operated in a non-equilibrium mode and the FT-IR E/T spectroscopy allows
determination of the particle temperature.

An example of the product yield, temperature, and residence time data
obtained for a Montana Rosebud coal from the heated tube reactor (HTR) is
shown in Fig. 2. The solid lines in Fig. 2a are predictions of a
previously described Functional Group (FG) model for coal pyrolysis[4,7-
9,13], but with adjustments of the previously reported rates which were
based on less accurate knowledge of particle temperatures. Similar
agreement was obtained between the theory and data from experiments on Zap
North Dakota lignite and Illinois #6 bituminous coal, using the same rates.
The experiments provide a validated set of kinetic rates for primary
pyrolysis.

The HTR was used to develop validated submodels for particle heat
capacity, emissivity, and heat of reaction as a function of temperature and
extent of pyrolysis. The upper and lower solid lines in Fig. 2b are the
temperature and residence time predictions, respectively. In general, the
agreement of theory and experiment is quite good. The apparent
overprediction of temperature is due to cooling of the particles at the
outer edge of the stream as they exit the tube and come into contact with
the cold ambient air. This point is discussed further in Ref. 10.

The next step was to use the validated temperature measurement
technique and heat transfer submodels to improve the prediction of particle
time-temperature histories in previously reported entrained flow reactor
experiments.[7-9,13] The remaining unknown was the effect of mixing of the
coal with the preheated gas near the injector. An adjustable mixing
parameter was used to match the measured particle temperatures at the
optical window.[23,24] A comparison of the calculated and measured particle
temperatures for EFR experiments with Zap lignite at a maximum furnace
temperature of 1300°C is given in Fig. 3a.

In general, the model predictions are in reasonable agreement with the
measured particle temperatures, especially at the shorter injection
distances. Similar results were obtained at 1100 and 1600°C (i.e. good
agreement for short distances and slightly high predictions at long
distances).[24] The discrepancy at long distances can be explained as
follows. The model assumes the particles to be in a central core while the
FT-IR temperature measurement reflects the presence of colder particles
closer to the walls.

The agreement is not a complete validation of the heat transfer model
because the particle temperature as a function of distance from the
injector varies with the injector position, and the temperature is measured
at only one position in the reactor. However, validation of submodels in
the HTR and agreement of the model for the EFR with data at several
different injector positions and at several temperatures[23,24] help to
establish confidence in the model. The calculated time-position histories
for each injector height are given in Fig. 3b.

The validated particle time-temperature model was subsequently used to
predict the results of material balance experiments in combination with
the Functional Group model as shown for a Zap, North Dakota lignite in
Fig. 3c. In general, good agreement was obtained for a wide range of
temperatures and coals (Kentucky #9 and Pittsburgh bituminous, Gillette and
Rosebud subbituminous).[23,24]

The final step in this phase of our pyrolysis research effort was to
use the validated particle temperature models for the HTR and EFR to
develop a set of kinetic parameters which is independent of reactor type,
coal rank, temperature (350 to 1600°C) and heating rate (0.5-20,000 K/s).
In order to obtain data at low temperatures and heating rates, a new

132

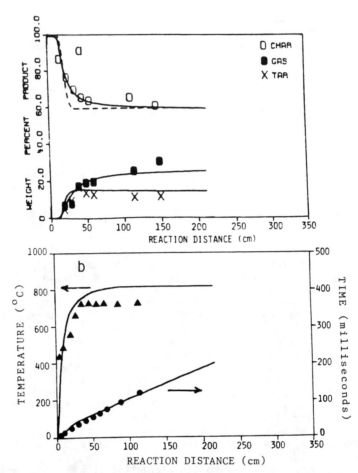

Figure 2.   Pyrolysis Results for Montana Rosebud, Case a.   a) Product
Distribution Symbols are Experimental Data.   Solid Lines are for the
Functional Group Model.[4,7-9,13]   Dashed Line is a Single, First Order Rate
Model with k = 4.28 x 10^{14} exp(-54,570/RT)s^{-1} and Percent VM (daf) =   40.
b) Time and Temperature of Particle vs. Distance. Lines are Calculations;
Symbols are Experimental Data.

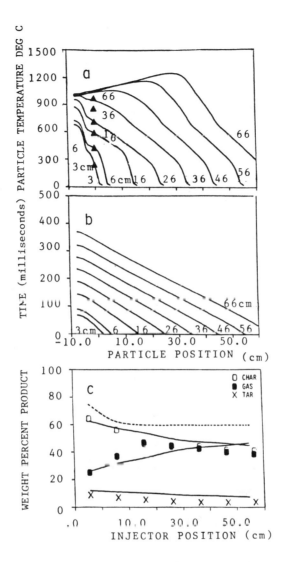

Figure 3.   Predictions of Particle Temperature and Functional Group Models
for 1300°C EFR Experiments. a) Comparison of Measured and Predicted
Particle Temperatures.   Solid Lines are Calculations.   (▲) FT-IR
determined Temperatures at Optical Port.   b) Predicted Cumulative Particle
Residence Time vs. Position. The Number Next to each Data Point or Theory
Line is the Injection Height in cm Relative to the Optical Port.   c)
Comparison of EFR Data for North Dakota Lignite, 1300°C, with Predictions
of FG Model (solid lines) and Single, First Order Model (dashed
lines) with k = 4.28 x 10^14 exp(-54,570/RT)s^-1 and Percent VM (daf) = 40.

134

thermogravimetric analyzer with evolved gas analysis (TGA/EGA) was developed.[25]

The TGA/EGA Apparatus consists of a sample suspended in a gas stream within a furnace. As the sample is heated, the evolving tars and gases are carried out of the furnace directly into a gas cell for analysis by FT-IR. The heating rate is slow enough (0.5 K/s) that the particles are close to equilibrium with the ambient gas temperature, which is measured with a thermocouple. Some representative data for Zap North Dakota lignite from the TGA/EGA compared with predictions of the FG model are shown in Fig. 4. Similar results have been obtained with a variety of other coals.

Using data from the above three reactors, a "universal" set of kinetic rates (i.e., independent of coal rank) was developed for the FG model.[23,24] This required some adjustments in previously published rate parameters because of the improved accuracy in the particle time-temperature histories. The results are in reasonable agreement with the predictions of a single first order model for primary pyrolysis weight loss which uses a rate constant $k = 4.28 \times 10^{14} \exp(-54,570/RT)s^{-1}$ (dashed lines in Figs. 2a, 3c and 4a). This is approximately one-half the k_{tar} rate indicated in Fig. 1 and Table I and suggests that the rate of primary pyrolysis is much higher at elevated temperatures than what is predicted by most of the results in Fig. 1. The rest of this paper will consider the reasons for these discrepancies.

DISCUSSION

There are a number of factors which could cause the wide variations observed in Fig. 1. The figure summarizes measurements on different coals, using different reactors, a variety of measurement techniques and different models to interpret the results. This section considers the possible factors which could cause the variations. They include: variations with coal rank; inaccuracies in determining the weight loss or the reaction time; mass transfer limitations; heat transfer assumptions; inaccuracies in particle temperature measurements; and differences due to the model used to interpret the data. The first three factors, which are considered to result in only small variations in the reported data, are discussed first. The last three factors, which are believed to be the primary reason for the wide variations in reported rates, are discussed last.

Variations in Rates with Coal Rank
One possible explanation for the wide variation in rates is that the rate-limiting step is a chemical kinetic rate and the observed variation is due to differences in the coal type studied. However, this is unlikely, because experiments in which the coal type alone was varied (i.e., all other conditions were held constant) typically show little variation with coal rank. For example, the rates reported by Anthony et al.[1] for lignite and bituminous coals are within a factor of 3, while Kobayashi reported a single rate for the same coals. Rates reported by Badzioch and Hawksley for anthracite and bituminous coals differ by less than a factor of 10 and rates reported by Solomon and coworkers[4,7-11] were insensitive to coal type, when coal type alone was varied over a wide variety of coals. A review of a number of other pyrolysis experiments shows no more than a five-fold variation in rate with coal type.[9]

Variations in Rates Due to Inaccuracies in Weight Loss or the Determination of Residence Time
The weight loss rate is determined by measuring a weight loss as a

135

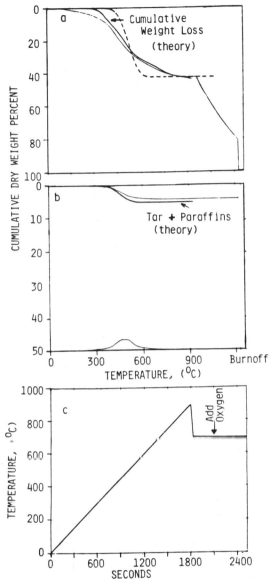

Figure 4. Pyrolysis Data and Theory Predictions for North Dakota Lignite Heated at 0.5 K/s in the TGA/EGA. Upper Solid Lines in 4a and 4b are Cumulative Evolution Data and Predictions of the Functional Group Model for Cumulative Product Evolution. The Theory Lines are Indicated with Arrows. Dashed Line in a) is Weight Loss Prediction of a Single, First Order Model with $k = 4.28 \times 10^{14} \exp(-54,570/RT)s^{-1}$ and Percent VM (daf) = 40. The Lower Solid Lines in 4a and 4b are Mass Evolution Rate Data (arbitrary scale). After the Coal Reaches 900°C, it is Quenched to 700°C and the Char is Oxidized for Elemental Analysis. Figure 4c is the Time-Temperature History of the Sample.

function of residence time. It is unlikely that inaccuracies in the determination of the pyrolysis weight loss or time could lead to the variation in the reported rates in Fig. 1. The weight loss can be easily measured to within 10%.

The time is accurately known for the heated grid experiments[1,3,4,6] and TGA/EGA experiment.[25] It is less precisely known for the entrained flow reactor experiments[2,5,7-9,13,14] and the laser experiment,[18] but estimates are unlikely to be off by more than a factor of two. The transit time was directly measured in the heated tube experiment.[10,11] Inaccuracies in the pyrolysis time can, therefore, account for at most a factor of two variation in the reported rates.

Mass Transfer Limitations

The role of mass transfer can be viewed from two different perspectives: 1) its effect on the product yields, 2) its effect on product evolution rates. The effect of mass transfer on pyrolysis has been addressed in recent reviews by Howard,[26] Gavalas,[27], and Suuberg.[28] The consensus is that, in principal, one cannot interpret coal pyrolysis yields without considering the coupling of mass transfer and kinetics. A key role is played by the tar species which can be converted to char and light gas via secondary reactions if they are hindered in their escape from the particle. The escape of tars may be hindered by either intraparticle pore transport in non-softening coals and evaporation in the case of softening coals. The relative importance of tar secondary reactions is greater in the latter case since softening coals typically produce more tar.

Models for coupled reaction and external transport of tar suitable for softening coals have been developed by Unger and Suuberg,[29] Solomon and King[30] and Squire et al.[31] According to these models, external transport at 1 atmosphere pressure and typical pyrolysis temperatures will not limit the transport of tar fragments of molecular weights of less than about 300 amu, even at heating rates of 20,000 K/sec. The transport limitation on larger molecules can lead to their cracking before leaving the coal particle. Because the larger molecules can crack to smaller molecules which are not mass transfer limited, the mass transport limitations will affect product molecular weight distribution, but have only a limited effect on the evolution rate.

A model for coupled reaction and internal transport of tars suitable for non-softening coals has been developed by Gavalas and Wilks.[32] Their model makes predictions of trends for pressure and particle size effects in terms of the intraparticle tar concentration which are in agreement with experiment.

Using a similar approach, Russel et al.[33] have presented estimates of the time scales for both diffusion and bulk flow as 10^{-3} sec in the case of 100 μm particles over the temperature range of 600 - 1000°C, at 1 atm pressure. This analysis was done for coals which retain their pore structure, although it has been applied successfully to those which do not. The time scale for mass transfer is, therefore, less, than the measured reaction time for almost all experiments. The time scales for mass transfer do, however, approach the time scales which have been measured in the HTR for kinetics of rapid pyrolysis.[10,11] Consequently, one might expect some mass transfer effect. However, for these experiments the analysis is made long after the heating cycle is complete, leaving ample time for removal of light pyrolysis products from the coal particles. For tars, however, molecules which are not volatile enough to escape the coal particles while they are hot may recondense upon cooling. This problem can be eliminated by extracting the tar with a solvent.

137

There are very few data or models available to assess the role of mass transfer on real-time measurements of product evolution. One exception is the work of Arendt and Van Heek[34] who investigated the effects of pressure up to 70 atm on the kinetics and yields of various products in a thermobalance at low heating rates (3 K/min) with on-line gas analysis. They found little effect on product evolution kinetics over this range of temperature even though significant changes in yields were observed.

Our conclusion is that while mass transfer limitations may affect product distributions and may have a limited affect on the kinetics for the higher heating rate experiments, it is not a major factor in explaining the wide variation in rates for experiments discussed in this paper. However, it should be noted that in many practical applications of pyrolysis kinetic models, such as in making predictions of ignition in combustion systems, a consideration of mass transport limitations will be important.

Influence of Heat Transfer Model Assumptions on Reported Rates in Entrained Flow Reactors

Heat transfer calculations have been performed for several of the entrained flow reactor experiments[2,7,8,14,23] as well as the heated tube reactor experiment.[10] As discussed by Maloney and Jenkins,[14] the accuracy of these calculations is open to question. For the entrained flow reactor experiments, the most difficult factor to determine is the effect of mixing of the injected coal and carrier gas with the preheated gas. In reference 23, the mixing was treated using a parameter which was adjusted to fit the temperature measurements. In the absence of such measurements, the effect of mixing is very difficult to predict. In addition to mixing, there are coal physical properties, the values of which have differed among models. These are the heat capacity and the emissivity.

For the heat capacity, the room temperature value has typically been used.[2,7,8,13,14] Data of Lee[35] and a model and data reported by Merrick,[36] however, indicate that the heat capacity increases by about a factor of 2.5 in going from room temperature to 773 K. The model developed in Refs. 10, 23 and 24 uses Merrick's model for the heat capacity.

To calculate the absorption of radiation by coal particles, coal has typically been assumed to be a gray body with emissivity values between 0.8 and 1.0.[2,7,8,12,13,14] Recent measurements[20-22] have shown, however, that while char particles are gray bodies with emissivities in this range, small coal particles typical of pulverized combustion are not gray and have spectral emittances which are dependent on particle size, coal rank and the extent of pyrolysis. Figure 5 compares the emitted radiation from char and coal particles to that emitted by a black or gray body. As can be seen in Fig. 5b, coal particles emit (and absorb) much less radiation than a black body. The average emittance at a typical furnace temperature for the −200, +325 mesh fraction of lignite shown in Fig. 5b would be about 0.4. The model of Refs. 10, 23, and 24 employs a spectral emittance which varies with particle size, the extent of pyrolysis and the temperature of the furnace.

A sensitivity analysis was done to examine the importance of the various submodels in prediction of the particle temperature. A series of 5 cases was examined for two temperature levels (800 and 1600°C): 1) C_p = 0.3 cal/g K; ε = 1.0; zero heat of pyrolysis; constant mass; 2) add particle mass (kinetic) submodel; 3) Add heat capacity submodel; 4) Add emissivity submodel; 5) Single particle in infinite gas. For successive cases the previous changes were retained. Results are presented for 200 x 325 mesh North Dakota lignite in Figs. 6a and 6b for 800°C and 1600°C experiments in the EFR, respectively.

138

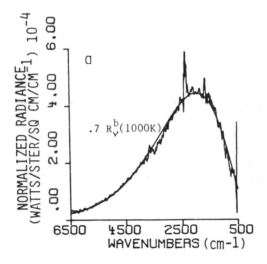

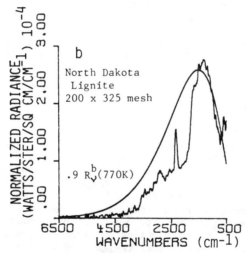

Figure 5.   Normalized Emission Spectra Compared with Theoretical Gray-Body
Curves   for a) North Dakota Lignite Char Previously formed at 1573   K.   b)
North Dakota Lignite Coal, 200 x 325 mesh.

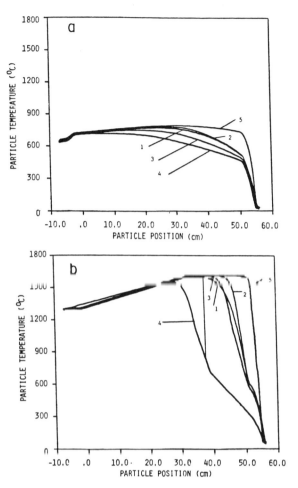

Figure 6. Effect of Various Model Assumptions on Predicted Temperatures in the Entrained Flow Reactor for North Dakota Lignite (200 x 325 mesh) Injected 56 cm above the Optical Port. Case 1: C_p = 0.3 cal/g K; ϵ = 1.0; Zero Heat of Pyrolysis; Constant Mass. Case 2: Add Particle Mass (kinetic) Submodel. Case 3: Add Heat Capacity Submodel. Case 4: Add Emissivity Submodel. Case 5: Single Particle in Infinite Gas. a) 800°C EFR Experiment in Nitrogen and b) 1600°C EFR Experiment in Nitrogen.

At 800°C, the particle temperature predictions are most sensitive to the variation in heat capacity with temperature and to a lesser extent on the emissivity assumptions. These make a difference of 50 – 100°C in the maximum computed temperature. At 1600°C (Fig. 1b) the predictions are extremely sensitive to the emissivity as well as the heat capacity models. For this case, the predicted particle temperature during pyrolysis is 800°C lower using the new models[23,24] than previous assumptions.[2,7,8,12,13,14] In addition, since primary pyrolysis occurs further away from the equilibrium temperature, the variation in particle mass becomes more important.

The net pyrolysis reaction heat was estimated to be between 60 and 80 cal/g (endothermic) for the 800°C and 1600°C experiments, respectively.[23,24] In both cases, the reaction heat had a negligible effect on the calculated particle temperature.

For both temperature levels, the heating rate of a single particle introduced without any cold gas, case 5, is significantly different than the more typical conditions of finite coal and carrier gas rates. This illustrates the sensitivity of particle temperature models to the assumptions concerning mixing and particle loading.

It therefore appears that, given the difficulty of accounting for mixing, possible underestimates of the heat capacity and possible overestimates of the emissivity, a factor of two inaccuracy in the particle heating rate is likely for the entrained flow reactor calculations. An inaccuracy of this magnitude in the heating rate can lead to errors of hundreds of degrees in the calculated particle temperature during pyrolysis. Such may be the case for Refs. 2,5,7,8,13 and 14. The direct measurement of temperature in entrained flow reactors is essential.

Influence of Heat Transfer Model Assumptions on Reported Rates in Heated Grid Reactors

It has been assumed that the coal particle temperature follows that of the thermocouple bead attached to or close to the screen. This assumption should be validated over the range of conditions of pressure, temperature and heating rates employed. However, no models for the particle temperature in heated grids have been published. Such models are needed to more rigorously evaluate the temperature data, and hence the reported rates for these reactors.

Inaccuracies in Measuring Particle Temperature

Among the papers reporting pyrolysis rates, particle temperatures were directly measured only for Refs. 10, 11, 23 and 24 by FT-IR E/T spectroscopy and Ref. 18 by three-color pyrometry. A possible problem of the temperature measurement in pyrolysis is temperature gradients within the particle. Since in both experiments the particles heat from the outside in, the surface could be hotter than the interior. In Refs. 10 and 24, the heating rates were on the order of 20,000 K/sec and 10,000 K/sec, respectively. The temperature difference ΔT from the center to the surface of the particle was estimated by a standard method.[37] Using a value of 1.2×10^{-3} cm^2 s^{-1} for the thermal diffusivity α at 700 K,[38] ΔT would be less than 30 K for heating rates less than 25,000 K s^{-1}. At higher temperatures, α increases sharply and ΔT would be even lower. Consequently, the assumption of uniform particle temperature would appear to be valid for Refs. 10,11,23,24.

For the laser heating experiment,[18] the heating rates are on the order of 10^6 K/sec. When the surface temperature has reached 2000 K, the calculated ΔT for a 65 μm diameter particle is ∿ 1500 K. The particle temperatures in Ref. 18 may, therefore, be highly non-uniform. This is in agreement with the observation that the particles undergo jet propelled motion in the direction of the laser beam when heated only from one side.

Soot formation from the evolving tars can also influence the temperature measurement. This problem was recently discussed by Grosshandler.[39] When a hot soot cloud surrounds the particle, the infrared particle temperature can be as much as 100% higher than the actual particle temperature. This can effect the temperature measurements in Ref. 18 where the peak temperatures in excess of 2000 K will promote soot formation. The temperatures during primary pyrolysis of Refs. 10, 23 and 24 will not be affected since no soot is produced in these experiments until after primary pyrolysis is complete.

The non-gray body emittance of coal (discussed above) can also affect the three color pyrometer temperature measurement which assumes a gray body. The spectral emittance has been determined for the FT-IR E/T measurement[20-22] and is used to determine the temperature.[10,24]

The temperature measurement is therefore a reasonable indication of the overall particle temperature in Refs. 10, 11, 23, and 24, but not in Ref. 18. For Ref. 18, a non-isothermal model is required to obtain kinetic rates.

The same caution on non-isothermal particle temperatures, soot, and non-gray body emittance must be applied to several other measurements of coal particle temperatures under very rapid heating, high temperature conditions which have been reported.[40-42]

Influence of Model Assumptions on the Reported Rates

Several different models have been used to describe the weight loss or tar loss during pyrolysis. The rates for Refs. 1,2,3,5,12, and 14 employ a single first order rate expression to describe weight loss. Reference 18 uses a two parallel reaction model. The two rates from Ref. 18, however, are close enough to be compared with the single rate model. References 4, 6-11 13, 23 and 24 report the rates for tar evolution. For bituminous coals, and lignites below 800°C, the tar (and products such as aliphatic gases which are released at rates similar to the tar) account for the major portion of the weight loss and so can be compared with the total weight loss. As can be seen in Table I, the single first order rate to describe weight loss for Refs. 10 and 23 is close to the tar rate and is between those for tar loss and aliphatic gas loss. For lignites at high temperatures, the slower evolution of CO can make the rate for overall weight loss appear substantially lower than the tar loss rate. Consequently, some of the discrepancy in reported rates can be attributed to comparing models which account only for rapid primary pyrolysis weight loss and those which also include the slower gas evolution from char formation.

The most significant difference among models appears to be between the single rate models and the distributed rate models employed in Refs. 1, 6, 7-11, 13, 23, and 24. The data and analysis of Anthony et al.[1] illustrate this problem. They presented two kinetic interpretations for their data: a single first-order process and a set of parallel processes with a Gaussian distribution of activation energies. Both interpretations fit the data using Arrhenius expressions for kinetic rates. For the bituminous coal, the single first-order process, which uses an activation energy of 13.3 kcal mol^{-1}, requires two parameters, while the distributed-rate model, which uses a mean activation energy of 50.7 kcal mol^{-1}, requires a third parameter to describe the spread in rates. While both models match the experimental data of Anthony et al[1], extrapolation to higher temperatures would give very different results for the two models. Similar results were observed by Suuberg et al.[6]

Part of the difficulty in the interpretation of the heated grid data of Anthony et al.[1] using a single first-order process is that the resultant low activation energy makes the analysis sensitive to possible pyrolysis occurring during the cooling period. As noted by Anthony in his thesis[1], this results in a lower apparent kinetic rate at a given temperature than if cooling is neglected (compare lines 1a and 1b in Fig. 1). Conversely, the multiple reaction model, which is centered around a high activation energy, is not much influenced by possible pyrolysis occurring during cooling. When the two models are used for a time-temperature history other than that used in Anthony's experiment, the predictions can be significantly different. This point has recently been discussed in a paper by Truelove and Jamaluddin.[43]

Another example of the variability in reported kinetic rate constants which can result from the assumptions of the analysis is illustrated in Fig. 7 (from Ref. 9), which shows numerical fits to the pyrolysis data of Campbell[44] for CO_2. The data are for the evolution rate for CO_2 measured as the coal sample was heated at a constant rate. A single first order process would produce a single peak. Campbell assumed the observed double peaks were due to two distinct sources (which we have designated loose and tight) and fitted each source (Fig. 7a) as a simple first order process using a rate constant which follows the Arrhenius temperature dependence. The low activation energy indicated in Fig. 7a is required because the peaks are wide. But suppose the wide peaks are caused by a distribution in activation energy. The simulations of Figs. 7b and c are performed assuming the Gaussian distributed activation energy model of Anthony et al.[1] where σ is the width of the distribution. Reasonable fits to the data can be obtained with arbitrary values of σ. The assumed value of changes the average activation energy and frequency factor substantially.

It is therefore obvious that the experimental data from experiments over a limited range of heating rates is insufficient to uniquely determine the rate constants. Such ambiguities can lead to wide variations in slopes of the lines in Fig. 1. Extrapolation of the rate outside of the original conditions may be very inaccurate. Eliminating such ambiguities can only be accomplished by considering additional experiments which provide sufficient variations in heating rate, reaction time, and final temperature. The multiple rate models appear to be applicable over a wider range of conditions than do the single rate models.

DETERMINATION OF CHEMICAL KINETIC RATE

Based on the above discussion, it is possible to choose appropriate experiments to define chemical kinetic rates. The HTR experiment allows the measurement of both time and particle temperature and does not require rapid mass transfer. The TGA/EGA experiment allows measurement of the time and the heating rate is slow enough so that the sample and nearby thermocouple are at the same temperature. Mass transport limitations are not a problem.

Each HTR experiment was analyzed to obtain the time to evolve 63% of the tar. The reciprocal of this time was plotted in Fig. 8 at the average temperature during pyrolysis. The HTR data are plotted as solid circles for 3 coals (North Dakota lignite, Montana Rosebud subbituminous and an Illinois #6 bituminous) at several conditions. The TGA/EGA data are plotted as squares for several coals (North Dakota lignite, Montana Rosebud subbituminous, Illinois #6 bituminous and Pittsburgh No. 8 bituminous.

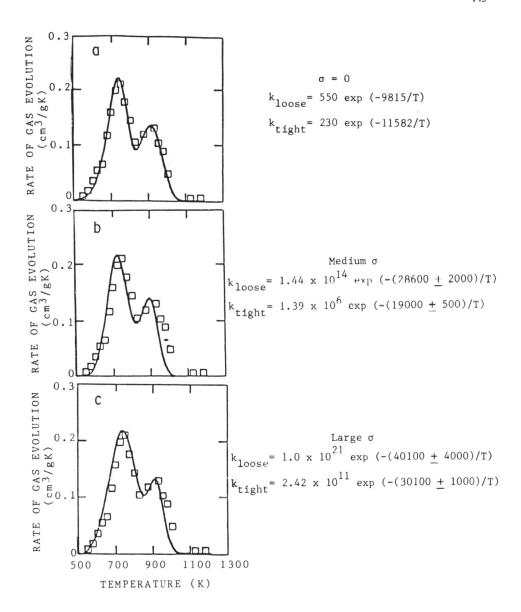

Figure 7. Effect of Variation of Distribution Parameter, σ on the Kinetic Rates. (Data of Campbell Ref. 44 for a heating rate of .055 K/sec).

144

Apparent rates were obtained using analysis of Juntgen and Van Heek[45] and an assumed prexponential factor of 8.57×10^{14}. The calculated rate is plotted at the temperature for the peak in the tar evolution rate curve. A kinetic rate was then defined using a Gaussian distribution of activation energies which fits both sets of data. The same procedure was followed with other pyrolysis species and the resulting rates were tested and sometimes adjusted by comparing to data from the EFR.[23,24]

Reactor independent rates using a Gaussian distribution of activation energies for weight loss, tar evolution and aliphatic gas evolutions have thus been established (see Table I) from three experiments[10,11,23,24] with heating rates of 0.5 K/sec (slow heating rate), 3000-12,000 K/sec (medium heating rate), and 20,000 K/sec (high heating rate). Particle temperatures were measured for the medium and high heating rate experiments and can be accurately determined from thermocouples in the surrounding gas for the slow heating rate experiment. Pyrolysis times were accurately determined for the slow and high heating rate experiments.

The rates from the three experiments are the highest which have been reported at 800°C, supporting the idea that for these experiments there is less effect from heat transfer limitations than has previously prevailed. On the basis of the above discussions, the rates should reflect the chemical kinetic processes in the coal. While the rank independent rates provide a good fit to the data, some improvements can be made with rates which vary systematically with rank.

INTERPRETATION OF TAR EVOLUTION RATES

Assuming that the rate for tar evolution is a chemical kinetic rate, how can it be interpreted? The activation energy of 54.6 kcal/mole is close to what is expected from thermochemical kinetics for ethylene bridges between aromatic rings[27] and agrees with pyrolysis rates for model compounds[46] and polymers[30,31] where these bonds are the weak links. The rates for four of these model compounds are plotted in Fig. 8. They are in good agreement with the coal tar evolution rate.

CONCLUSIONS

- There is a wide variation in the rates for pyrolysis weight loss or tar loss reported in the literature.

- It is unlikely that variations with coal rank, inaccuracies in measuring weight loss, pyrolysis times or mass transfer limitations can account for the wide variations in rates.

- The determinations of particle temperature appear to be a major source of the variation in rates reported from entrained flow reactor experiments. Differences in the assumptions employed to describe mixing and the values assumed for coal particle heat capacity and emissivity can easily lead to errors of factors of two or more in the heating rate. This can lead to hundreds of degrees variation in particle temperatures during pyrolysis. Direct measurement of particle temperatures is essential.

- Particle temperatures have been measured by three color pyrometry for Ref. 18. For the high heating rates ($\sim 10^6$ K/sec) and high temperatures (> 2000 K) of this experiment, it is likely that surface temperatures and the temperatures of soot surrounding the

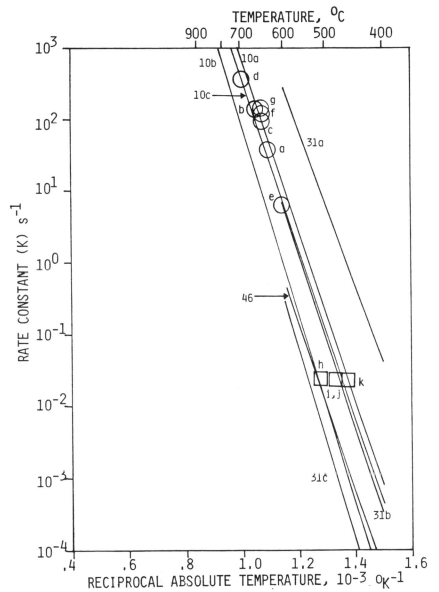

Figure 8. Comparison of Kinetic Rates for Tar Evolution from Coal,
Polymers, and Model Compounds. The Numbers Next to Each Line Refer to a
Reference and Case in Table I. The Data Points are Rates for Evolution
from Experiments in the HTR (O) and TGA/EGA (□). The Letters Next to
Each Data Point Indicate the Coal Type and Experimental Conditions. The
Experimental Conditions for the HTR Experiments are given in Ref. 10. The
Experimental conditions for the TGA/EGA Experiments were 0.5 K/s to 900°C.
Coal Types were as Follows: North Dakota (Zap) Lignite (b,c,d,e,k);
Montana Rosebud Subbituminous (a,f,i); Illinois No. 6 Bituminous (g,j);
Pittsburgh No. 8 Bituminous (h).

particles can exceed interior temperatures by hundreds of degrees. These effects will yield low pyrolysis rates at an "apparent" high temperature. Proper interpretation of these data would require a non-isothermal particle model.

- Particle temperatures were measured by FT-IR E/T spectroscopy for Refs. 10, 11, 23, and 24. Heating rates are equal to or less than 20,000 K/sec and temperatures during primary pyrolysis are less than 1073 K. For these conditions, temperature differences within the particles are calculated to be less than 30 K and soot is not formed. The measured temperatures are therefore representative of the temperature throughout the particle and an isothermal model of pyrolysis is applicable.

- In programmed heating experiments, such as the heated grid, there is less uncertainty in the knowledge of temperature and time than for entrained flow reactors. The assumption that the measured thermocouple is a good indication of the coal particle is probably adequate for experiments in helium at above 1 atm pressure and below 1000 K/s heating rate, although additional modeling will be required to firmly establish this. However, even in those cases, large variations in the rate constants can be obtained which depend on the pyrolysis model assumptions. Single first order rate models yield low activation energies, which can give disproportionate emphasis to possible pyrolysis occurring during cooling and lead to low apparent rates. Models employing a distribution of activation energies can yield much higher values for the average activation energy, depending on the assumed width of the spread in the distribution. Experiments with large variations in particle heating rates are required to determine the proper width in the distribution.

- Reactor independent rates for weight loss and tar evolution have been established from three experiments[10,11,23,24] with heating rates between 0.5 K/sec and 20,000 K/sec. Particle temperatures were measured for the high heating rate experiments and can be accurately determined from thermocouples in the surrounding gas for the 0.5 K/sec experiment.

- The rate for tar loss which fits the three reactors for lignite and bituminous coals using a Gaussian distribution of activation energies model is $k = 8.57 \times 10^{14} \exp(-54,570/RT)sec^{-1}$ with $\sigma = 3.0$ kcal/mole. A single first order rate which is a reasonable approximation for the initial rapid weight loss of primary pyrolysis is $k = 4.28 \times 10^{14} \exp(-54,570/RT)sec^{-1}$.

- While the rates from the three experiments are the highest which have been reported at 800°C, the rates are in good agreement with the chemical kinetic rates for bond breaking of an ethylene bridge between aromatic rings in polymers and model compounds.

- While mass transfer limitations do not affect the determination of rates where products are collected long after pyrolysis occurs, such limitations might well affect processes such as ignition where products are rapidly consumed. Such limitations must be considered in combustion or gasification models.

ACKNOWLEDGEMENT

Support for this work was provided by the Morgantown Energy Technology Center under contracts DE-AC21-81FE05122, DE-AC21-84MC21004 and DE-AC21-85MC22050.

REFERENCES

1. Anthony, D.B., Howard, J.B., Hottel, H.C. and Meissner, H.P., 15th Symposium (Int) on Combustion, The Combustion Institute, Pittsburgh, PA pg. 1303 (1975) and Anthony, D.B., Sc.D. Thesis, M.I.T., Dept. of Chemical Engineering, Cambridge, MA (1974).
2. Kobayashi, H., Howard, J.B. and Sarofim, A.F., 16th Symposium (Int) on Combustion, The Combustion Institute, Pittsburgh, PA, pg. 411, (1977). and Kobayashi, H., Ph.D. Thesis, M.I.T. Dept. of Mechanical Eng., Cambridge, MA (1976).
3. Niksa, S., Heyd, L.E., Russel, W.B. and Saville, D.A., 20th Symposium (Int) on Combustion, The Combustion Institute, Pittsburgh, Pa, pg. 1445, (1984).
4. Solomon, P.R. and Colket, M.B., 17th Symposium (Int) on Combustion, The Combustion Institute, Pittsburgh, PA pg. 131 (1978).
5. Badzioch, S. and Hawksley, P.G.W., Ind. Eng. Chem. Process Design Develop., **9**, 521, (1970).
6. Suuberg, E.M., Peters, W.A. and Howard, J.B., 17th Symposium (Int.) on Combustion p. 131, The Combustion Institute, Pittsburgh, PA (1979).
7. Solomon, P.R. and Hamblen, D.G., EPRI Final Report for Project RP 1654-8 (1983).
8. Solomon, P.R. and Hamblen, D.G., Chemistry of Coal Conversion, Editor, R.H. Schlosberg, Plenum Press, NY, Chapter 5, pg. 121, (1985).
9. Solomon, P.R. and Hamblen, D.G., Progress Energy Combustion Science, **9**, 323 (1984).
10. Solomon, P.R., Serio, M.A., Carangelo, R.M., and Markham, J.R., "Very Rapid Coal Pyrolysis", Fuel, **65**, 182 (1986).
11. Solomon, P.R., Serio, M.A., Carangelo, R.M., and Markham, J.R., ACS Div. of Fuel Chem. Preprints, **30**, 1, 266, (1985).
12. Freihaut, J.D., Ph.D. Thesis, Pennsylvania State University, (1980).
13. Solomon, P.R., Hamblen, D.G., Carangelo, R.M. and Krause, J.L., 19th Symposium (Int) on Combustion, The Combustion Institute, Pittsburgh, PA, pg. 1139 (1982).
14. Maloney, D.J. and Jenkins, R.G., 20th Symposium (Int) on Combustion, The Combustion Institute, Pittsburgh, PA, 1435, (1984).
15. Lockwood, F.C., Rizvi, S.M.A., Lee, G.K., and Whaley, H., 20th Symposium (Int) on Combustion, The Combustion Institute, Pittsburgh, PA, pg. 513, (1984).
16. Truelove, J.S., 20th Symposium (Int) on Combustion, The Combustion Institute, Pittsburgh, PA, 523, (1984).
17. Jost, M., Leslie, I., Kruger, C., 20th Symposium (Int) on Combustion, The Combustion Institute, Pittsburgh, PA, 1531, (1984).
18. Witte, A.B. and Gat., N., "Effect of Rapid Heating on coal Nitrogen and Sulfur Release", Presented at the DOE Direct Utilization AR & TD Contractors' Meeting, Pittsburgh, PA (1983).
19. Solomon, P.R., Best, P.E., Carangelo, R.M., Markham, J.R., Chien, P., Santoro, R.J. and Semerjian, H.G., "FT-IR Emission/Transmission Spectroscopy for In-Situ Combustion Diagnostics", Accepted for the 21st Symposium (Int.) on Combustion, (1986).
20. Best, P.E., Carangelo, R.M., Markham, J.R., and Solomon, P.R., "FT-IR

148

Emission - Absorption Measurements from Coal and Char: Temperature and Emittance", Combustion and Flame, to be published (1986).

21. Solomon, P.R., Carangelo, R.M., Best, P.E., Markham, J.R., and Hamblen, D.G., "Analysis of Particle Composition, Size and Temperature by FT-IR Emission/Transmission Spectroscopy", ACS Div. of Fuel Chem. Preprints, **31**, 141, (1986).

22. Solomon, P.R., Carangelo, R.M., Best, P.E., ,Markham, J.R., and Hamblen, D.G., "The Spectral Emittance of Pulverized Coal and Char", Accepted for 21st Symposium on Combustion, Munich, Germany, August, 1986.

23. Serio, M.A., Solomon, P.R., Hamblen, D.G., Markham, J.R. and Carangelo, R.M., "Coal Pyrolysis Kinetics and Heat Transfer in Three Reactors", Poster Session for Combustion Symposium, Munich, Germany, August, 1986.

24. Solomon, P.R. Serio, M.A., Hamblen, D.G., Squire, K.R., Carangelo, R.M., Markham, J.R. and Heninger, S.G., "Coal Gasification Reactions with On-Line In-Situ FT-IR Analysis", Final Report for U.S. Department of Energy, Morgantown Energy Technology Center, Contract No. DE-AC21-81FE051222, December, 1985.

25. Carangelo, R.M., Solomon, P.R., and Gerson, D.J., ACS Fuel Chemistry Preprints, **31**, No. 1, p. 152, April 13-18 Meeting, New York (1986).

26. Howard, J.R., in "Chemistry of Coal Utilization", M.A. Elliot, (Ed.), Wiley, New York, Chapter 12, (1981).

27. Gavalas, G.R., "Coal Pyrolysis", Elsevier, New York, Chapter 5, (1982).

28. Suuberg, E.M., in "Chemistry of Coal Conversion", Plenum Press, New York, R.H. Schlosberg, (Ed.), Chapter 4, (1985).

29. Unger, P.E. and Suuberg, E.M., 18th Symposium (Int) on Combustion, The Combustion Institute, Pittsburgh, PA, 1203, (1981).

30. Solomon, P.R. and King, H.H., Fuel, **63**, 1302, (1984).

31. Squire, K.R., Solomon, P.R., Carangelo, R.M. and DiTaranto, M.B., Fuel, **65**, 833 (1986).

32. Gavalas, G.R. and Wilks, K.A., AIChE J., **26**, 201, (1980).

33. Russel, W.B., Saville, P.A., and Greene, M.I., AIChE J. **25**, 65 (1979).

34. Arendt, P., and Van Heek, K.H., Fuel, **60**, 779 (1981).

35. Lee, A.L., ACS Div. Fuel Chem. Preprints, **12**, (3), 19, (1968).

36. Merrick, D., Fuel, **62**, 540, (1983).

37. Carlsaw, H.S. and Jaeger, J.C., "Conduction of Heat in Solids", 2nd Ed., Clarendon Press, Oxford (1959).

38. Badzioch, S., BCURA Mon. Bull., **31**, 193 (1967).

39. Grosshandler, W.L., Combustion and Flame, **55**, 59, (1984).

40. Ballantyne, A., Chou, H.P., Orozvo, N. and Stickler, D. "Volatile Production During Rapid Coal Heating", Paper to US Dept. of Energy Direct Utilization AR & TD Contractors Review Meeting, Pittsburgh, 1983.

41. Seeker, W.R., Samulesen, G.S., Heap, M.P. and Trolinger, J.D., in 18th Symposium (Int) on Combustion, Combustion Institute, 1981, p. 1213.

42. Midkiff, K.C., Altenkirch, R.A. and Peck, R.E., Combut. Flame, **64**, 253 (1986).

43. Truelove, J.J., Jamaluddin, A.S., Combustion and Flame, **64**, 369 (1986).

44. Campbell, J.H., Fuel, **57**, 217 (1978).

45. Jüntgen, H. and Van Heek, K.H., Fuel, **47**, 103 (1968) and "Progress Made in the Research of Bituminous Coal," paper given at the annual meeting of the DGMK, Salzburg, 1968, Translated by Belov and Associates, Denver, CO, APTIC-TR-0779 (1970).

46. Stein, S.E., "New Approaches in Coal Chemistry", (Ed. B.D. Blaustein, B.C. Bockrath and S. Friedman), ACS Symposium Series, **169**, ACS, Washington, D.C., 208 (1981).

DISCUSSION

H. Juntgen

1) I could not get the meaning of rate constant reported on : overall rate constant ? rate constants of tar formation ? rate of CH_4 formation ?

2) What kinetic law do you use ? First order kinetics ?

3) Our experience is that experimental curves cannot exactly be described using a simple kinetic law with first order, so that measurements are not properly evaluated due to the incorrect assumption of the very simple kinetic law. See e.g. publications of Hanbada, Juntgen, Peters or Juntgen, Van Heek. Good coincidence between experimental curves and calculation is only achieved, if the reaction order is between 1 and 2 or if activation energy distribution is used.

4) I have some doubt on your conclusion, that kinetics of pyrolysis generally is independent on coal rank.

P.R. Solomon

Comparison using two different models were made in the paper. The most accurate model is our "Functional Group (FG) model". This model is described in several references given in the paper. It describes pyrolysis as the parallel evolution of individual species, i.e. tar, aliphatic gases, CO, CO_2, H_2O, CH_4 etc. Each species "i" is modeled as a single first order process

$$\frac{dW_i}{dt} = - k_i W_i$$

where a gaussian distribution of activation energies model is used for each. So k_i is described by three parameters, k_{oi}, the pre-exponential factor, Eo_i, the activation energy and σ_{oi}, the width in the distribution. In the paper we have reported the rate constants for tar and aliphatic gases and compared these to the weight loss rate constants. The justification for this is that tar and aliphatic gases dominate "primary pyrolysis" especially for bituminous coals.

We use one set of rate constants for all coals and find that good agreement to the data can be obtained. Only the amounts of each species were varied with rank. There are some variations in rate with rank, about a factor of 5 increase in the rate for tar loss in going from high rank bituminous coals to lignites. This factor of 5 is small, however, when compared to the factors of 100 to 1000 variations among rate constants reported by different investigators.

In addition to the "F.G." model, we have also used a single first order model for weight loss. This simple model does not fit the data as well as the F.G. model. It varies more sharply with temperature. It does, however, match the

data at the center of the distribution (i.e. at half the
weight loss) over a wide range of heating rates (0.5 K/s to
20.000 K/s).

A. Williams

Values of A & E derived from heated grid experiments nu-
merally form a plane having a mathematical surface with elongated contours which makes them difficult to separate with
accuracy, and may thus lead to "high" or "low" activation
energies. Your experiments give direct experimental information on rate constants but your experimental technique has the
disadvantage that tar may be lost from coal particles by
disruptive bubble bursting. Consequently your rate constants
may be too high by 10 to 20 % because we have some evidence
that this amount of tar may be released by disruptive
ejection. Please can you comment.

P.R. Solomon

You may be correct. Considering the factors of 100 to
1000 variation in rate constants reported in the literature we
would be overjoyed with an accuracy of 20 %.

M. Hertzberg

I generally agree with your viewpoint that there is a
major uncertain in particle temperature (or temperature profile) associated with heat transport limitations. In addition to
the higher heat capacity and lower emissivity values that you
mentioned, it is also necessary to modify the Nusselt number
used to calculate the heating flux to the particle. Once there
is any significant amount of volatiles emitted from the
particle there is a drastic decline in the temperature
gradient in the gas phase and the Nusselt number will drop by
an order of magnitude. The simple Nusselt heat transfer model
for an inert particle assumes that the high temperature gas is
displaced about one particle radius from the surface and that
the boundary layer through which heat is conducted has a
thickness of only one radius. However, as soon as any mass is
emitted from the particle that boundary layer advances with
the flow of pyrolysis products to distances that can be an
order of magnitude larger. The heat transfer rate would then
drop by orders of magnitude.

R.P. Solomon

This is a good point which requires consideration. The
situation you describe, where the volatiles are evolved in a
homogeneous cloud will prevail for CO_2 and H_2O prior to coal
melting. This evolution will decrease heat transfer. However
if the coal melts, and volatiles are evolved in a jet, the
particle can move more rapidly relative to the surrounding gas
or increase the mixing with the surrounding gas and the heat
transfer could increase.

MODELLING OF COAL DEVOLATILIZATION WITH A NON-LINEAR HEATING RATE.

G. PRADO, S. CORBEL AND J. LAHAYE
CENTRE DE RECHERCHES SUR LA PHYSICO-CHIMIE DES SURFACES
SOLIDES - 24 Avenue Kennedy, 68200 MULHOUSE (France)
and
ECOLE NATIONALE SUPERIEURE DE CHIMIE
3 Rue A. Werner, 68200 MULHOUSE (France)

The paper of J.B. Howard in this book reviews the different models of coal devolatilization which have emerged from experimental studies.

Among these models, the multiple reactions model appears to represent, with a sufficient accuracy, the experimental results obtained mainly with the heated grid technique, under isothermal conditions and for linear heating rates.

In practical systems, the coal is submitted to non-linear heating rates, and it is important to assess the effect of practical heating rates on the kinetics of coal devolatilization. The objective of this note is to present a computation of coal devolatilization under the heating rate which corresponds to the drop tube furnace experiments developed in our laboratory, and to compare computed results with experimental measurements of combustion times of volatile matter.

MODELLING OF COAL DEVOLATILIZATION

We have used the model of Anthony and Howard (1), which represents coal devolatilization with a large number of parallel and independent reactions. The rate of production of volatile matter due to reaction i is :

$$\frac{dV_i}{dt} = k_i \ (V_i^* - V_i)$$

k_i : kinetic (Arrhenius) constant for reaction i.

V_i^* : maximum total amount of volatile matter produced through reaction i.

The kinetic constants k have all the same pre-exponential factor, k_0, and differ only by their activation energy E_i

$$k_i = k_0 \ \exp(-E_i/RT)$$

Furthermore, the number of reactions is assumed to be large enough to describe the activation energies E as a continuous function f(E). It follows that :

$$\frac{dV}{dt} = k_0 \ \exp(-E_i/RT) \ x \ (V^*F(E)dE - V_i)$$

Anthony and Howard represent f(E) with a Gaussian distribution of activation energy :

$$F(E) = 1/\sigma \sqrt{2\pi} \ \exp(-(E-E_0)^2 / 2\sigma^2)$$

For a bituminous coal, the values proposed by Anthony and Howard are :

$k_0 = 1.67 \times 10^{13} \ s^{-1}$

$E_0 = 54.8 \ kcal.mole^{-1}$

$\sigma = 17.2 \ kcal.mole^{-1}$

$V* = 57.2\%$

We have integrated numerically the system, with an increment $\Delta E = 1$ kcal.mole-1. Two ranges of activation energy were used :

$E_A = 0 - 61 \ kcal.mole^{-1}$ for primary reactions, and

$E_A = 61 - 90 \ kcal.mole^{-1}$ for secondary reactions.

Primary reactions form non-reactive species which escape from the coal. Secondary reactions form reactive species which are pyrolysed inside or at the surface of the grain.

The numerical method used is a GEAR method, included in the code LSODE, from Sandia National Laboratory.

COMPARISON WITH EXPERIMENTAL RESULTS

The experimental equipment is described elsewhere in this book (2). Briefly a single particle is introduced into a drop tube furnace, electrically heated up to 2000 K.

The heating rate is computed from a precise energy balance, taking into account convective, conductive and radiative heat transfer. The particle temperature versus time is plotted on fig. 1 for three furnace temperatures (1400 - 1600 and 1800 K) for a 80 μm diameter coal grain.

These three curves were fitted with an exponential law :

$T = A + B \exp (C.t)$

with

A (K)	B	C
1400	− 875	− 205
1600	− 1075	− 205
1800	− 1275	− 205

Temperature distribution inside the particle were also computed, using the general equation

$$\frac{\partial^2 T}{\partial r} + \frac{2}{r} \ \frac{\partial T}{\partial r} + \frac{1}{a} \ \frac{\partial T}{\partial t}$$

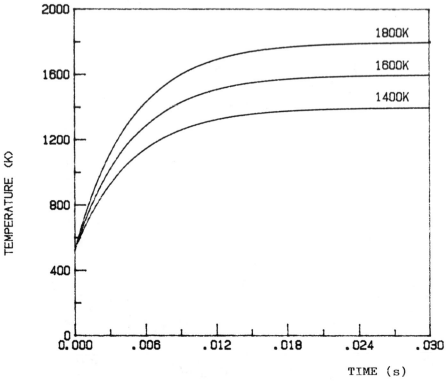

Figure 1 : Particle Heating Rate

a : thermal diffusivity = $\dfrac{\lambda}{\rho C_p}$

λ : thermal conductivity (1.25 W/(m.K))

ρ : particle density (860 kg/m³)

Cp : specific heat (711 J/kg.K)

These values are for a bituminous coal, and are supposed constant as a first approximation.

The general equation was solved numerically. Thirty dis-crete points inside the particle were considered and the re-sulting matrix solved with the Gauss method.

For a 80 µm particle, the maximum difference between surfa-

ce and core temperature is 9 K, for a furnace temperature of
1400 K. Consequently, temperature gradients inside coal grains
were neglected for devolatilization kinetics calculation.

The computed particle weight losses versus time for the
three exponential heating rates described above are compared
with a linear heating rate of 10^5 K.s $^{-1}$ on fig. 2.

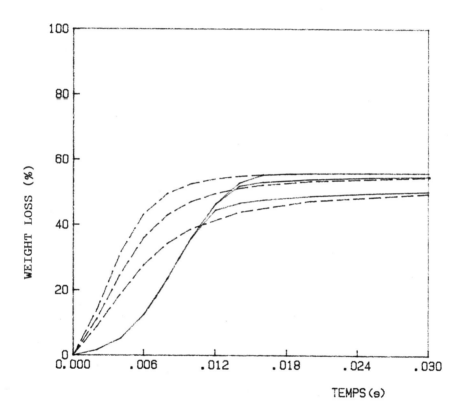

Figure 2 : Volatile Matter versus Time
——————— : linear heating rate
- - - - : exponential heating rate

156

The maximum particle temperatures are, for both cases, 1400, 1600 and 1800 K. The parameters used are those proposed by Anthony and Howard (see above). With the exponential heating rate one obtains a devolatilization duration of approximately 8 ms, whereas with the linear heating rate the devolatilization duration is about 14 ms.

Experimentally, we do not measure directly the devolatilization rate but only the duration of combustion of volatile matter around the particle.

These are measured either with a fast movie camera, or by recording the luminous flame around the grain with a two color pyrometer. They are certainly somewhat shorter than devolatilization time, but should not be too different, especially as in our equipment, combustion occurs in a quasi stagnant atmosphere.

We have measured the volatile matter combustion duration for approximately 100 particles, with initial diameters in the 70-90 μm range.

The combustion times range from 1 ms to 17 ms, with a maximum at 5 ms. The complete histograms is plotted on fig. 3.

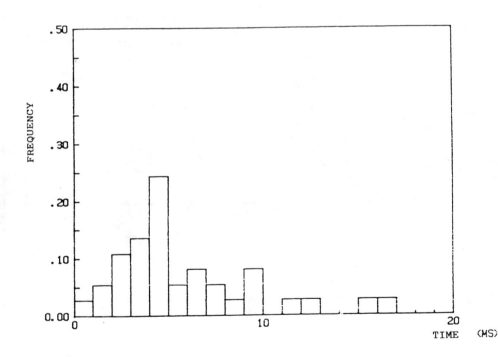

Figure 3 : Histogram of Time of Combustion of Volatile Matter.

These values are in excellent agreement with the predicted devolatilization rate.

CONCLUSION

The model of Anthony and Howard for the kinetics of devolatilization of a bituminous coal, coupled with an experimental non-linear heating rate, predicts very closely the duration of combustion of the volatile matter escaping from a burning coal particle. For 80 µm particles, heated at approximately 10^5 K/s, volatile matter combustion occurs in approximately 5 ms. The temperature gradient inside the particle is negligible.

REFERENCES

(1) Anthony D.B., Howard J.B., Hottel M.C. and Meismer H.P.
    15th Symp. (Int.) on Comb. (The Comb. Inst.)
    p. 1303 (1974).

(2) Prado G., Froclich D. and Lahaye J.
    Fundamentals of the Physical-Chemistry of Pulverized Coal
    combustion (J. Lahaye and G. Prado Ed.) Martinus Nijhoff
    Publ. (1987).

158

DISCUSSION

I.W. Smith

What is the effect of mass flux of volatiles on heat transfer? If circumferentially uniform rates flux then is Nu <<2 ?
If local jets of volatiles, is Nu >> 2 ?

G. Prado

We have not investigated these effects. In our furnace, we
use a quasi-stagnant atmosphere (very small slipping velocity), and the radiative heat transfert dominates, so then effects should be negligible.

Van Heeks

During the last years we have made a systematic study into
the transition from chemical reactions controlled to transport
controlled kinetics for the pyrolysis of coal particles. Thereby grain size ranged from 60 μ to 1 mm, heating rates were
up to 3000 K/s and pressure varied from 1 to 100 bar. the atmosphere was either N_2 or H_2. The results have been published
in FUEL 64, 1985 p. 571.

POROUS MORPHOLOGY OF COAL.
ITS TRANSFORMATION DURING PYROLYSIS.

J. LAHAYE AND G. PRADO
CENTRE DE RECHERCHES SUR LA PHYSICO-CHIMIE DES SURFACES
SOLIDES - 24 Avenue Kennedy, 68200 MULHOUSE (France)
and
LABORATOIRE D'ENERGETIQUE ET COMBUSTION
ECOLE NATIONALE SUPERIEURE DE CHIMIE DE MULHOUSE
2 Rue A. Werner, 68200 MULHOUSE

In the combustion of pulverized coal, devolatilization and gasification steps are dependent upon the porosity of initial coal and its development during these steps. Reciprocally, morphological transformation of coal (porosity, swelling) are strongly affected by devolatilization and gasification conditions.

The morphological evolution of pulverized coal particles during heat treatment is indeed important. It must be kept in mind, however, that the kinetics result from three main factors :

. diffusion processes
. surface areas involved
. intrinsic reactivity of surfaces.

The initial porosity of coal and its evolution during devolatilization will be successively examined. Pore evolution during heterogeneous combustion will be just mentioned only in the conclusion.

Some background on porosity and methods to characterize it will be given in the first part.

I. CHARACTERIZATION OF POROSITY AND SURFACE AREAS

When speaking of porosity, one often thinks of cylinders or of canals with circular cross sections. This oversimplification which is difficult to avoid may be misleading particularly for polyphasic materials such as coals.

Pores in coal may result either from the chemistry of transformation of kerogens into coal or from the mechanical formation of fissures between the different phases or macerals. The chemistry is expected to produce pores of small dimensions either closed or with circular apertures; the mechanical fractures produce fissures with very dissymetrical apertures. For conveniency, in this last case, the pore "diameter" is usually considered to be equal to the width of the fissures. Therefore, the traditional description of cylindrical pores for coal is acceptable for small pores but very idealized for large pores.

In 1962, I.U.P.A.C. classified pores as :
. micropores : diameters below 2 nm
. mesopores : diameters between 2 and 50 nm
. macropores : diameters larger than 50 nm.

Actually, the different techniques developed to characterize coal porosity as well as the different pore domains
corresponding to the processes responsible for devolatization
and gasification of coal lead to a slightly different classification. To avoid confusion with the I.U.P.A.C. definition,
pores in coal will be classified as follows :
Small pores : diameters below 1.2 nm
Medium pores : diameters between 1.2 and 30 nm
Large pores : diameters larger than 30 nm.
A comprehensive description of the analytical procedures
developed for characterizing the morphology of porous materials is of course out of the scope of the review. The main
methods used for coal characterization and their domains of
application will be summarized.

I.1. Pore volumes and pore size distributions
Three main techniques are used :
. helium pycnometry
. mercury pycnometry and porosimetry
. gas adsorption.
- In helium pycnometry, helium is assumed to penetrate the
entire open porosity. The density is quoted d
- In mercury porosimetry, mercury does not wet solids so
that a pressure must be applied for mercury to penetrate the
pore network. For non interconnected cylindrical pores and a
contact angle of mercury on solids equal to 141°, the classical Washburn equation (1), derived from Laplace equation can
be used :

$$P = \frac{1.5}{d} \times 10^5$$

with P in Pascal
     d in μm
Corrections have to be made to take into account the compres-
sibility of solid and the penetration of mercury into the
interparticle voids in the case of pulverized materials. These
corrections are particularly significant for small pore
volumes and for pore diameters smaller than 50 nm.
At atmospheric pressure (mercury pycnometry) pores with
diameters equal or larger than 15 μm are penetrated; the corresponding density is quoted d_{Hg}
The open pore volume for pore sizes below 15 μm in diameter is equal to

$$\frac{1}{d_{Hg}} - \frac{1}{d_{He}}$$

In assuming the value of the true density d_{tr} of the
solid i.e. the density of the solid material itself, the volume of the closed porosity is equal to

$$\frac{1}{d_{He}} - \frac{1}{d_{tr}}$$

- gas adsorption will be examined when describing the measurements of specific surface areas.

I.2. Surface areas
Two main methods are used to characterize coal surface
areas :
. nitrogen adsorption at the temperature of liquid nitrogen
. carbon dioxide at 0°C or at room temperature.

- Nitrogen adsorption

The adsorption isotherm of nitrogen at low temperature gives the area of the pores with diameters larger than ca. 5Å. The pore volume distribution can be computed from the adsorption or the desorption isotherm. From the adsorption branch, using the method of Cranston and Inkley (2), Walker et al. (3) considered that pore size distributions in the diameter range of 1.2 to 30 nm could be computed. Mercury and nitrogen porosimetrics overlap in the 10-30 nm domain, which is quite convenient for cross-examining experimental results.

- Carbon dioxide adsorption

Areas computed form carbon dioxide adsorption at 298 K are usually considered to be the closest approximation to the total surface area of coals because this measurement includes the surface area of the micropores which constitutes the majority of the surface. Using the Dubinin-Raduchkevich equation (4), the accessible pore volume also can be determined.

This equation can be written

$$\log V - \log V_o \quad \frac{BT}{\beta^2} \quad (\log \frac{P_o}{P})^2$$

where V = volume of gas adsorbed per gram of adsorbent at the

relative pressure $\frac{p}{Po}$ and the temperature T(K).

V_o = maximum volume of micropores accessible to the adsorbed phase.

β = coefficient of affinity of the adsorbent.

B = structural constant of adsorbent (correlated to average diameter of micropore).

P = pressure of absorbent.

P_o = vapor pressure of adsorbent at temperature T.

The difference between N_2 adsorption and CO_2 adsorption is

not correlated to their molecular sizes ($\sigma_{N_2}^{77K}$ = 0.162 nm²;

$\sigma_{CO_2}^{273K}$ = 0.207 nm²). It results from activated processes.

However, pores with diameters below ca. 3Å are not accessible and one must be careful when attempting a correlation with the helium density.

II. POROSITY AND SURFACE AREAS OF COALS

During two decades after W.W.II, a huge amount of data dealing with coal porosity and surface areas were produced (e.g. P. Chiche at the C.E.R.C.H.A.R. in France).

The work of P.L. Walker Jr. et al. (3) on widly used american coals will be employed to introduce the nature of coal porosity. Twenty seven samples encompassing all the ranks from lignite to anthracite were selected; their carbon contents ranged from 63.3 % to 91.2 % (% dry, ash free basis).

In figure 1, surface areas have been plotted against carbon content.

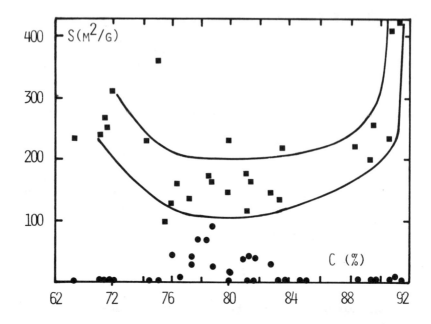

FIGURE 1. Variations of nitrogen and carbon dioxide surface areas of coals with carbon content
o N_2 □ CO_2 (3).

It is seen that coals with more than 10 m^2/g nitrogen surface area fall in the carbon content range 75.5 to 81.5 %. On both sides of this range, nitrogen surface areas are smaller than 1 m^2/g (except for anthracites). The CO_2 surface areas vary as the inverse of the nitrogen areas. This means that coals exhibiting the lowest total internal surface area (as measured by CO_2) and therefore the lowest micropore volume have the largest volume of pores accessible to nitrogen i.e. larger than 5 Å.

The CO_2 and N_2 surface areas of lignites must be critically examined. Indeed, prior to surface characterization, the sample is heated at 130°C under low pressure; the low nitrogen surface area (below 1 m^2/g) may result from irreversible transformation of lignite during pretreatment.

Let us examine the porosity of coals. On some of the samples, helium and mercury densities were determined. In table I the helium and mercury densities are given.

TABLE I. Open and close porosities of american coals (3).

Sample	Carbon (% dry, free ash)	Helium density* (g/cm³)	Mercury density* (g/cm³)	Mass volume (cm³) ($\frac{1}{d_{Hg}}$)
PSOC 80	90.8	1.53	1.37	0.730
127	89.5	1.33	1.25	0.800
135	88.3	1.32	1.25	0.800
135	83.8	1.28	1.23	0.813
105 A	81.3	1.27	1.07	0.935
Rand	79.9	1.25	1.14	0.877
PSOC 26	77.2	1.27	1.06	0.943
197	76.5	1.29	1.13	0.885
190	75.5	1.30	1.00	0.00
141	71.7	1.35	1.17	0.833
87	71.2	1.40	1.22	0.820
88	63.3	1.45	1.31	0.763

* Mineral-matter free basis.

Sample	Open pore volume (m³/g) ($\frac{1}{d_{Hg}} - \frac{1}{d_{He}}$)	Closed pore volume (m³/g) ($\frac{1}{d_{He}} - \frac{1}{d_{tr}}$)	Open porosity (% volume)	Closed porosity (% volume)
PSOC 80	0.076	0	10.4	0
127	0.052	0.011	6.5	1
135	0.042	0.017	5.3	2
4	0.033	0.040	4.1	5
105 A	0.144	0.047	15.5	5
Rand	0.083	0.060	9.5	7
PSOC 26	0.158	0.047	16.7	5
197	0.105	0.034	11.9	4
190	0.232	0.028	23.2	3
141	0.114	0.051	13.3	6
87	0.105	0.025	12.8	3
88	0.073	0	9.6	0

These data were used by Walker et al. to compute the open porosity with respect to the volume of the coal samples. Also included in this table is our evaluation of the closed porosity. The vitrinite content of these different samples is between 60 and 97.5 % with a high proportion of samples between

164

75 and 80 %. Their true densities were approximated using the corresponding vitrinite true densities (6) namely 1.45 for lignite, 1.35 for bituminous coal and 1.5 for anthracite.

Open porosities lie between 4.1 and 23.2 %. The order of magnitude for closed porosity is less than 7 %. Though less significant than open porosity, the contribution of the closed porosity in processes where closed pores are opened (devolatilization and heterogeneous combustion) may not be negligible.

Harris and Yust (7) pointed out that micropores, mesopores and macropores are favoured by certain macerals, namely vitrinite, inertinite and exinite. Thus micropores are favoured and porosity decreases in high rank coals (vitrinite rich). Similar conclusions have also been reached using mercury porosimetry (8).

Walker et al. (3), for the same series of coal, calculated the volumes of large, medium and small pores and quoted them as V_1, V_2 and V_3 respectively. The total open pore volume for pores accessible to helium at 305.5 K was quoted as V_t. The values of V were obtained by difference i.e. $V_T - (V_1 + V_2)$.

Results in percent of the total volume (V_T) are plotted on table II.

TABLE II. Gross pore distributions in coals (3).

Sample	Rank	V_T (cm³/g)	V_1 (%)	V_2 (%)	V_3 (%)
PSOC-80	Anthracite	0.076	11.9	13.1	75.0
PSOC-127	LV Bit.	0.052	27.0	nil	73.0
PSOC-135	MV Bit.	0.042	38.1	nil	61.9
PSOC-4	HVA Bit.	0.033	51.5	nil	48.5
PSOC-105A	HVB Bit.	0.144	25.0	45.1	29.9
Rand	HVC Bit.	0.083	20.5	32.5	47.0
PSOC-26	HVC Bit.	0.158	19.6	38.6	41.8
POC-197	HVB Bit.	0.105	20.9	12.4	66.7
PSOC-190	HVC Bit.	0.232	17.2	52.6	30.2
PSOC-141	Lignite	0.114	77.2	3.5	19.3
PSOC-87	Lignite	0.105	59.1	nil	40.9
PSOC-89	Lignite	0.073	87.7	nil	12.3

The influence of rank on the proportion of the different pores volumes is not monotonic and it is difficult to draw clear conclusions. It appears, however, that only H.V. bituminous coals have a significant pore volume between 1.2 and 30 nm. Whatever the rank, the internal surface is composed of a network of interconnected pores of different diameters. For lignite, there is an apparent anomaly : small pores (d < 1.2. nm) represent 12.3 % of the total porosity ; these small volumes obtained by difference $V_T - (V_1 + V_2)$ are in contradiction with their high CO_2 surface areas.

It is of primary importance to quantitatively describe the volume proportion of the different classes of pores in initial coal as done by Walker et al. However, for a given pore distribution, the topology of porosity has a significant effect on the behaviour of the material during devolatilization and subsequent heterogeneous combustion. The topology of porosity is defined as the steric distribution of pores (e.g. for the same pore distribution, a particle may have a totally different reactivity depending upon whether the large pores or the small pores emerge from the surface).

The topology of porosity may be represented by the two following images :
. a random repartition of pores of different sizes in the bulk of the material (9).
. the "pore tree" structure proposed by Simons (10-11).

Each pore that reaches the external surface of the coal particle has been depicted as a trunk of a tree (cylindrical large pore). Pores of decreasing diameters located inside the particle are described as branches and leaves In Simons work the number of pores within the volume V whose pore radius is between r_p and $r_p + d_{r_p}$ is denoted by $Vf(r)dr$. The porous volume V is expressed as :

$$V = \int_{r_{min}}^{r_{max}} \pi r_p^2 l_p \, Vf(r_p) dr_p$$

where l_p is the pore length.

Assuming all pores cylindrical and their lengths l_p proportional to radius r_p the result is that $f(r_p)$ is proportional to r_p^{-3}.

No morphological evidence of these structures has been provided by these authors or by other laboratories. These structures must be considered as models useful to describe semi-quantitatively the texture and its evolution during devolatilization or gasification.

A more realistic description of pore topology in coal particles might be achieved by a more systematic use of some complementary methods such as optical microscopy, electron microscopy and, eventually fractal approach. Image analysis of optical micrography will be illustrated (12) in chapter concerned with coal devolatilization.

<u>Characterization by electron microscopy</u>

A method used to characterize the textural evolution of coking coals during cokefaction (13) might be useful to characterize medium and large pores topology. It is based on the phase contrast observation of the matrix (interferences produced by superposition of the transmitted beam and a diffracted beam).

Each pore wall is made of basic structural units less than 1 nm in size ; these units (a few parallel aromatic molecules) give a phase contrast; they are associated edge to edge, parallel to the wall and form larger wrinkled layer stacks. The local molecular orientation can be correlated to the size of pores. In the work of A. Oberlin pores were classified into 11 stages (stage 1, 5-10 nm in size; stage 11 >> 1 μm).

Such a method applied to cuts of coal and coke might determine what model (Gavalas, Simons or other model) is the most relevant to coal and coke texture.

Fractal approach

Surface-science has in the past tended to deal with the problem of description of geometrically irregular surfaces by treating irregularity as a deviation from well-defined Euclidian shapes. An entirely different approach is possible (14-15) if the surface is (statistically) invariant over a certain range of scale transformation i.e. if the surface is self-similar upon changes in resolution power. For such surfaces, the degree of irregularity can be given by one number : the fractal dimension D (16) of the surface where 2 < D < 3.

Pfeifer et al. using molecular probes of different sizes determined the fractal dimension of porous solids and deduced pore distribution laws. This approach can be applied to coal and coke provided precautions are taken to avoid misleading conclusions due to activated diffusion of molecular probes in the pore network.

In a recent Ph.D. dissertation thesis (17) on the "Study of the evolution of the texture of combustible solids during incandescence" it was shown using 4 molecular probes (N_2 and Ar at 77 K, CO_2 at 195 K and butane at 273 K) that the fractal dimension of a char ex poly (phénol formaldehyde) is close to 3 (29). Using the fractal theory, Pfeifer et al. showed that

$$- \frac{dV}{dr_p} \, \alpha \, r_p^{2-D}$$

For a fractal dimension of 3, the above equation is equivalent to the distribution function theoritically computed by Simons on coal and coke.

Bale et al. (18) in the case of lignite correlated submicroscopic porosity determined by small angle X ray scattering with fractal properties of the system. A fractal dimension of 2.56 was found; its meaning is not clear.

Fractal approach is becoming a fashionable way to examine solids. It is too early to know whether such a concept applied to coal and coke will be fruitful. However, the fractal approach by itself is not expected to give topological informations and has to be combined with optical and electronic microscopies.

III. PORE EVOLUTION DURING DEVOLATILIZATION

The mechanisms responsible for pore evolution have been listed by Simons (19) as follows :

1. Bulk growth : Growth of existing pores due to the breaking of organic bonds in bulk along the walls of the pores.
2. Combination : Apparent destruction (generation) of small (large) pores due to the growth of large pores into the material containing the small pores.
3. Exposure : Formation of small pores due to the exposure of previously closed pores to the open pore structure.
4. Generation : Formation of small pores (several angstroms)

due to the breaking of individual organic bonds.

These four mechanisms are involved in pore evolution whatever model is adopted to describe porous texture. Of course,
Simons did the computations for the pore tree theory.

The overall result of all these processes leads to an increase in porosity. For lignite (20) the pore volume goes
through a maximum at 1300 K over a large range of heating
rates (1 to 2 000 Ks $^{-1}$).

Simons demonstrated theoritically that the distribution
fonction in r_p^{-3} is approximately self-preserved. This was
confirmed by Koranyi et al. (21) for two british bituminous
coals. From other experimental results (3), enlargement of
large pore is greater and this effect increases with heating
rate.

Besides the four processes mentioned above, shift in pore
diameter distribution may results in the two following phenomena :
. Secondary reactions of tars are promoted by pressure. Since
the pressure in small pores of non swelling coal can reach
1000 atm or even more the overall rate of gas evolution during
pyrolysis is expected to be smaller in small pores. Suuberg et
al. (22) found that at 1000 atm the calculated gas phase equilibrium agreed well, with their observed product composition
for 75 μm Montana lignite heated to 1073 to 1273 K at 10 3 KS-1.
. It has been observed on activated carbons, at the temperature of devolatilization that the mobility of carbon atoms in
the matrix is high enough to close micropores. A curve of
closed porosity versus temperature would be useful to check
the possibility of the occurence of this phenomenon in coal.

Phenomenon of pore closing can be illustrated by table III
and figure 2.

TABLE III. Effect of heat treatment on properties of glassy
carbon (23).

HTT, °C	Surface Area, m^2/g	
	CO_2	N_2
700	830	420
900	994	282
1100	513	92
1300	154	81
1400	72	72
1500	52	67

In table III (23), when increasing the temperature of treatment from 700 to 1500°C the surface areas (N_2 and CO_2 adsorption) dramatically decrease. At 1400°C and above CO_2 and N_2
surface areas are equivalent which means that all pores with
diameters smaller than 5 Å are closed.

Influence of ashes is illustrated in figure 2 (24).

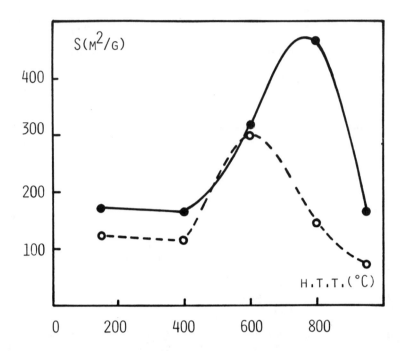

FIGURE 2. CO_2 Specific Surface Area (S_{CO_2}) versus Temperature of Treatment (HTT)
o Initial coal (Algerian Didi Coal)
o Ash free coal (24)

The CO_2 surface area of an algerian low rank coal with 13 %
ashes content (Didi Coal) is plotted versus the temperature of
treatment. In the presence of ashes, the maximum occurs at
600°C; for the ash free sample the maximum is at ca.800°C.

Experimental results of pore evolution during coking (a few
Kelvin per minute) are not relevant to the pore evolution of
pulverized coal.

To illustrate what is happening at ca. 10^5 Ks^{-1} , we selected a recent paper of Jones et al. (12). A comparative study
of rapid pyrolysis of different coals as well as the rapid
pyrolysis of the extracted vitrinite and inertinite from these
coals was carried out under conditions representative of pulverized coal combustion.

The coal particles were pyrolysed by passing through a H_2/O_2
flame. Rapid pyrolysis of the coal particles occurs in the
flame but burn out is prevented by the absence of free oxygen
behind the flame front. Coal residence time was 34 ms. The
flame temperature was controlled at 1773 K by adding nitrogen
into the combustion gases. The calculated heating rate of a
50 μm particle is 10^5 K/s.

In table IV are plotted the carbon contents of the coal macerals used by Jones et al.

TABLE IV. Analysis (% C) of coal macerals (12)

Coal code	Coal origin	Vitrinite C (%)	Inertinite C (%)
A	Southern Africa	81.2	86.0
B	Southern Africa	81.7	84.1
C	Australia	88.7	91.4
D	Australia	82.1	87.6
E	Australia	80.7	84.5
F	Australia	88.0	92.4
G	Canada	87.7	91.8
H	Britain	81.6	84.8
I	Britain	90.9	
J	Britain	92.7	
K	Southern Africa	80.5	

The quantitative measurements of char morphology were achieved by techniques of image analysis of the optical micrograph (Leitz TAS + image analyser). Images of char can be analysed to measure particle volume, macroporosity and pore size distribution. By this technique, only pores with diameter larger than 2 µm are visible.

The good agreement (table V) among the values of porosity determined by mercury density and image analysis confirms that the submicron pores do not contribute greatly to the total char porosity.

TABLE V. Average macroporosity of chars from whole coals (12).

Method	Macroporosity (%)	
	Mercury density	Image analysis
Particle size	75-100 m	50-63 m
Coal A	50	49
G	49	47
I	48	57
K	70	73

The authors examined the influence of macerals on the development of the morphology of char. This point is particularly important as microscopic analysis of pulverized coal indicates that maceral disproportionation can occur during grinding. With a high-inertinite, low-rank bituminous coals, it was found that the largest particles in pulverized coal (> 100 µm) were enriched in vitrinite, compared with the parent coal, by 10-20%. These particles therefore have a higher volatile yield than the average for the coal and will yield less char on pyrolysis. Inertinite is concentrated in the smaller particles.

The internal morphology of the pyrolysed coals was investigated by reflected light microscopy. The char structures were classified into one of three types, as follows :

Cenospheric chars. Hollow rounded chars with walls of variable thickness and few perforations, enclosing a small number of large, rounded pores. These chars derive from vitrinite and their morphology alters with coal rank.

Honeycomb chars . Irregularly shaped particles, often elongated, with a large number of long, narrow, parallel pores. Their morphology indicates a limited fluidity during pyrolysis.

Unfused chars . Irregularly shaped particles with little or no porosity. No softening is apparent during pyrolysis and they are recognizable as inertinite remains by comparison with the parent coal. Many of the apparent "pores" contain ashes.

The yield of cenospheric chars was compared with that calculated assuming they derive exclusively from vitrinite. The yield of cenospheric char was calculated from the maceral composition and the intact char yield, correcting for the small difference in maceral density. There was good agreement between calculated and observed values. Cenospheric chars therefore derive from vitrinite and honeycomb and unfused chars from inertinite.

Typical examples of these chars are shown in Figures 3-5.

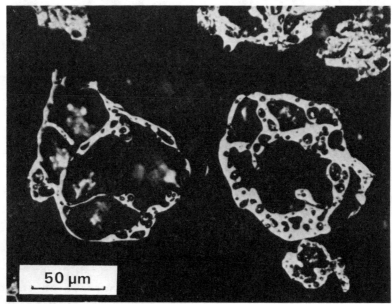

50 µm

FIGURE 3. Cenospheric chars derived from vitrinite (12).

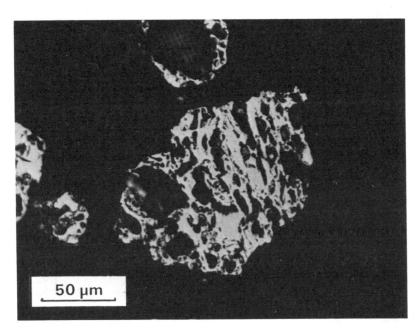

FIGURE 4. Honeycomb chars derived from low-reflectance inertinite (12).

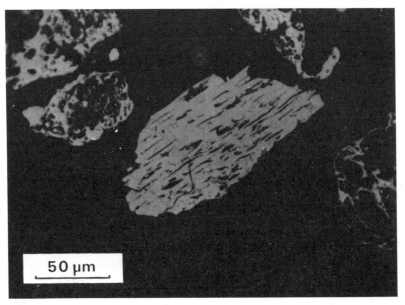

FIGURE 5. Unfused chars derived from high-reflectance inertinite (12).

It was shown that both char porosity and pore size decrease in the order cenosphere > honeycomb >/unfused char.

The morphology of inertinite derived chars does not change significantly in the range of rank studied. In contrast, the porosity of vitrinite derived chars decreases with increasing rank.

The type of char and the corresponding porosity are both strongly affected by swelling which accompanies devolatilization. As mentioned by Mulcahy and Smith (25), it has been found both by microscopic examination of rapidly devolatilized pulverized coal and by cinephotography of single particles subjected to rapid heating that the swelling of the particles increases with heating rate. Shibaoka(26) recorded increases in diameter up to 800% and Littlejohn (27) found the upper size limit of char from a bituminous coal to be 2-5 times greater than that of the original coal. The expanded particles are highly porous and under the microscope burning can be seen to occur within the pores (26).

D. Froelich (28) has studied the devolatilization and heterogeneous combustion of particles of a high volatile bituminous coal (Freyming) in a drop tube furnace. The histogramme of particle diameters before and after heat treatment was drawn from optical micrographs of the samples. The diameters of 1,000 particles were measured for each sample. Though the swelling index of the coal is 3-5, no swelling was observed during devolatilization either in nitrogen or in air. The flux of volatile matter leaving the particle during the plastic phase does not produce swelling of the carbon matrix.

These different observations demonstrate the necessity to carefully examine the relationship between swelling and rate of devolatilization.

High speed cinematography as well as histogram of size distribution before and after treatment have to be systematically made to have a clear description of the relation between swelling, type and granulometry of coal and heat treatment conditions.

A new approach by W.S.Fong et al. (29) is promising. The plastic behaviour of coal is determined under conditions pertinent to modern high temperature coal conversion process, using a specially developed fast-response plastometer. A mathematical model of the kinetics of coal plasticity is under development.

Swelling is a key problem in pore development during devolatilization and it must be clarified. Resolidification and eventual contraction after swelling may be responsible for the formation of fracture (large pores) and this aspect must also be considered.

CONCLUSION

A significant part of carbon in pulverized coal combustion is gasified by oxygen (heterogeneous combustion). This step largely determines the time scale of the process and therefore the dimensioning of pulverized coal combustion facilities.

Oxidation step, pore evolution in particular is dependent on the initial porosity of the char and on the kinetical and diffusional aspects. Many useful informations have been publi-

shed by I. Smith (25), R. Essenhigh (30), G.R. Gavalas (9),
G.A. Simons (10-11), N.M. Laurendeau (31) etc. However, a
critical evaluation of pores enlargement according to pore
models, would require comparing predicted and measured variations of particle size and density with burn-off. Unfortu-
natly very few data are available.

Therefore, for the moment it can be considered that no description of pore enlargement during heterogeneous combustion
is accurate enough for a physical-chemical description of the
procedure even if models may approximate observations.

A systematic study of pore evolution of well characterized
char (produced in experimental conditions relevant to industrial facilities) using the different techniques mentioned in
the paper must be carried out in extreme conditions i.e. kinetically controlled (low temperature and pressure) and diffusion controlled burn off. Such a study would permit to
validate models of pore structure in coal and char as well as
to quantitatively describe heterogeneous combustion.

REFERENCES

1. Washburn E.W.
   Proc. Math. Acad. Sci. U.S.A. 7, 115, 1921.
2. Cranston R.W. and Inkley, F.A. "Advances in Catalysis"
   Vol. 9, Academic Press, New York p. 143, 1957.
3. Gan H., Nandi S.P. and Walker P.L.,Jr Fuel 51, 272, 1972.
4. Dubinin M.M.
   Chemistry and Physics of Carbon (Ed. P.L. Walker, M.
   Dekker Inc. New York) 7, 51, 1966.
5. Anderson J.R., Structure of Metallic Catalyst (Academic
   Press), 1975.
6. Van Krevelen D.W., Coal Typology, Chemistry, Physics and
   Constitution (Elsevier Publishing Company), 314, 1961.
7. Harris L.R. and Yust C.S. Fuel 55, 233, 1976.
8. Orenbakh M.S. Reaction Surface during Heterogeneous
   Combustion, Novosibirsk, 1973.
9. Gavalas G.R. Combustion Science and Technology 24, 197,
   1981.
10. Simons G.A.
    . 19th Symposium (Int.) on Combust. : The Combustion Ins-
    titute, 1067, 1982.
11. Simons G.A.
    . Prog. Energy Combust. Sci. 9, 269, 1983.
12. Jones R.B., Mc Court C.B., Morley C., and King K.
    Fuel 64, 1460, 1985.
13. Bensaïd F., Ehrburger P., Lahaye J. and Oberlin A.
    16th Biennial Conference on Carbon. Extended abstracts
    (American Chemical Society), 68, 1983.
14. Avnir D., Farin D. and Pfeifer P.
    Nature 308 (5956), 261, 1984.
15. Pfeifer P.
    Appl. Surf. Science 18, 146, 1984.
16. Mandelbrot B.B. The fractal geometry of nature (Freeman,
    San Francisco, 1982).
17. Jabkhiro El Hassan "Etude de l'Evolution de la texture de
    solides combustibles au cours de l'incandescence chimique.
    Dissertation thesis (thèse de 3e cycle) Lille (France),
    1986.
18. Bale H.D. and Schmidt P.W.
    Physical Review Letters 53 (6), 596, 1984.
19. Simons G.A. Combustion and Flame 53, 83, 1983.
20. Simons G.R., Kothandaraman G., Schertzer S.P. and
    Palm M.J.
    Effect of Preignition on Pulverized Coal Combustion
    DOE/PC/30293-8/PSITR-329 Final Report 1982.
21. De Koranyi A. and Balek V.
    Thermochimica Acta 93, 737, 1985.
22. Suuberg E.M., Peters W.A. and Howard J.B.
    17th Symp. (Int.) on Combust. The Combust. Inst., 117,
    1979.
23. P.L. Walker Jr Proceedings of the 5th London Internatio-
    nal Carbon and Graphite Conference (Society of Chemical
    Industry) Vol. I, 427, 1978.
24. F. Addoun. Caractérisation des matières minérales présen-
    tes dans le charbon. Leur rôle dans le développement de la

porosité des cokes.
Dissertation thesis (Docteur d'Etat es Sciences Physiques, Université de Haute-Alsace, 1985).

25. Mulcahy M.F.R. and Smith I.W.
Rev. Pure and Appl. Chem. 19, 81, 1969.

26. Shibaoka M.J.   J. Inst. Fuel 42, 59, 1969.

27. Littlejohn R.F. J. Inst. Fuel 40, 128, 1967.

28. Froelich D. Etude Expérimentale et Modélisation de la Combustion d'un grain de charbon.
Dissertation Thesis (Docteur en Chimie Physique) Université de Haute-Alsace, 1985.

29. Fong W.S., Khalil Y.F., Peters W.A. and Howard J.B.
Fuel 65, 195, 1986.

30. Essenhigh R.H. Chemistry of Coal Utilisation (Second Supplementary Volume (Ed. Elliot M.A.) John Wiley and Sons, 1153, 1981.

31. Laurendeau N.M.
Prog. Energy Combust. Sc. 4, 221, 1978.

176

DISCUSSION

I.W. Smith

It is important to have knowledge of pore structure effective at combustion temperatures, i.e. what are values of pore diffusion coefficients and available surface areas that apply to the hot char. Are they any techniques that can measure these properties at elevated temperatures ?

J. Lahaye

I am not aware of in situ techniques to characterize surface area and porosity of hot chars. One could eventually take into account the dilatation of the material between the temperature at which the determination has been carried out and those of the hot char. I do not expect the correction to be significant.

E.M. Suuberg

A general question about the application of CO_2 to surface area determination in low rank coals. It seems reasonably well established in chars and high rank materials that CO_2 sorption and sorption of other species give comparable results. But there is still a concern about the quadrapole moment of CO_2 in connection with the polar surfaces of low rank materials. CO_2 has been shown to swell coals, which is another sign of specific interactions.

J. Lahaye

For determining the specific surface areas of solids it is very important to avoid adsorbates exhibiting specific interactions with the substrate such as the interaction between the quadrupole moment of CO_2(of nitrogen too) with the polar groups of the substrate. These specific interactions are quite low and at temperature equal or higher than 273 K there can be neglected.

H. Jüntgen

1) Supplement remarks to micropores in coals : You did not mention the concept of "bottle neck pores" - pores with diameters of pore mouth in the order of gas molecules and their characterisation by measurement of activated diffusion using methane (van Krevelen) and higher hydrocarbons (Handaba, Jüntgen, Peters). According to the new finding concerning the model structure of coal macromolecule and its presentation by steric models (Kosky) it has to be expected a correlation between bottle neck pores, detemined by activated diffusion on the one hand, and steric presentation of coal macromolecule, on the other hand, because the gases diffuse into the holes between the aromatic clusters of aromatic-hydroaromatic structures. This correlation could help in future research to define pore morphology in the region of microporosity.

2) Question to your determination of closed pores by
using the equation

$$V_c = \frac{1}{d_{He}} - \frac{1}{d_{tr}}$$

d_{tr} determined by van Krevelen was based on measurements of d_{He}
by R. Franklin. Are they differences by the determination of
d_{He} (P.L. Walter jr) and d_{tr} (R. Franklin).

3) I would like to stress the statement given in my paper that for combustion the porosity of parent coal has less
significance than the porosity of char created by the pyrolysis dependent on heating rate, particle size during pyrolysis
and rank (and especially swelling properties) of parent coals.

J. Lahaye

1) I thank you for having emphasized that point. A significant part of the micropore volume can correspond to
bottle neck pores. If, during devolatilisation or combustion
the pore mouth is opened then the corresponding pore becomes a
medium or large pore with a totally different reactivity : in
the devolatilisation step the diffusion of primary tars increases and therefore secondary reactions become less significant ; in the combustion step diffusion to and from the pores
in also increased so that the gross gasification rate
increases.

2) You are right though the determination of helium
densities by R. Franklin were made on isolated vitrinite and
not on the entire coal. It might be better to use X ray
scattering measurements.

R. Cypres

Concerning the reactivity of chars, we made determinations on chars prepared at low heating rate, under different
gas atmospheres (N_2, H_2, H_e) and pressures (up to 70 bars).
Always the most important parameters of air reactivities
under isothermal conditions, was the residual gas ambient of
the char. This fits well with the general view mechanism of
coal combustion.

THE ROLE OF VOLATILES IN COAL COMBUSTION

ROBERT H. ESSENHIGH
E. G. Bailey Professor of Energy
Conversion
The Ohio State University
Columbus, Ohio   43210

ERIC M. SUUBERG
Associate Professor of Engineering
Brown University
Providence, Rhode Island   02912

1. INTRODUCTION

Our knowledge of the role of volatiles in coal combustion ranges at this time from the self-evident to the ambiguous. The clearest point on which all agree is that pyrolysis will occur during the total coal combustion process, and the volatile matter products can then burn completely or partially in the vapor phase, depending on a variety of factors. The significance of the pyrolysis and VM combustion process to the overall combustion process is, however, open to argument.

Flames have many properties. These include: flame speeds and limits, temperature and concentration profiles, peak temperatures, ignition temperatures or times, reaction times, and so forth. The focal question then is: What flame properties depend on or are critically influenced by the existence of volatiles in the burning coal. In asking this question, the crucial word is "dependence"; there is often an apparent dependence, revealed by correlations, where the functional dependencies, or mechanisms of any dependence are less clear or even unknown.

The most general illustration of this situation is the acknowledged rank-dependence of coal "flammability" in both use and in safety. In use, it is generally believed that anthracites or low-volatile fuels cannot be fired in boilers designed for bituminous coals; the so-called U-flame and W-flame designs must be used instead. This differentiation is paralleled by safety regulations in mining where there are rock (stone) dusting or barrier requirements for deep mining of bituminous coal to prevent or attemperate explosions; but these regulations are not usually required or are much less stringent in anthracite mining because the lower VM content of anthracites is believed to make them less prone to accidental ignition.

Thus the phenomenological (correlational) influence of coal rank is clearly evident. The mechanistic (causal) dependencies, however, are not. In characterizing coals to determine both use and safety requirements, the VM content is the most commonly used characteristic. It is almost inevitable, then, that the most common explanation invoked for the demonstrated flammability/rank correlations is the intuitively appealing assumption that it is due to release and combustion of the volatiles. However, this final step in logic still represents an intuitive leap rather than a conclusion that has been satisfactorily confirmed. Although the general correlation of flammability with VM content is clear, we do not know whether this is specifically due to the flammability characteristics of the volatiles, _per se_; or whether it is due to the VM content being just a general index of coal rank that accurately or broadly represents some other significant reactivity factor. Moreover, the conclusions may very well change from one flame environment to another.

This point identifies the focus of this paper. The intuitive explanation for the rank/flammability correlations may be correct, although there is increasing evidence against it. However, if "volatiles flammability" were to be eliminated as an explanation, we would then have a mechanistic void. Identification and discussion of possible alternatives is, thus, a major objective of this paper.

The problem of determining the mechanistic role of the volatiles is compounded by the lack of data on their combustion rates. At this time there is no direct information available on even the global kinetics of combustion of volatiles. Attempts to calculate the VM combustion rates have to depend on indirect methods, primarily lumped-parameter types of assumptions about the principal VM constituents and/or use of "slow-step" assumptions--usually that the terminal step is CO reaction. Such approximations may be reasonable, but this applicability has never yet been tested.

2. BACKGROUND
2.1. The Problem Formalized

In the most general terms we can say that "flame properties" depend on coal rank, and the task is to determine the appropriate functional dependence. Flames have many properties. These include: flame speed, ignition temperature, ignition time, reaction time, concentration and temperature profiles, peak temperature, flame emissivity, radiative properties, and so forth. If the flame properties are designated as FP, then FP depends in any given situation on the coal properties (CP), the particular experimental system (ES), and the firing conditions (FC). For any particular flame property FP(i), this can then be represented, we assume, as a function of CP, ES, and FC, thus

$$FP(i) = f_i(CP, ES, FC) \qquad (1)$$

In principle, the function f_i is usually obtained as a solution to an appropriate differential equation or set of DE's.

The Firing Conditions (FC) include such factors as: particle size and size distribution, stoichiometry, firing rate, and similar factors that are independent variables. An important firing condition is coal particle concentration which can lie between the extremes of dilute streams and non-dilute streams.

The use of a particular coal in a particular system, of course, sets the CP and ES functions as constants, and experiments in recent years have generatlly focused on the functional structure of the FC component, as discussed later in this article.

The Experimental Systems (ES) can be broadly classified in four groups: single particles; fixed beds; fluid beds; and entrained flow. In principle, all systems can be operated in either time-independent or time-dependent modes. Single particle studies, by the nature of the system, must be carried out in a time-dependent mode, but a quasi-steady method of analysis is often adequate. The fixed, fluid, and entrained flow systems are all multiple particle, and in principle can be designed to operate as 1, 2, or 3-dimensional reactors. For experimental/analytical purposes, to understand and to test kinetic components of theoretical analyses, one-dimensional systems are preferred. This substantially or totally designs-out the aerodynamic flow factors. Heat transfer is then a dependent parameter in concentrated particle flows, but is a partially controllable (independent) parameter in dilute flows as, for example, in drop tube experiments.

These general characteristics of the experimental systems are of essentially secondary importance to us in this article, but they are necessary for context.

The Coal Properties (CP) are the focus of concern in this article, and more complete specification is necessary. We have
- Physical properties: density, internal surface, porosity, pore structure, ...

- Chemical Properties: The classical analyses: Elemental (C,H, O,N,S percentages), Proximate (VM%, moisture, ash, fixed carbon), Maceral (Vitrinite, exinite, inertinite, etc.). In recent years, other factors have also been found to be important. For example, heating rate has been found to influence the amount of volatile matter, its composition, and the reactivity of the char left behind. The composition of mineral matter in coal also affects the nature of the volatile matter and the reactivity of the char. These same factors may play a role in determining the physical properties.

Thus CP is nominally an N-parameter set of variables where we are not even certain of the value of N. However, we can reasonably suppose that many of these properties are inter-dependent, so that we should have additional constitutive and/or auxilliary relationships that will reduce the number of independent variables although few such relationships are known at this time. This is the problem of coal systematics and constitution. There is also the question of which of the properties should be considered independent, and which are dependent. The components of the elemental analyses and particularly the C, H, O percentages are the best starting candidates as the independent parameters. This may best be seen by noting that most coals fall in a relatively narrow band in C-H space (see Figure 1). If all other coal properties were functions of C,H, and O (neglecting N and S), and of some unspecified or unknown parameters, x_i to x_n, we would have for any coal property CP [say CP(i)]

$$CP(i) = F_i(C,H,O,x_i \ . \ . \ x_n) \tag{2}$$

which is a function of at least 3 variables.

Of course, we also have the crude relationship between C,H, and O embodied in Figure 1. Note that because it is possible to represent an oxygen content axis in the same plane as the carbon and hydrogen axes, this means that the oxygen content is, crudely, a function only of the carbon and hydrogen content of the coal. This means that CP(i) can actually be expressed as a function of 2 compositional variables and the x_i; note that any coal can, in theory, be represented by its C and H coordinates alone in Figure 1.

Figure 1 was constructed so as to include a large percentage of all coals in the narrow band shown in C-H space. If as a further approximation it is assumed that all coals actually lie on a single line that is enclosed within the band, then it would be possible to define a typical coal in terms of only one elemental composition parameter (e.g., the carbon content). This is, in fact, the basis for correlating coal properties with rank as measured by carbon content, a very common procedure in discussing the properties of coal. In theory, it should be possible to use any of the three main compositional axes to define a "point" on the coal line, but carbon content is generally preferred because there is too little systematic variation of hydrogen in coals of low carbon content and too little systematic variation of oxygen in coals of high carbon content.

Further refinements of this type of analysis have been performed, leading to empirical correlations in terms of elemental ratios found in typical coals, e.g.,

$$O/H = (O/C)/(H/C) = [1 + B(O/C)]/A$$

or

$$H/C = A(O/C)[1 + B(O/C)]$$

with $A = 80\pm6$, $B = 92$, with a correlation coefficient of 0.9889. Thus, coal properties are sometimes presented as functions of atomic ratios, rather than as a function of carbon content alone. For example, the specific volume (v) may be correlated by an expression of form:

$$v/C = A'(O/C)/[1 + B'(O/C)]$$

where A' and B' are empirical coefficients. v/C is the specific volume per unit of carbon.

To review all attempts at correlating coal chemical and physical properties with rank or elemental ratios is far beyond the scope of this paper (the reader is referred to the excellent reference works by Van Krevelen, Elliott, and Meyers, to name but a few).

Still, when it comes to parameters necessary to represent properties important in combustion, simple correlations with rank, represented by carbon content or elemental ratios are often inadequate. Other parameters x_i must be specified.

This is implied, in essence, by the typical methods used for classifying coals. ASTM classification uses VM% and calorific value (which is calculable with some accuracy from elemental analysis). See Figure 2 for a rudimentary coal classification diagram, based upon these parameters. Note that lignites are often treated separately from "true coals." They would fall to the right of subbituminous coals on the diagram of Figure 2. The U.K. National Coal Board system uses VM% and Gray-King coke type. The International Classification of Hard Coals uses VM% and caking properties (with Gray-King, Dilatometer, and other parameters as alternates). The VM% correlates quite well, but not quite uniquely--i.e., not with sufficiently small error--with C% (or H/C or O/C); however, the additional parameter(s) and how to incorporate it (them) into an appropriate monotonic function to be able to calculate VM% with sufficient accuracy, is still unknown. This is further discussed below.

When these considerations are applied to combustion properties, the same conclusion can be expected to apply: no one parameter alone will be sufficient to describe all properties of interest. It is reasonable to expect that any combustion property will be a function of coal rank, but an appropriate specification will probably require the use of at least two parameters. In view of the complexity of coal (and the lack of fundamental knowledge of coal constitution) it would not be surprising that only two coal parameters may still not suffice for the prediction of coal combustion properties. In short, the demonstration that there is a general, but not unique, monotonic trend of any combustion property with VM% alone is not a priori surprising; the reasonable expectation would be that at least two parameters would be required to accurately specify the combustion properties. By the same token, the existence of any unique trend for any combustion property with any single parameter then becomes, conversely, of major significance.

2.2. Historical: Theories and Experiments

In the process of turning the function(s) f_i in Eq. 1 into mathematical/quantitative expressions, a number of qualitative (physical) concepts have been developed over the years. These concepts have included: the process of pyrolysis; the process of ignition; the process of VM combustion or particle combustion with cracking and soot formation; and so forth. The necessary concepts were originally derived from the results of empirical experimentation over many decades, where the experiments reported by Faraday and Lyell (1845) can be taken as a reasonable starting point. It is then a quirk of history that the development of the concepts happened to depend to a significant extent on the direction of particular interest at the time. As this shaped views on the role of coal rank in combustion, it is of some importance to summarize the elements of that history.

Original interest in the scientific study of coal combustion derived totally from considerations of safety and explosion in underground coal mines. Although pulverized coal combustion was entering the commercial market in the 1880's, it would appear that an interest in applying fundamental combustion concepts to fixed beds (grates) or pulverized coal combustion (p.c.) firing did not develop until after 1910 or 1920. The intervening period of 7 or 8 decades from the 1840's was dominated by the development of bench-scale ignition-testing devices in the interest of safety (with the development of the large scale--up to 1000 ft.--explosion galleries about the turn of the century.

The combination of the focus of interest on safety and the small-scale of the ignition testers allowed the examination of very large numbers of coals where "large" in this context could mean 40 or 50 or more. By comparison, documented testing in the large explosion galleries and, more particularly, in boilers and furnaces was usually limited to a few coals (perhaps half a dozen), and frequently only one. The difference in the number of coals tested can then lead to different inferences regarding possible mechanistic influences.

Comparison of the bench ignition tests with the gallery experiments brings out two points. First, comparing the correlation coefficients between flammability and VM% for the two methods of experiment, they are so sufficiently different that it is reasonable to conclude that the two methods of experiment do not correlate; this was, in fact, the conclusion of Mason and Wheeler that the bench scale test was only a rough predictor of gallery behavior, and there was the further implication that the mechanisms of ignition and of combustion or propagation, and the influence of the volatiles on these, could be quite different. The second conclusion of the comparison is that apparent trends with rank can be misleading or false if only a few coals (half a dozen or so) are used; either it is necessary to use 40 or 50 to obtain "high density" plots, or the correlation coefficients between the combustion and rank parameters must be very high (possibly in the region of 0.97 or 0.98).

It is also clear that a broad or close correlation between some coal properties or combustion parameters with VM% supports ab initio no particular mechanistic interpretation. In particular, the idea of coal flammability being conferred by volatiles flammability has no support from phenomenological correlations. The problem, at bottom, is still the lack of an adequate model of coal constitution that can be used to provide a mechanistic basis for accurately modeling the effect of coal rank. At the same time, however, observed behavior does imply some simple regularities with a dependence on only one rank parameter (VM), which is a surprise, so that the problem may be inherently much simpler than it is often thought to be, if only we have the wit to ask the right questions.

3. IGNITION AND EXTINCTION

All flames start with ignition but the mechanisms and parameter values of ignition are uncertain to the point of controversy. Extinction is generally of importance where safety is a major concern, notably in underground mining, but also in grinding, and in boiler explosions.

Parameter scope - properties include
- the flammability limits
- ignition time
- ignition temperatures
- ignition energies

- extinction temperatures
- extinction concentrations
- extinction by additives

Theoretical basis - is Thermal Explosion Theory (TET), to the extent that volatiles are not the primary combustibles involved at the moment of ignition and extinction: again a controversial point.

Relationships - If an ignition temperature can be defined, the mathematical relationships to ignition times and energies can nominally be written down. If the variation of ignition temperature with coal concentration can be established, flammability limits are also established (in principle). Then the role of coal rank (and influence of VM%) follows.

Problem: define ignition temperature.

Ignition temperature: methods of definition
    - operational (experimental)
    - theoretical: TET or other

Experimental definitions of ignition temperature:
* Bench-scale batch methods
        - single captive particles (Tretyakov, Bandyopadhyay and Bhadhuri, Karcz et al. . . .)
        - "single" particles dispersed in hot air (gas) (Cassel and Liebman, Chen et al. . . .)
        - dispersion of batches of particles (G-G test, Hartman, Wheeler, etc.)
        - TGA (Tognotti et al.)
        - Crossing point (ASME)
        - heated grid (de Soete, Juntgen and Van Heek)
*Flame measurements
        - determination of flame temperature at flame front (Howard and Essenhigh . . .)

Theoretical definitions - All experiments have been evaluated in terms of TET, implying heterogeneous ignition. Only Annamalai and Durbetaki (19877) have developed alternative theoretical predictions for homogeneous and heterogeneous ignition.

Scope of Results - mixed and somewhat confusing.

Ignition Temperatures: range from 300C to 1000C. Crossing point and "Single" dispersed particles show lowest values: Chen and Tognotti are in the region of 350 to 450C. Higher values are believed to be in error due to experimental method used. Lower values agree with crossing point; also make sense with temperature limits on air-swept mills in coal grinding (exhaust temperatures not to exceed about 150F). Theory shows that ignition temperatures should drop with increasing coal concentration so 400C would be about upper limit for clouds. In these experiments, the particles heat at their "natural" rate for cold particles plunged into hot gas with gas temperature set at the critical ambient temperature for TET ignition.

Flames: ignition temperatures are generally (much) higher. Howard and Essenhigh (1967) give 900 to 1000C. Heating is now being driven at rates above "natural" heating rate, and final ambient gas temperature (which is rising as the particles temperatures rise) is much higher than the critical ambient value required in TET. Thus, this difference from the "single" particle results appears to be the difference in ambient temperature, and difference in heating rate of particles. These differences have not been adequately accommodated in theoretical developments.

Batch clouds (G-G test, Hartman, Wheeler, Carpenter). Temperatures are much higher: 500 to 1000C. Reason: measuring different things. Temperatures reported are usually Ignition Source Temperatures (IST), not particle temperatures, or gas temperatures. Relation of IST to gas or particle temperatures is unknown. Alternatively, the process involved is not heterogeneous ignition and a different process is taking place. In the G-G test, Godbert measured combustion times (flash duration). They are values to be expected from volatiles combustion--or from extinction of char as whole coal burning is extinguished: we don't know. Temperatures required in this test may very probably be determined by the requirements of heating up cold air that replaces hot air swept out of the furnace as the coal sample is injected.

Extinction - Major large scale experimental studies using explosion galleries; very extensive bench scale studies (using G-G test furnace by adding increasing amounts of rock dust to coal sample: Godbert; Godbert and Greenwald, etc.): studies of very little value.

Ubhyakar and Williams (1976) determined rate constants of burning char particles from particle residues on extinction: ignited by laser in cold oxygen-enriched air. Waibel and Essenhigh (see Essenhigh, 1979) calculated extinction conditions in a boiler using TET on a flame ball model, and D. Stickler, et al., did the same using a fluid mechanics model.

There are also numerous determinations of lower limits, notably by Hertzberg (1981, 1982, 1984), and some of upper limits (e.g., Deguingand and Galant, 1981).

Interpretations - Most experiments that have been interpreted have been in terms of TET assuming heterogeneous ignition. Support: ignition temperatures are close to or even below the pyrolysis temperatures at low heating rates, and are even below estimated pyrolysis temperatures at high heating rates when ignition is at 1000C and pyrolysis onset is at 1200C (most direct experiments on this are Howard and Essenhigh (1967) in flames). Juntgen and Van Heek (1979) have noted the particle size and heating rate dependence of the ignition process, and seen ignition temperatures in the range 780-820°C for heterogeneous ignition. G. G. de Soete (1985) has most recently obtained similar results with a heated grid in very extensive expriments: particles ignite--at low temperatures about 350-450C, and below pyrolysis temperature.

At very high heating rates (10^5 and 10^6 deg C/sec) particles still probably ignite heterogeneously, but then start to pyrolyze so soon after that the heterogeneous ignition is of no practical importance. Thus, in high intensity combustion (Goldberg and Essenhigh, 1979, Farzan et al., 1982) at fuel rich conditions (equivalence ratio of 1.4), combustion involves (essentially) all volatiles. This also is probably the case in the G-G test furnace.

Influence of Coal Rank - Very little: second order at most. Example: Chen's experiments show very little difference in ignition temperatures between anthracite, bituminous coal, and chars. In TET analysis, the materials igniting are carbon to solid hydrocarbons. There can/will be different values of kinetic constants, but evidently the variation only accounts for second order differences in ignition temperature: typically, it would seem, as activation energy E increases, the frequency factor rises to offset it.

A broad band variation of furnace temperature (IST) is found in such batch experiments as the G-G test. This may be due to the requirements for heating the coal particles and pyrolyzing off sufficient volatiles to burn. However, very much better correlations are found with other factors such as reactivity to oxygen at 100C or to pyridine extract (Essenhigh and Howard, 1971).

One-dimensional flames show variation in ignition times, but this has been identified as due to the optical properties of the flames downstream of the flame front, not to variations in reactivity. Ignition times can then be calculated quite exactly assuming 1000C ignition temperature and all particles have properties independent of coal rank. Measurements show that temperatures at the flame front are all close to 1000C for coals, chars, petroleum cokes. Allowing for the difference in heating rates, this is consistent with the "single" particles measurements and the de Soete hot grid measurements.

4. PROPAGATION AND COMBUSTION

The mechanism of flame propagation is predominantly thermal exchange. Propagation properties are clearly rank dependent; but role of volatiles is still ambiguous.

Question - Are volatiles effects (if any) due to the volatiles per se: or due to volatiles as a (relatively poor) index of coal rank?

Arguments are from different flame types and systems. The question is whether it is possible to draw general conclusions that apply to all flame systems.

Flame Systems
* Stabilized
    * Single burner: experimental and many process furnaces.
    Can be defined by dimension
                [0] - high-intensity, jet mixed approximating
                      to perfectly stirred.
                [1] - Types I and II (see Table 1 and below)
            [2], [3] - jet flames
                    - with and without axial symmetry
                    - with and without swirl
* Moving - many batch testing devices
            - large experimental explosion galleries and pipes
            - unwanted explosions -- confined
                                  -- unconfined

Flame Types - One-dimensional flames can be classified into several types. Table 1 provides a classification into 4 types. Predominant difference between types is the dominant mode of energy exchange.

Other flame types: energy exchange
[0] - high-intensity mixing provides for thermal exchange by this mixing process (high speed convection).

[2], [3] - jet flames ignite by thermal feedback of hot combustion products to the incoming (cold) jet. This is a convection-driven system with heat then diffusing to the center of the cold jet by thermal diffusion or turbulent exchange. Thus, particles can be represented as being in an ambient gas of increasing temperture where the temperature of the ambient gas is determined by the crossjet mixing and by the temperature of the combustion products being recirculated. This applies to either internal or external recirculation.

Role of Volatiles - This most probably varies from system to system. There are many ambiguities and/or uncertainties.

Most broadly, volatiles can affect propagation and/or combustion either
        - intrinsically, on account of their ability to ignite the char
          (touchpaper mechanism); or
        - extrinsically, on account of the independent combustion characteristics
          of the volatiles removal has on the char combustion properties (for
          example, production of internal surface).

Zero Dimensional Flames - Propagation is not an issue in these flames; only stability. Goldberg and Essenhigh (1979) have shown that at equivalence ratios of 1.4, temperatures peaked and combustion was for all practical purposes 100% volatiles, with a residence time of 10 to 20 msec. However, as the concentration was leaned out, char combustion increasingly participated (with temperatures falling), and stable flames were maintained to as low as 800°C. Q-factors were about 1.8.

The conclusion is that: at an equivalence ratio of 1.4, volatiles are produced by pyrolysis and consume available oxygen at a rate faster than competitive reaction of char/oxygen. As the coal concentration drops, the system becomes lean with respect to volatiles concentration, and the char reaction participates increasing in about the same residence time, so that competitive reaction is determined more by volatiles concentration (availability) than by actual reaction rates.

The inference is that as the volatiles content of a coal is reduced, flame temperatures would probably drop; the equivalence ratio for peak temperature would change to suit; and low limits might rise on account of lower char reactivity to the extent that reactivity is dependent on accessible internal surface. This would be worth testing experimentally. The role of volatiles in this case would seem to be dominant to controlling. There is a possibility that coals below a certain volatile content would not be able to support flames in this device. If this is the case, determination of that limit would be of considerable significance.

One-Dimensional Flames: Type I (Smoot et al., 1977) - As these are volatiles flames, the role of the volatiles is controlling. Without volatiles, there would seem to be no possibility of stabilizing flames in the systems used unless the propagation mechanism changes. As with the zero order systems, there may be a VM content below which flames cannot be obtained. Again, this would be of value to determine.

One-Dimensional Flames: Type II (Howard and Essenhigh, 1967; Cogoli and Essenhigh, 1977) - Experimentally, these flames showed no particular dependence on VM%. Fuel type only affected the ignition distance. Xieu et al. (1981) showed that this could be explained by the different optical properties of the flames, with the optical properties being controlled by the mode of reaction (relative importance of internal and external reaction). Otherwise, flame temperatures and flame speeds were similar, and with no evident effect of fuel type; in experiments by Gee et al. (1984), the highest flame temperatures were obtained with petroleum coke. In the modeling of the flames, using experimental temperature profiles, the same kinetic constants were used in all cases for coal, chars, and petroleum cokes, with the char model able to describe coal flames with no pyrolysis component in the model. In these radiation-stabilized flames, there is no evidence of any significant direct contribution of volatiles. Coal or fuel rank is important to the extent that it controls the rate of reaction of the solid, but not on account of the volatiles.

One-Dimensional Flames: Type III - These are propagating subsonic (deflagration) explosion flames. There is good evidence that they are rank dependent, notably from the Mason and Wheeler experiments from which it was established that an "Index of Inflammability," determined by the quantity of stonedust required to prevent propagation, was proportional to the VM content. The mechanism of propagation is paraxial turbulent, so that it may be presumed that the speed of propagation is determined by the flame temperature, and thus by the rate of combustion. The key factor here is that, for bituminous coals, on pyrolysis and combustion of pyrolysis products, typically 1/3 to 2/3 of the coal mass is burned in about 1/10 of the total combustion time (depending on the local stoichiometry). Radiation helps to suppress temperature excursions that would otherwise occur, but faster combustion rates should result in faster propagation. The critical factor then is the speed of pyrolysis compared with the speed of combustion. To the extent that the jet stirred reactor mimics the head of an explosion flame (see Goldberg and Essenhigh, 1979), the inferential conclusion is that the coal does pyrolyze fast enough in the explosion flame for the volatiles to be important. The importance depends on the rapid rate of heat release that produces higher temperatures in the flame and thus increased exchange of heat for ignition. There is then the dilemma that the correlation between Mason and Wheeler's Inflammability and VM% in the experimental gallery is very good when the bench scale experiments show rather poor correlation, and also show better correlation with other reactivity parameters. A partial explanation for the poor correlation in the small scab tests is very likely to be found in the composition of the volatiles. As Hertzberg et al. (1981) have shown, the combustible volatile contents at the lean limits in small scale experiments were roughly constant at 40-60 mg/l. Further experimental study of this point is warranted. This, of course, does not explain why a good correlation was obtained in the large scale tests. This point also warrants further examination.

One-Dimensional Flames: Type IV - These are supersonic, detonation flames with heat transferred by propagation of the associated shock wave. However, the energy released into the shock to maintain it again depends on reaction rates. Most of the observations made on the Type III flames apply here. Again, the key and unknown factor is the relative rates of pyrolysis and combustion under the conditions of shock heating. Nettleton and Stirling have some relevant data, but the interpretation in relation to detonation waves is still uncertain.

Two- and Three-Dimensional Flames - As described above, flame stability depends on the rate of recirculation of hot gases from downstream to the incoming cold jet upstream. The intensity of that recirculation governs the rate of rise of the jet temperatures, and thus of the incoming coal particles. This has similar characteristics, therefore, to the turbulent exchange flames (Type III). The magnitude of the thermal flux feeding back will depend on the temperature of the hot gases downstream, and again, if there is rapid release and combustion of the volatiles, this should provide higher temperatures downstream that will be reflected in faster ignition and improved flame stability. By the same token, reduction of the volatile content of the coal should reduce the temperatures, speed of ignition, and reduce flame stability, possibly to the point that a coal of VM content below a certain value cannot be used in a particular system.

Overall Evaluation - With the exception of the One-Dimensional, radiation-stabilized flame (Type II), this argument strongly suggests that flame behavior is significantly dependent on the volatiles on account of the combustion behavior, and notably as the result of the generation of local volumes of high temperature gases in the flame that are moved upstream by different mechanisms to ignite the fresh fuel. This argument still allows heterogeneous ignition--which is important as a mechanism from the point of view of correct analysis of the ignition process; the heterogeneous ignition is then irrelevant to the subsequent behavior on account of the rapid onset of pyrolysis soon after ignition (a few hundredths of a second later, or sooner).

What is still an issue, however, is the question of the quantity of volatiles that are released, and the speed of release. As a rule of thumb, it is commonly stated that the VM combustion is about 1/10 of the total combustion time. Also a rule of thumb, the combustion time in a boiler is generally said to be between 1 and 2 seconds, giving a volatiles combustion time of about 1/10 sec. In pyrolysis experiments, however, the pyrolysis times are shown to be as low as 1/100 sec., or 10 times faster. If these values are generally correct, it implies that the slow step in the flame is the combustion, not the pyrolysis, and that the volatiles are not burning as they are evolved. This is a factor considered separately (below).

Another confusing factor is that the effects of the volatiles may be determined as much by the mixing conditions as by the VM content. Thus, the discussion above follows very much the conventional path, with the conventional conclusion that VM content and release determine the role of the volatiles. It would then seem to follow that the maximum temperature would be obtained at a coal concentration that would provide a stoichiometric concentration of volatiles. On that argument, boilers of high rank coals have been designed on the assumptions that: the flame speed is low, and that the primary stream has to be very rich. The result has been the U and W flame boiler designs with downward injection of a rich primary stream, and supply of secondary air through front wall ports. With this configuration, the gases in the boiler circulate in an "O." In the early 1960's, however, the basic assumption was challenged. When the air was increased in the primary stream to a value from about 20% of requirements to 80%, the flames were found to be hotter, more compact, and giving higher burn-out. As the result of modeling, this was attributed in part to the higher momentum of the primary jet that penetrated the secondary air more efficiently. This again suggests that the behavior of the volatiles may still be secondary to other considerations.

A factor of final concern is the relationship between VM content and the heat of combustion. Suppose, for example, that the volatiles were a large fraction of the coal but the heat of combustion of those volatiles were very low. This would be partially reflected, of course, in the total heat of combustion, but if a large fraciton of the coal produces little heat, it will have proportionately less effect on the flame and propagation speeds. The question to be asked, then, is what relationship exists between the VM contents and their heats of combustion. This does not seem to have been established at any time. Ergun (1979) shows that the total heat of combustion of whole coals (not volatiles) becomes almost independent of VM% between about 30 and 50% VM. That is, VM is a poor predictor of the heating value of coals; and thus by inference, of the heating value of volatiles. If this is added to the concept of rapid and slow release of volatiles (the origin of the Two-Component Hypothesis of Coal Constitution), this may partly explain the poor correlations found in most cases between flammability and VM%. The issue of time dependence of volatiles heating value has been considered by Suuberg et al. (1979) and will be discussed further below.

5. THE NATURE OF COAL VOLATILES AND THE FLAMMABILITY CHARACTERISTICS OF MIXED HYDROCARBON GASES

5.1 Compositions and Kinetics of Release of Coal Volatiles

A complete review of this topic will not be presented here. Rather, only important highlights will be provided, and the reader is referred to other recent, extensive reviews for more details (Anthony and Howard, 1976; Howard, 1981; Howard, et al., 1981; Gavalas, 1982; Solomon and Hamblen, 1983, 1985). The viewpoints concerning many details of the pyrolysis process vary widely among these different reviewers, and it is fair to say that this is an area requiring considerably more attention. This observation pertains not only to the very complicated chemistry of devolatilization, but to the complex nature of mass transfer processes involved in escape of volatiles from the particle as well (Suuberg, 1985).

The most comprehensive of the reviews, in terms of providing an indication of experimental techniques for studying devolatilization, are those by Howard and coworkers. An enormous variety of techniques has been used to study the phenomenon, and it has been this fact which has led to some confusion concerning the rates and compositions of the products. It is, however, amply clear from most recent experimental studies that the standard proximate volatile yield is only a very crude characterization of devolatilization behavior under pulverized coal combustion conditions. Table 2 illustrates this point, if one compares the entries in the VM column (standard proximate values) with those in the weight loss column. The point to be made is that there is no easy way to predict by how much the devolatilization yield of any particular coal will exceed the proximate volatile matter content. The observation that the volatile matter release during rapid heating typically exceeds the proximate volatile matter has led to attempts at empirical correlations with coal type (e.g., the "Q factor" of Badzioch and Hawksley). The relationship between actual volatile yields and proximate volatile matter has again recently been examined (Desypris et al., 1982). However, correlations based on total volatile matter release convey no information concerning the actual composition of volatiles, which will be very important in understanding the combustion behavior of volatiles.

The variables which determine the yield and composition of devolatilization products include:

1. The chemical composition of the coal.
2. The temperature history to which the coal is subjected.
3. The pressure and composition of the gaseous environment surrounding the particles of coal.
4. The particle size.

These factors are listed, crudely, in decreasing order of importance. Proximate volatile matter itself is one of the two key rank classification criteria. This naturally implies a strong correlation between elemental composition and volatile matter. But it must be recalled that coals with widely differing compositions (e.g., a Pittsburgh high volatile bituminous, #14, and a Montana lignite, #8 in Table 2) may exhibit rather similar proximate volatile matter yields and yet be entirely different in combustion behavior because of different compositions of volatiles.

The compositional (or composition related) variables which play a role in determining the total yield of volatile matter and their composition have been given as (Solomon and Hamblen, 1985):

- Elemental Composition
- Functional Group Composition
- Molecular Weight of Ring Clusters
- Plasticity
- Variations in Bridging Material between Clusters
- Porosity
- Hydrogen Bonding
- Catalytic (Mineral) Constituents
- "Guest" Constituents such as Methane

The relative importance of these factors may be open to debate, and some of the factors may be viewed as compositionally dependent rather than independent (e.g., plasticity). Still, it is undeniable that compositional variables are the most important in determining volatile yields.

There have been numerous attempts to correlate volatile matter yields with elemental composition. A review of coal characteristics has shown a crude relation between volatile matter and the H/C atomic ratio of a wide range of coals (Ergun, 1979). For most coals of interest, any attempted correlation would be no better than ± 20%, and worse yet for a few "special cases." This re-emphasizes the point that the search for significant correlations must be made with a large number of coals; a sample of a half dozen or so is simply inadequate. Still, there is undeniably a correlation between composition and volatile matter, a fact recognized by Mott (1948) who used a variant of Seyler's coal classification chart, which had as the orthogonal axes volatile matter and calorific value, and showed elemental composition as "derived parameters." A very recent attempt at correlation of proximate volatile matter with elemental composition has been somewhat more successful than some of the earlier attempts (Neavel et al., 1986). The regression equation was found to be:

$$\%VM = -0.408[C] + 11.25[H] + [O] + 1.3[S]$$

where the quantities in brackets are all mass percentages on a dry basis (oxygen by difference on an ASTM ash basis). A standard deviation of 1.5% is noted, implying 95% confidence limits that at 40% V.M. would be ±8%, relative to the volatile matter estimate. While the results of this painstaking study are of great interest in coal classification work, it should be remembered that the proximate volatile matter that is predicted is not the volatile matter yield of relevance in combustion work.

The modest success achieved in correlating standard volatile matter yields with simple elemental composition parameters has led to other correlative approaches in recent years. In one study, (Neoh and Gannon, 1984), volatile yields were measured on 13 coals heated under pulverized coal heating conditions (10^5 K/s) to high temperature (1600-2400 K) in steam. It was found possible to correlate the volatiles yields by:

$$VM_{1600} \; K \; = \; 48.1 \left[\frac{[H] + 2[O]}{[C] + [S]} \right] - 1.41$$

and

$$VM_{2400} \; K \; = \; 52.6 \left[\frac{[H] + 2[O]}{[C] + [S]} \right] + 6.89$$

where the compositional ratios are on an atomic basis. Correlation coefficients of 0.95-0.96 were obtained, but a spread of $\pm 10\%$ (relative to volatiles) was seen, for what must be viewed as a relatively modest number of samples. It should be noted also that the 1600K yields were slightly in excess and the 2400K yields were greatly in excess of the ASTM proximate volatile yields.

Another correlation was developed between volatile yields and elemental composition (Solomon et al., 1981) in which the devolatilization was performed at heating rates of about 1000 K/s to a temperature in excess of 1000K in vacuum. The correlation was developed for 27 coals, and is expressed as:

$$VM = 0.86 \; [O] + 12.6 \; [Hal]$$

In this case, the elemental compositions are taken as weight percents (dry, mineral free basis) and [Hal] represents aliphatic hydrogen. This correlation depends upon a more sophisticated analysis; in order to obtain [Hal] a spectroscopic measurement is needed. The correlation coefficient is 0.84, and deviations of ~10% (relative to volatiles) are seen.

It is not surprising that simple elemental analyses are not sufficient for very accurately correlating volatile matter. For example, it has been shown that not only does mineral matter yield some "volatile matter" (Scholz, 1980), but perhaps more importantly, it has an enormous catalytic effect on certain pyrolysis reactions (Tyler and Schafer, 1980; Franklin et al., 1983; Morgan and Jenkins, 1986). Generally, the effect involves a substantial decrease in volatiles with an increase in cation content (e.g., a 20% reduction in the case of a lignite studied by Morgan and Jenkins).

Consequently, it must be concluded that prediction of volatile yields to better than about $\pm 10\%$ from easily obtained elemental composition data is not likely to be possible, although estimates of the accuracy cited above have been possible for some time. It must be asked whether it is worth considerable additional effort to correlate simple weight loss with the more sophisticated compositional information that is now becoming available on coals. It is not likely that the effort is warranted, unless the effort goes hand-in-hand with development of correlations for significant compositional properties of volatiles as well. Specifically, the additional features of interest are:

- The heating value of the volatiles.
- The elemental composition of the volatiles.

- The molecular weight distribution of the volatiles.
The time variation of these properties is also of interest. The reasons for the interest in these properties are likely to be apparent; whether the volatiles are combustible (e.g., tars, Ch_4) or incombustible (e.g., H_2O and CO_2) will have an important bearing on flammability characteristics. The partitioning of nitrogen between char and volatiles will have an important impact on NO_x formation in many situations. The concentration of high molecular weight species may tend to determine the propensity towards soot formation (Seeker et al., 1981, McLean and Hardesty, 1981).

In recognition of these considerations, increasing emphasis has been placed in recent years on measurement of actual product compositions and the kinetics of formation of individual products. The present review of this information is biased towards results from high heating rate experiments ($\gtrsim 1000K/s$) with fine particles, as would be most relevant to pulverized coal combustion.

5.2. Measurements of Volatile Compositions and Evolution Kinetics at High Heating Rates

The major experimental techniques used to study high heating rate pyrolysis include:

- Captive solid sample techniques, including the electrically heated wire mesh, fluidized beds, and shock tubes.
- Entrained flow reactors.

The most widely used is the electrically heated wire mesh, apparently developed first by Loison and Chauvin (1964) and now employed in numerous research laboratories. The technique involves heating of a thin layer of particles (normally in a size range between 50 and 1000 μm) in the folds of a wire mesh which acts as a resistance element in a circuit. Heating rates up to 10,000 K/s can be achieved, and temperatures up to about 1300K are attained with stainless steel meshes; temperatures up to about 1900K are attainable with tungsten meshes. An important advantage of this technique is that any volatiles that leave the particles, and thus the mesh, are usually quickly quenched by contact with the unheated surroundings of the mesh (either coal gas or cold walls), provided that the reactor vessel is large enough. Thus, primary products of pyrolysis are observed in the gas phase.

Shock tube studies of pyrolysis have become more common in recent years, as have laser pyrolysis studies. Higher heating rates are available to these devices, but product collection has been difficult (particularly the condensed phase products).

The entrained flow experiments also have the advantage of higher heating rates (10^5 - 10^6K/s), more similar to those thought relevant in pulverized coal combustion. The difficulty in interpreting the results from these experiments is that direct temperature measurement has historically been difficult (in the past, calculated temperatures have generally been employed). Recently, however, optical temperature measurement techniques seem to be developing to the point of reliability in such systems (Solomon et al., 1986; Tichenor et al., 1984). A remaining disadvantage of entrained flow systems is the coupling of gas and particle residence times. This means that primary volatile products of pyrolysis travel concurrently with the particles down an entrained flow reactor (though not necessarily at the same velocity) and therefore have an opportunity to undergo further reaction prior to sampling.

The compositional data shown in Table 1 have been assembled from the above types of reactors. The common thread is that they have all been obtained at the high rates of heating of relevance to pulverized coal combustion (>100K/s). The key to references is CL (Cliff et al., 1984), SOL (Solomon and Hamblen, 1983), SU (Suuberg, 1977), SC (Scelza, 1979), SO (Solomon and Hamblen, 1985),

LC (Loison and Chauvin, 1964), VH (Arendt and Van Heek, 1981), LA (Lander, 1979), FR (Franklin et al., 1981), FH (Freihaut et al., 1982), UN (Unger, 1983), and TY (Tyler, 1980). More detailed data on nirogenous products of pyrolysis are available from Blair et al. (1976), Pohl and Sarofim (1976). An enormous variety of compositional data is seen in Table 2. The variations are in some cases enormous, depending upon the particular conditions employed. In addition to providing "typical" compositions of volatiles, Table 2 has been designed to emphasize the following:

1. The products of pyrolysis vary from oxygen dominated to hydrocarbon dominated products as rank is increased from lignites to bituminous coals. It is apparent that there are no simple correlations of product yields with elemental analysis, however. This conclusion is reached even if data from systems in which cracking is likely to be a problem (Sec. 3. below) are excluded. Clearly, if prediction of yields of key products is a goal, we are rather far from it at present.

2. In considering the relevance of any set of data to combustion modeling, obviously preference should be given to studies in which the difference between the column marked "Total" and the "Wt. Loss" is small, indicating relatively complete accounting of all products. In fairness to some authors whose data are displayed with numerous "ND" entries, the measurements may have been made, but merely not reported in the "readily available" literature. In light of this, it is suggested that future studies concerning products of devolatilization endeavor to routinely present mass balance closures.

3. Data in rows #1 through 4 are all in a similar type of coal, indicating how not only temperature but gas phase residence time can affect measured products. Comparison of #11 and #12 and comparison of #13 and #14 show that fluid beds tend to promote cracking of tars, and favor light gas formation. Likewise, #18 and #20 show how entrained flow reactors favor cracking of tars to light gases. Thus, data from heated grids are probably "cleanest" in terms of indicating what the primary products are.

4. Comparison of #24 and #25 reaffirms that importance of cations in cracking of tars, as these two samples differed only in added cation content.

5. Comparison of #8 and #9, comparison of #14, #15, and #16, suggest volatile product compostions are not overly sensitive to heating rate in the range from a few hundred to 10^4 K/S heating rates. This is supported by a comparison of total volatile yield in #1 vs. wt. loss in #4. The data in #6 and #7 or #18 and #19 seem to contradict this conclusion, but in both cases, there seems to be evidence of high light gas yields, suggesting cracking, at the lower heating rates, possibly because of a small reactor volume. In examining the gaseous volatiles only, Blair et al. (1976) found little effect of heating rate on light gaseous yields above 500 K/sec. Data on mass loss alone (Niksa et al., 1982) seem to confirm that above a few hundred degree per second heating rate, there is minimal effect of heating rate on volatiles yields.

No distinction is made in Table 2 between volatile yields measured at vacuum and atmospheric pressure conditions. There are, however, important differences, particularly as pertains to tar yields. It has been shown (Suuberg et al., 1979) that the yield of tars from a bituminous coal (#15 in Table 2) is decreased from 36.5% (MAF) at vacuum to the indicated 29.2% at atmospheric pressure. In a study on a wider range of coals, Arendt and Van Heek (1981) have shown the same trends for four other bituminous coals. The decrease in tar yield is partially compensated for by an increase in light gaseous yields (Suuberg et al., 1979; Arendt and Van Heek, 1981). The nature of the tar has also been shown to change with changes in pressure, increasing in molecular weight with decreased pressure (Unger and Suuberg, 1984). This is consistent with a partially vaporization-rate limited process of tar escape from pyrolyzing particles (Unger and Suuberg, 1981; Suuberg et al., 1985). Thus, devolatilization measurements performed at pressures different from those of interest in combustion may not give a true picture of volatiles compositions.

It is considerably more difficult to find data on the effect of particle size on devolatilization yields. It has been shown that increasing particle size seems to depress tar yield and increase gas yield, somewhat as does increasing pressure (Suuberg et al., 1979). However, it is difficult to get reliable data on high tar yield coals, since these tend to soften and change morphology (occasionally forming balloon-like cenospheres).

The kinetics of devolatilization are to be considered in a separate presentation at this conference, and so will not be reviewed extensively here. It has been suggested that one quasi-universal set of kinetic parameters can describe the evolution of the main products of pyrolysis from a wide variety of coals (Solomon and Hamblen, 1985). These constants are shown in Table 3. Also shown for comparison are constants from other studies, including some at low heating rates (Campbell and Stephens, 1976; and Weimer and Ngan, 1979). The constants of Solomon and Hamblen are presented as reasonably "universal" to all ranks, while the remaining workers have presented their constants as reasonably specific to particular coals (although some similarity may exist between the constants for different ranks of coal).

Not shown in Table 3 are the kinetic constants for tar degradation. Serio (1984) has determined that secondary reactions of Pittsburgh high volatile bituminous tar can be modeled by a distributed activation energy process, with a pre-exponential of 3.96×10^9 sec^{-1}, a mean activation energy of 49.5 kcal/mol and a standard derivation of activation energies of 8.1 kcal/mol. The gaseous products are similar to those of primary pyrolysis. Doolan et al. (1985) have reached the same conclusion in work on Millmerran subbituminous coal tar, except that surprisingly little H_2O and CO_2 are formed during tar decomposition. Also, little acetylene is formed below 1300K.

There remains considerable debate as to the values of kinetic constants for the devolatilization process. Some of the debate stems from poor knowledge of temperature (experiments involving calculated temperatures suffer from this). Some of the debate arises due to attempts to fit processes involving distributed activation energies by single activation energies, which invariably leads to unrealistically low activation energies. Evidence of this is seen in Table 3. This is an area which requires careful re-evaluation of existing data, and further measurements to verify newly developing models. This is particularly true of the combined chemical kinetic-mass transfer models which seek to further explain the effect of pressure and particle size on the yields and compositions of volatiles.

In short, while there exist a large number of data on devolatilization, they paint a rather confusing picture both in terms of compositions of volatiles and kinetics of release. In designing future experiments, the following questions should be addressed:

1. How important is heating rate in determining key properties of interest? It appears as though heating rate variation in the range from a few hundred to 10^4 K/s has little effect on total volatile yields, composition and kinetics of release. Selection of a heating rate in this range enables use of the heated grid, which is perhaps the simplest of the devices commonly used. The number of data obtained at higher heating rates ($>10^5$K/s) is limited, and some suggest somewhat higher yields of volatiles are possible; but these are normally seen in experiments also involving

temperatures in excess of about 1600K (see, for example, the review by Howard, 1981). Unless the experiments to be undertaken involve striving for such high temperatures, efforts to ensure "realistic" heating rates of $>10^5$K/s are likely to be unnecessary, in terms of providing a "realistic" picture of volatiles release.

2. How important are secondary reactions? Again, the heated grid, in a <u>large</u> reactor volume, provides volatiles compositions that are most free of secondary cracking reactions. But is this an important goal in providing data for pulverized combustion modeling? How do secondary reactions (e.g., cracking of tars) affect the end use to which the data are put?

6. KINETICS OF COMBUSTION OF COAL VOLATILES

There is unfortunately little fundamental flammability data on coal volatiles. This is relatively easy to understand in terms of the difficulties in performing reliable experiments on coal volatiles as such. It must be remembered that coal volatiles are produced at temperatures ranging from a few hundred to, at sufficiently high heating rates, temperatures over 1000°C. As has been pointed out already, many of the heavy, tarry volatiles are marginally volatile even in such a high temperature range. This means that classical flammability limit tests, as performed in room temperature tubes, would involve a rather messy mixture of flammable vapor and condensed phase, complete with all the difficulties inherent in droplet flammability tests (e.g., gravitational settling, droplet size dependence of limits, loss of droplets to the walls by collision). Of course, classical limit tests of the flammability tube type are not of great relevance to practical applications, except as they might provide data which perhaps could be analyzable in terms of global kinetic constants for combustion of the volatiles. Thus, few efforts have been made to pursue these measurements with real coal volatiles.

Instead, the flammability of volatiles has been characterized indirectly, by measurement of ignition temperatures (Darbetaki et al., 1980), or measurements of stable operating limits in swirl type burner devices such as that in use at Ohio State University (Essenhigh, 1986). The problem of feeding heterogeneous mixtures to flat flame type devices makes use of these devices, which are a bit more suitable for fundamental studies of flame speed, somewhat tricky.

Some data are available on flammability limits of coal gas mixtures and mixtures of hydrocarbon gases. These data are summarized in the well-known reports of the U. S. Bureau of Mines (Coward and Jones, 1952; Zabetakis, 1965), and demonstrate the applicability of LeChatelier's law of mixtures:

$$L = 1/\Sigma p_i/N_i \qquad (6.1)$$

where p_i is the mole fraction of gas i in the mixture, N_i is the flammability limit of pure i in air, and L is the limit for the mixture in air. Normally, this law is applied with N_i and L as lower limits, but Coward and Jones (1952) suggest its applicability to upper limits as well. It may be applied to liquid mixtures, if Raoult's law or some similar vapor-liquid relation is known to apply (Zabetakis, 1965). The method does fail occasionally, however, particularly if one component is prone to cool flame formation (Coward and Jones, 1952).

A graphical approach which embodies LeChatelier's law and extends it to include non-flammable diluents has recently been presented (Wierzba and Karim, 1983). While the approach discussed the "special" problem of dilution by CO_2 (presumably mainly because of the effects on mixture heat capacity), the technique is otherwise identical to the use of (4.1), except that the summation is performed only over the <u>flammable</u> species i, noting that in the case of a mixture containing diluents $\Sigma p_i < 1$.

Again, the major problem involved in application of this law to coal volatiles mixtures is the handling of the heavy, tarry fractions for which there are no real data available. Then too, there are the problems associated with extrapolation of data obtained at standard ambient conditions of temperature, pressure, and oxidizer fraction to conditions of actual interest in any particular application.

The same difficulties that apply to the measurement of flammability limits would apply to measurements of global kinetics of combustion. The literature on the global kinetics of hydrocarbon gas combustion is limited in any case. The review by Westbrook & Dryer (1981) provides a relatively comprehensive set of single-step, two-step and quasi-global reaction constants for <u>pure</u> hydrocarbon species. These constants are summarized in Table 4. Also shown are quasi-global kinetic constants for combustion of carbon monoxide, hydrogen, and ammonia.

For some applications, a single pseudo chemical reaction step may be sufficient (Coffee et al., 1983). This may be described by the general equation for reaction:

$$Fuel + n_1 O_2 \longrightarrow n_2 CO_2 + n_3 H_2 O$$

the n_i are determined from the choice of fuel. The appropriate rate law is

$$k_{ov} = AT^n \exp(-E_A/RT)[Fuel]^a[O_2]^b \qquad (6.2)$$

where the symbols in square brackets represent concentrations, in units of mole/cm^3, R is the gas constant in cal/mol-K, T is the temperature in K, and the other symbols are as shown in Table 4. Note that a negative value of "a" suggests that the fuel acts as an inhibitor, which may create numerical problems as its concentrations approach zero. In this case, the rate expressions may be arbitrarily truncated at low values of concentration.

The two-step reaction mechanism may be represented by:

$$C_n H_m + \left[\frac{n}{2} + \frac{m}{4} \right] O_2 \longrightarrow nCO + \frac{m}{2} H_2 O$$

followed by:

$$CO + 1/2\ O_2 \longrightarrow CO_2$$

This recognizes that the single step reaction overpredicts the heat of combustion by assuming that all carbon in the fuel is burned to CO_2 and all hydrogen in the fuel is burned to H_2O. At flame temperatures, significant amounts of CO, H_2 and dissociated species may exist. As a compromise, partial combustion to CO is allowed in the two-step mechanism, although all hydrogen is still assumed to burn to H_2O. The rate constants for the first step are shown in Table 4, utilizing the nomenclature of equation 6.2, but recognizing that the products are now CO and H_2O. The values of a and b are unchanged.

Finally, the H_2 - H_2O equilibrium may be accounted for by assuming that the fuel is broken down to CO and H_2 in a first step, followed by a detailed mechanism for combustion of CO and H_2, involving key elementary reactions which have been well characterized (see Table 5). Thus, the breakdown of the fuel is represented by (Edelman and Fortune, 1969):

$$C_n H_m + \frac{n}{2} O_2 \longrightarrow n\ CO + \frac{m}{2} H_2$$

This is followed by all the reactions of Table 5 (and their reverse reactions as well). The quasi-global model has also been applied to aggregated species, and not just the pure species shown in Table 4. The rate expressions for quasi-global breakdown of long chain aliphatics to CO and H_2 has been modeled by (Edelman, 1977):

$$\frac{d[C_nH_m]}{dt} = -6 \times 10^4 \, TP^{0.3} \left[C_nH_m\right]^{1/2} \left[O_2\right] \exp\left[-\frac{24,400}{RT}\right]$$

and for cyclic hydrocarbons:

$$\frac{d[C_nH_m]}{dt} = -2.08 \times 10^7 \, TP^{0.3} \left[C_nH_m\right]^{1/2} \left[O_2\right] \exp\left[-\frac{39,000}{RT}\right]$$

where the units are pressure in atmospheres, T in K, E in cal/mol, and concentrations in mol/cm^3.

Recently, Hautman et al. (1981) have proposed a sequence of global processes for aliphatic hydrocarbon combustion, involving ethylene and H_2 intermediates. The sequence involves a series of four reactions, involving pyrolysis of the aliphatic to ethylene and hydrogen, followed by combustion of these species. This scheme will not be outlined here, noting that except in a few cases, coal does not yield a great deal of long chain aliphatic products.

In addition to the above reactions, more recently Edelman et al. (1983) have also considered emissions from combustion of synthetic fuels components, and have included a global reaction for soot formation from various components of a toluene-iso-octane global model. The global rate expression employed was of form:

$$d[soot]/dt = 4.66 \times 10^{14} \, T^{-1.94} \, [HC]^{1.81} \, [O_2]^{-0.5} \exp\left[-\frac{32,000}{RT}\right]$$

where [HC] is the concentration of toluene plus acetylene in this case. Naturally, a soot burnout term was also included. A meaningful treatment of coal volatiles combustion will doubtless require an allowance for soot formation from volatiles also, since recent coal flame work has established that at least some soot seems to be produced directly from coal volatiles (McLean et al., 1981; Seeker et al., 1981).

7. COMPARATIVE TIME SCALES FOR DEVOLATILIZATION AND COMBUSTION

Having now assembled data on rates of production and destruction of volatile species, it is instructive to reconsider the issue of comparative time scales for the various processes involved. Such a comparison is presented in Table 6, based upon:

1. Isothermal processes, at the indicated temperatures (no temperature gradients considered).
2. A somewhat arbitrary loading of coal of about 250 g/m^3, shown to give near peak temperature in bench scale flammability tests with Pittsburgh high volatile bituminous coal (Cashdollar and Hertzberg, 1982). The air surrounding the particles is assumed at atmospheric pressure. The yield of tar is 30%, and the tar molecular weight is 400.
3. The time scale for tar and H_2 release is calculated from the distributed rate model of Solomon and Hamblen, with conversion of 50%.

4. .The time scale for tar cracking is calculated from the constants presented by Serio, for a conversion of 50%.
5. The time scale for cyclic hydrocarbon combustion is calculated from the expression for cyclic hydrocarbons by Edelman; that for benzene from the data in Table 4. Both these rates are for the quasi-global process, for a conversion of 50%.

It may be noted immediately from Table 6 that there is a wide disparity in the time scales for release of volatiles. Release of tars occurs on a much shorter time scale than does release of H_2. Obviously, homogeneous cracking of the tar, in the absence of oxygen, is a slow process compared to its release from the coal, but still a rapid process in comparison to the overall time scale of pyrolysis as defined by H_2 release.

It is instructive to note that the predicted time scales for tar cracking and combustion are both long in comparison to the time scale for tar release at almost all temperatures of relevance. It is, however, noteworthy that if hydrogen release defines the time scale for devolatilization, then the combustion and devolatilization time scales are comparable, or devolatilization is even a bit longer. This illustrates the importance of specifying which species are actually of interest in defining the time scale. It is true that on a mass basis, tar will normally be present in far greater quantities than will hydrogen, but the latter on a molar or volumetric basis is the more "important" product.

8. THE HEATING VALUE OF VOLATILE PRODUCTS

Only limited data are available on the heating value contents of volatile products. This is again not surprising because obtaining the data is difficult. Still, as shown by Hertzberg et al. (1981), the limits of flammability of coal dusts seem to be predictable from an assumption of a constant limit mixture heat of combustion based on volatiles. The limit mixtures seemed to all fall in the range of 10 to 13 kcal/mol of flammable mixture, based on volatiles. Clearly, the testing of samples should be carried on to lignites, whose volatiles are dominated by oxygenated species and, therefore, would be expected to not be nearly as flammable as the bituminous coals previously studied.

There have been only limited attempts to calculate the heat of combustion of actual rapid devolatilization product mixtures. Some such data are shown in Figures 3 and 4, for a lignite and bituminous coal, respectively. It is immediately apparent that the lignite evolves much less of its total heating value into the vapor phase than does the bituminous coal (about one quarter vs. about one half). The difference is in the relative yields of tar, mainly. The consequences for combustion behavior were examined in a separate paper (Suuberg et al., 1979). The key results are summarized in Figures 5 and 6, which were calculated from the data as shown in Tables 2 and 3, assuming a heating rate of 10^4 °C/sec.

What is immediately apparent in comparing Figures 5 and 6 is that the rates of devolatilization and total mass loss from the two coals are quite comparable. What is very different is the heating value of the volatiles, represented in the bottom panels of both figures as "heat of combustion fluxes" (for particles of 74 μm). The vapor surrounding the bituminous particle is clearly "richer" in fuel. These calculations were used to infer that the flux of volatiles from a 55μm lignite particle might at no time be great enough to shield the particle surface from direct attack of oxygen. Not only does this support the picture of heterogeneous ignition for small particles, but also makes the point that volatile combustion may well occur simultaneously with combustion of what might be termed fixed carbon. Obviously, there will be difficulty in predicting and modeling such important behavior without adequate knowledge of time resolved volatiles composition.

9. CONCLUSIONS

In this brief review, we have attempted to summarize the evidence concerning the role of volatiles in coal combustion processes. It has been noted that, in some cases, the volatiles may play a significant role, whereas in others, they seem to play little role in determining the nature of the combustion process. It is fair to say that there remains a great deal of uncertainty as to mechanistic details in a great many processes.

The information concerning composition of volatiles and the kinetics of their release remains spotty. Despite a large number of investigations in this area, a legitimate case may still be made in favor of a more systematic study of a large number of coals. Likewise, little systematic information is available on the heating value of volatiles, their flammability characteristics or global kinetics of combustion.

In short, while the role of volatiles in coal combustion has been acknowledged for more than a century, we remain in a situation in which it is still only possible to make semiquantitative guesses about the details of their participation in the processes.

REFERENCES

K. Annamalai and P. Durbetaki, Comb. and Flame, 29, 193 (1977).

D. B. Anthony and J. B. Howard, AIChEJ. 22, 625 (1976).

P. Arendt and K.-H. Van Heek, Fuel, 60, 779 (1981), also K.-H. Van Heek, "Druckpyrolyse von Steinkohlen," VDI Forschungsheft Nr. 612/1982, Verlag GmbH, Dusseldorf (1982).

S. Badzioch and P. G. W. Hawksley, I&EC Proc. Des. Dev., 9, 521 (1970).

S. Bandyopadhyay and D. B. Bhaduri, Comb. and Flame, 18, 411 (1972).

D. W. Blair, J. O. L. Wendt, and W. Bartok, 16th Symp. (Int.) on Combustion, p. 475, The Combustion Institute, 1976.

M. C. Branch and R. F. Sawyer, 14th Symp. (Int.) on Combustion, p. 967, The Combustion Institute (1973).

J. H. Campbell and D. R. Stephens, ACS Div. of Fuel Chem. Preprints, 21(&), 94 (1976).

D. L. Carpenter, Proc. S. Wales Inst. Engrs., 69 (1953-54); Comb. and Flame, 1, 63 (1957).

K. L. Cashdollar and M. Hertzberg, Temperature, 5, 453 (1982).

H. M. Cassel and I. Liebmann, Comb. and Flame, 3, 467 (1959); 6, 153 (1962); 7, 79 (1963).

M.-R. Chen, L.-S. Fan and R. H. Essenhigh, 20th Symp. (Int.) on Combustion, p. 1513, The Combustion Institute (1984).

D. L. Cliff, K. R. Doolan, J. C. Mackie, and R. J. Tyler, Fuel, 63, 394 (1984).

T. P. Coffee, A. J. Kotlar, and M. S. Miller, Comb. and Flame, 54, 155 (1983).

J. G. Cogoli and R. H. Essenhigh, Comb. Sci. and Tech., 16, 177 (1977).

H. F. Coward and G. W. Jones, "Limits of Flammability of Gases and Vapors," U. S. Bureau of Mines Bulletin No. 627 (1965).

R. Deguingand and S. Galant, Proc. 18th Symp. (Int.) on Comb., p. 705, The Combustion Institute (1981).

G. de Soete, Rev. de l'Institut Francais du Petrole, 40, 650 (1985).

J. Desypris, P. Murdoch, and A. Williams, Fuel, 61, 807 (1982).

K. R. Doolan and J. C. Mackie, 20th Symp. (Int.) on Comb., p. 1463, The Combustion Institute, 1984.

K. R. Doolan, J. C. Mackie, M. F. R. Mulcahy, and R. J. Tyler, 19th Symp. (Int.) on Comb., p. 1131, The Combustion Institute, 1982.

K. R. Doolan, J. C. Mackie, and R. J. Tyler, Proc. 1985 Int. Conf. on Coal Sci., p. 961, Pergamon (1985).

F. L. Dryer and I. Glassman, 14th Symp. (Int.) on Comb., p. 987, The Combustion Institute (1973).

P. Durbetaki, V. L. Wolfe, W. R. Arthur, G. H. McAuliffe, and R. T. Gibbs, "Preliminary Studies in the Pyrolysis of Coal and Wood," Paper WSS 80-23, Western States Meeting, Combustion Institute, April 1980.

R. B. Edelman in Alternative Hydrocarbon Fuels: Combustion and Chem. Kinetics (C. T. Bowman and J. Birkcland, Eds.), pp. 294-301, Vol. 62, Prog. Astronautics and Aeronautics, AIAA (1977).

R. B. Edelman, R. C. Farmer, and T.-S. Wang, in Combustion of Synthetic Fuels (W. Bartok, Ed.), American Chemical Society Symposium Series No. 217, Chapter 2 (1983).

R. B. Edelman and O. F. Fortune, "A Quasi-Global Chemical Kinetic Model for the Finite Rate Combustion of Hydrocarbon Fuels with Application to Turbulent Burning and Mixing in Hypersonic Engines and Nozzles," AIAA Paper 69-86 (1969).

M. A. Elliott, Ed., Chemistry of Coal Utilization, Second Suppl. Vol., Wiley, 1981.

S. Ergun in Coal Conversion Technology, (C. Y. Wen and E. S. Lee, Eds.,), pp. 1-56, Addison-Wesley, Reading, Massachusetts (1979).

R. H. Essenhigh, Personal Communication (1986).

R. H. Essenhigh, in Coal Conversion Technology (C. Y. Wen and E. S. Lee, Eds.), Chap. 3, Addison-Wesley, Reading, Mass., 1979.

R. H. Essenhigh and J. B. Howard, "Combustion Phenomena in Coal Dusts and the Two-Component Hypothesis of Coal Combustion," Penn State University Studies No. 31, Pennsylvania State University, 1971.

M. Faraday and C. Lyell, Phil. Mag., 26, 16 (1845).

H. Farzan and R. H. Essenhigh, Proc. 19th Symp. (Int.) on Comb., p. 1105, The Combustion Institute, 1982.

J. D. Freihaut, D. J. Seery, and M. F. Zabielski, "Investigation of the Coal Devolatilization: Kinetics and Product Characterization," United Technologies Research Center Report R82-955568-4, 1982.

H. O. Franklin, R. G. Cosway, W. A. Peters, and J. B. Howard, I&EC Proc. Des. Dev., 22, 39 (1983).

H. D. Franklin, W. A. Peters, F. Cariello, and J. B. Howard, I&EC Proc. Des. Dev., 20, 670 (1981).

G. R. Gavalas, Coal Pyrolysis, Elsevier, Amsterdam/New York (1982).

G. T. Gee, K. Han, E. Hardin, D. Liu, D. Purvis, D. Shaw, and R. H. Essenhigh, Paper #CSSCI-84-13, Spring Meeting, Central States Section, The Combustion Institute, 1984.

A. L. Godbert, Fuel, 9, 57 (1930) and Safety in Mines Research Establishment: Research Report No. 58, HMSO, London, 1952.

A. L. Godbert and H. P. Greenwald, U. S. Bureau of Mines Bull. No. 389 (1935).

P. Goldberg and R. H. Essenhigh, Proc. 17th Symp. (Int.) on Comb,, p. 145, The Combustion Institute (1979).

D. Hautman, F. Dryer, K. Schug, and I. Glassman, Comb. Sci. Tech., 25, 219 (1981).

M. Hertzberg, K. L. Cashdollar, D. L. Ng, and P. S. Conti, Proc. 19th Symp. (Int.) on Combustion, p. 1169, The Combustion Institute, 1982.

M. Hertzberg, K. L. Cashdollar, and C. P. Lazzara, Proc. 18th Symp. (Int.) on Combustion, p. 717, The Combustion Institute, 1981.

M. Hertzberg, R. S. Conti, and K. L. Cashdollar, Proc. 20th Symp. (Int.) on Combustion, p. 1681, The Combustion Institute, 1984.

J. B. Howard in: Chemistry of Coal Utilization. Second Supplementary Volume (M. A. Elliott, Ed.), pp. 625-784, Wiley, New York (1981).

J. B. Howard and R. H. Essenhigh, I&EC Proc. Des. and Dev., 6, 74 (1967); Proc. 11th Symp. (Int.) on Comb., p. 399, The Combustion Institute, 1967.

J. B. Howard, W. A. Peters, and M. A. Serio, "Coal Devolatilization Information for Reactor Modeling," Final Report, EPRI Project No. (1981).

J. B. Howard, G. C. Williams, and D. H. Fine, 14th Symp. (Int.) on Combustion, p. 975, The Combustion Institute (1973).

H. Juntgen and K. H. Van Heek, Fuel Proc. Tech., 2, 261 (1979).

H. Karcz, W. Kordylewski, and W. Rybak, Fuel, 59, 799 (1980).

E. P. Lander, M.S. Thesis, Department of Chemical Engineering, Carnegie-Mellon University, 1979.

A. Levy, H. A. Arbib, and E. L. Merryman, in Combustion of Synthetic Fuels (W. Bartok, Ed.), American Chemical Society Symposium Series No. 217, Chapter 6 (1983).

R. Loison and R. Chauvin, Chimie et Industrie, 91, 269 (1964).

T. N. Mason and R. V. Wheeler, Safety in Mines Research Board Papers No. 33 (1927), No. 48 (1928); No. 64 (1931); No. 29 (1933); No. 95 (1936); No. 96 H.M.S.O., London.

W. J. McLean, D. R. Hardesty, and J. H. Pohl, 18th Symp. (Int.) on Comb., p. 1239, The Combustion Institute (1981).

R. A. Meyers, Ed., Coal Structure, Academic Press, 1982.

M. E. Morgan and R. G. Jenkins, Fuel, 65, 757 (1986).

R. A. Mott, J. Inst. Fuel, 22, 2(1948).

R. C. Neavel, S. E. Smith, E. J. Hippo, and R. N. Miller, Fuel, 65, 312 (1986).

K. G. Neoh and R. E. Gannon, Fuel, 63, 1347 (1984).

M. A. Nettleton and R. Stirling, Proc. Roy Soc., A300, 62 (1967), A322, 207 (1971), and Comb. and Flame, 22, 407 (1974).

S. Niksa, W. B. Russel, and D. A. Saville, 19th Symp. (Int.) on Combustion, p. 1151, The Combustion Institute (1982).

J. H. Pohl and A. F. Sarofim, 16th Symp. (Int.) on Combustion, p. 491, The Combustion Institute, 1976.

S. T. Scelza, M.S. Thesis, Department of Chemical Engineering, Carnegie-Mellon University, 1979.

A. Scholz, Fuel, 59, 197 (1980).

W. W. Seeker, G. S. Samuelsen, M. P. Heap, and J. D. Trolinger, 18th Symp. (Int.) on Comb., p. 1213, The Combustion Institute (1981).

M. Serio, Ph.D. Thesis, Department of Chemical Engineering, M.I.T., 1984.

C. A. Seyler, Proc. S. Wales Inst. Engr., 53, 254-327 and 396-407 (1938).

L. D. Smoot, M. D. Horton, and G. A. Williams, Proc. 16th Symp. (Int.) on Comb., p. 375, The Combustion Institute, 1977.

P. R. Solomon, "The Evolution of Pollutants During the Rapid Devolatilization of Coal," United Technologies Research Center Report R77-952588-3, November 1977.

P. P. R. Solomon, R. M. Carangelo, P. E. Best, J. R. Markham, and D. G. Hamblen, ACS Div. Fuel Chem. Prepr., 31(1), 141 (1986).

P. R. Solomon and D. G. Hamblen in Chemistry of Coal Conversion (R. H. Schlosberg, Ed.), pp. 121-251, Plenum, New York (1985).

P. R. Solomon and D. G. Hamblen, Prog. Energy Comb. Sci., 9, 323 (1983). Also, ACS Dev. Fuel Chem. Prepr., 24(3), 154 (1979) and Proc. 5th EPA Fundamental Comb. Res. Workshop (1980).

P. R. Solomon and D. G. Hamblen, Progr. Energy and Combust. Sci., 323 (1983).

P. R. Solomon, R. H. Hobbs, D. G. Hamblen, W.-Y. Chen, A. LaCava, and R. S. Graff, Fuel, 60, 342 (1981).

D. Stickler, R. Gannon, L. Young, and K. Annamalai, Paper #20, AFRC International Symposium on Combustion Diagnostics, Akron, Ohio, 1983.

E. M. Suuberg, Sc.D. Thesis, Department of Chemical Engineering, M.I.T., 1977.

E. M. Suuberg, in Chemistry of Coal Conversion (R. H. Schlosberg, Ed.), pp. 67-119, Plenus, New York (1985).

E. M. Suuberg, W. A. Peters, and J. B. Howard, 17th Symp. (Int.) on Combustion, p. 117, The Combustion Institute (1979).

E. M. Suuberg, P. E. Unger, and W. D. Lilly, Fuel, 64, 956 (1985).

D. A. Tichenor, R. E. Mitchell, K. R. Hencken, and S. Niksa, 20th Symp. (Int.) on Combustion, p. 1213, The Combustion Institute (1984).

L. Tognotti, A. Malotti, I. Petarca, and S. Zanelli, Paper presented at Poster Session, 20th Symp. (Int.) on Comb., University of Michigan, Ann Arbor, 1984.

S. Tretyakov, cited in S. Bandyopadhyay and D. B. Bhaduri (1972).

R. J. Tyler, Fuel, 59, 218 (1980).

R. J. Tyler and H. N. S. Schafer, Fuel, 59, 487 (1980).

S. Ubhayakar and F. A. Williams, J. Electrochem. Soc., 123, 747 (1976).

P. E. Unger, Ph.D. Thesis, Department of Chemical Engineering, Carnegie-Mellon University, 1983.

P. E. Unger and E. M. Suuberg, Fuel, 63, 607 (1984).

P. E. Unger and E. M. Suuberg, 18th Symp. (Int.) on Combustion, p. 1203, The Combustion Institute, 1981.

D. W. Van Krevelen, Coal, Elsevier, 1961.

R. F. Weimer and D. Y. Ngan, ACS Div. of Fuel Chem. Preprints, 24(3), 129 (1979).

C. K. Westbrook and F. L. Dryer, Comb. Sci. and Tech., 27, 31 (1981).

R. V. Wheeler, J. Chem. Soc., 103, 1715 (1913).

I. Wierzba and G. A. Karim, J. Inst. Energy, p. 68, June 1983.

D. V. Xieu, T. Masuda, J. G. Cogoli, and R. H. Essenhigh, Proc. 18th Symp. (Int.) on Comb., p. 1461, The Combustion Institute, 1981.

M. G. Zabetakis, "Flammability Characteristics of Combustible Gases and Vapors," U. S. Bureau of Mines Bulletin No. 627 (1965).

TABLE 1

FLAME TYPES

on the Propagation of One-Dimensional Flames of
Particulate Coal, Char, and Coke

Type	Flame Speed m/sec	Heating Rate deg C/sec	Characteristics	Propagation Mechanism
I	< 0.3	air > 10^5 particulate < 10^4 -	Primarily volatiles flame. Flame thickness: 10 cm Volatiles released downstream diffuse upstream against incoming air.	Conduction/Diffusion
II	0.2 - 0.5	10^4	Heterogeneous ignition at 1000C-delayed pyrolysis. Flame thickness: > 1 m	Radiation
III	< 1000	10^6	Flame thickness: > 10 m Explosive deflagration	Turbulent Exchange
IV	> 1000	--------		Detonation

Table 2

#	Ref.	Reactor	Gas/Pressure (ATM)	Heating Rate (K/S)	Max. Temp. (K)	Rank	C (Wt% MAF)	H	O (by diff.)	H/C	O/C	V.M. (MAF)	Particle size (μm)	CO	O_2	(H_2O)[1]	H_2	C_2H_2	CH_4	C_2H_4	C_2H_6	Other HC Gas	Liquids and Tar	Total	Wt. Loss% (MAF)
1	CL	ST	Ar/>1	10^7	2343	Brown	65.9	4.6	27.6	0.83	0.31	53.5	<8	34.5	8.1	-6	0.8	6.5	2.9	2.3	0.2	>1.9	ND	>63	ND
2	↓	↓	↓	↓	1623	↓	↓	↓	↓	↓	↓	↓	↓	22.2	7.4	-8	0.8	1.0	1.5	3.3	0.6	>1.8	ND	>47	ND
3	↓	↓	↓	↓	1273	↓	↓	↓	↓	↓	↓	↓	↓	5.5	5.6	-4	0.2	0.4	0.4	1.0	0.2	>0.4	ND	>17	ND
4	↓	FB	N_2/1	10^4	1273	Brown	69.0	4.6	25.7	0.80	0.27	50.7	98	24	8.5	2.5	1.5	1.7	3.0	3.3	0.2	>2.0	3.6	>50	64
5	SOL	HG	VAC	2000	2073	LIG	70.9	3.9	23.5	0.66	0.25	ND	<150	19.1	2.7	3.5	3.5	1.8	0.9	ND	ND	ND	15	>47	67
6	↓	HG	VAC	2000	1273	↓	↓	↓	↓	↓	↓	↓	↓	9.3	4.4	5.1	2.2	<0.1	1.3	ND	ND	ND	15	>38	54
7	↓	HG	VAC	10^4	1173	↓	↓	↓	↓	↓	↓	↓	↓	20.8	4.2	5.1	1.7	ND	ND	ND	ND	ND	10	>37	47
8	SU	HG	He/1	10^4	1168-1378	LIG	71.2	4.6	21.8	0.78	0.23	44.3	53-88	8.1	0.0	ND	ND	ND	1.5	1.0	0.3	1.5	6.5	>29	44
9	SC	HG	He/1	1000	1098	LIG	73.4	4.0	21.5	0.65	0.22	44.6	55-74	11.4	.5	11.7	0.6	ND	1.6	0.7	0.2	0.8	6.5	45	45
10	↓	HG	He/1	1000	1173	SUBB	78.4	6.4	13.5	0.98	0.13	51.4	76-104	5.1	.2	(4.8)[2]	ND	ND	1.2	0.4	0.7	1.4	4.6	28	34
11	TY	FB	N_2/1	10^4	1213	↓	↓	↓	↓	↓	↓	↓	53-88	ND	ND	ND	ND	ND	5.4	8.6	0.7	>1.1	15.2[4]	-	66
12	SU	FB	N_2/1	1000	1173	HVB	77.5	5.5	9.2	0.85	0.09	44.6	76-104	7.0	2.5	3.6	ND	ND	3.7	2.9	0.7	3.7	35[4]	>59	59
13	TY	FB	N_2/1	10^4	1123-1273	↓	↓	↓	↓	↓	↓	↓	53-88	ND	0	-	ND	ND	7.8	5.6	1.1	>0.2	19.2	-	62
14	SU	HG	He/1	1.3×10^4	↓	↓	↓	↓	↓	↓	↓	↓	53-88	2.6	1.3	6.1	ND	ND	2.8	0.8	0.7	3.1	29.6	47[5]	52
15	SOL	HG	He/1	1000	2073	HVB	82.0	5.4	9.4	0.79	0.09	38.9	<150	2.8	.4	7.4	1.2	ND	2.9	1.0	0.6	3.0	29.2	50[5]	52
16	↓	↓	↓	400	↓	↓	↓	↓	↓	↓	↓	37.1	↓	2.8	.8	6.0	3.0	0.9	2.5	0.5	0.7	2.5	28.4	45[5]	51
17	↓	HG	VAC	2000	↓	↓	↓	↓	↓	↓	↓	33.1	↓	0.7	.9	0.6	1.1	0	1.0	ND	ND	ND	28	>36	65
18	SOL	HG	VAC	2000	1173	↓	↓	↓	↓	↓	↓	↓	↓	2.4	.9	2	3.3	ND	2.3	ND	ND	ND	40	>49	65
19	↓	HG	VAC	600	1373	↓	↓	↓	↓	↓	↓	↓	53-74	1.9	.2	0.5	0	1.3	ND	ND	0	15.2	34	>40	50
20	SO	EF	He/1	2×10^3	1323	HVB	80.6	5.2	12.0	0.77	0.11	38.9	50-80	8.3	.8	4.8	1.0	ND	7.3	2.6	-0	ND	13.4	55	60
21	LC	HG	N_2/1	1500	↓	HVB	85.9	5.3	7.2	0.74	0.06	37.1	↓	6.0	2.5	5.5	0.9	ND	2.2	ND	ND	ND	20	>45	41
22	↓	↓	↓	↓	↓	HVB	86.3	5.2	5.7	0.72	0.05	33.1	↓	2.5	2.5	4.4	1.0	ND	3.1	ND	ND	ND	24	>35	40
23	↓	↓	↓	↓	↓	HVB[6]	82.9	5.7	7.2	0.83	0.07	41.5	↓	4.0	.6	4.7	1.0	ND	2.5	ND	ND	2.2	27	>41	48
24	FR	HG	He/1	1000	1050-1300	HVB[6]	82.9	5.7	7.2	0.83	0.07	41.5	45-53	2.4	0	3.0	ND	ND	3.6	1.0	0.7	2.2	33.6	48	51
25	VH	HG	He/0.1	210	1273	HVB[3]	83.0	5.6	7.3	0.81	0.07	36.4	200-315	3.4	.5	ND	ND	ND	2.9	0.9	0.7	1.5	25.9	38	48
26	UN	HG	He/1	1000	1273	HVB	84.4	5.6	7.4	0.80	0.07	37.0	62-88	3.8	.8	0.4	0.6	ND	2.5	0.4	0.5	2.8	28	>37	41
27	VH	HG	He/0.1	210	1273	HVB	86.4	4.6	5.5	0.64	0.05	32.4	200-315	3.6	.5	ND	1.1	ND	3.3	0.7	0.5	1.6	29.1	41	44
28	LC	HG	He/0.1	1500	1323	HVB	88.0	5.0	4.9	0.68	0.04	29.6	50-80	2.4	.5	3.1	0.8	ND	3.4	0.6	0.7	ND	27	>36	40
29	↓	HG	N_2/1	1500	1323	MVB	88.1	4.7	4.5	0.64	0.04	22.8	50-80	2.5	.8	2.7	1.2	ND	3.1	ND	ND	ND	21	>32	38
30	VH	HG	He/0.1	210	1273	MVB	89.5	4.2	3.3	0.56	0.04	22.7	200-315	1.5	.2	ND	1.1	ND	2.9	0.4	0.6	ND	18	>27	30
31	↓	HG	↓	↓	↓	LVB	90.4	4.3	2.2	0.57	0.02	16.3	↓	1.1	.3	3.2	1.2	ND	3.8	0.3	0.5	ND	17	>26	28
32	LC	HG	N_2/VAC	1500	1323	LVB	88.7	4.5	4.2	0.61	0.04	19.6	50-80	1.1	.3	ND	1.3	ND	3.4	0.2	0.6	ND	10	>17	19
33	FH	HG	He/1	1000	1323	LVB	85.0	4.6	4.2	0.65	ND	19.6	53-150	1.0	.3	3.2	1.2	ND	3.0	0.3	0.6	ND	16	>25	25
34	LC	HG	N_2/1	1500	1273	LVB	90.8	4.5	4.0	0.59	0.03	19.1	55-74	1.2	.3	ND	(1.2)[2]	ND	3.1	ND	0.4	1.4	17	>22	25
35	LA	HG	N_2/1	1500	1323	ANT	91.5	3.7	2.9	0.49	0.02	10.2	55-74	1.2	.3	0.6	1.1	ND	3.4	ND	0.4	ND	11.8	>9	22
36	LC	HG	He/0.1	210	1273	ANT	94.4	3.4	0	0.43	0	5.9	200-315	0	0	1.8	0.9	ND	1.1	ND	ND	ND	2	2	15
37	VH	HG	He/0.1	210	1273	ANT	ND	ND	ND	ND	ND	ND	ND	0	ND	ND	ND	ND	0.8	ND	ND	ND	ND	—	6

NOTES:

ND	not determined or not reported
HG	heated grid
FB	fluidized bed
ST	shock tube
EF	entrained flow
VAC	vacuum

LIG	lignite
SUBB	subbituminous
HVB	high volatile bituminous
MVB	medium volatile bituminous
LVB	low volatile bituminous
ANT	anthracite

[1] Excludes moisture
[2] Back calculated from elemental analysis.s
[3] Treated to contain 5.9% by mass of $Ca(OH)_2$ + $CaCO_3$
[4] Back calculated assuming H_2 and C_2H_2 yields small
[5] Does not include roughly 1% H_2S
[6] Demineralized

TABLE 1
KINETIC CONSTANTS FOR DEVOLATILIZATION

	Solomon and Hamblen, 1983			Suuberg et al. 1978			Juntgen and Van Heek, 1972			Campbell & Stephani, 1976			Weimer and Ngan, 1972			Doolan et al. (1981)		
	A	E	α	A	E	α	A	E	α	A	E	α	A	E	α	A	E	α
CO_2 - extra loose	3.4×10^{14}	44.5	4.0	2.1×10^{11}	36.2	0	$7.5 \times 10^2 - 1.2 \times 10^4$	21.8 - 26.5	0	550	19.5	0						
CO_2 - loose	4.8×10^{14}	58.8	7.2	5.1×10^{13}	64.3	0	$1.7 \times 10^5 - 1.1 \times 10^7$	37.0 - 43.5	0	230	23.0	0	1.67×10^{13}	48.9 - 61.8	9.5 - 18.1	1.3×10^5	12.0	---
CO_2 - tight	1.1×10^{15}	76.1	7.2	5.5×10^6	42.0	0												
H_2O - loose	1.7×10^{14}	54.6	4.6	7.9×10^{13}	51.4	0	$1.0 \times 10^6 - 3.3 \times 10^6$	31.0 - 33.5	0									
H_2O - tight	1.7×10^{14}	65.0	6.6				$1.4 \times 10^5 - 5.8 \times 10^5$	33.0 - 34.0	0									
CO - loose	1.7×10^{11}	49.7	5.0	1.8×10^{12}	44.4	0	$1.7 \times 10^3 - 3.3 \times 10^3$	22.0 - 27.0	0	55	18.0	0	1.67×10^{13}	30.8 - 51.9	6.0 - 7.8			
CO - tight	1.0×10^{14}	80.5	11.9	2.6×10^{12}	59.5	0	$3.3 \times 10^4 - 2.0 \times 10^6$	33.0 - 41.0	0	2.5×10^3	30.2	0	1.67×10^{13}	66.7 - 72.7	5.1 - 13.4	1.6×10^5	16.7	
CO - extra tight				5.9×10^9	58.4	0												
HCN - loose	1.7×10^{12}	59.6	3.0															
HCN - tight	1.0×10^{14}	94.4	9.4															
NH_3	1.2×10^{12}	54.2	6.0															
CH_4 - aliphatic	1.7×10^{14}	59.6	3.0	1.7×10^{16}	70.1	2.5 - 3.5	$1.67 \times 10^{14*}$	56.5 - 62.0*		7.3×10^6	35.0*		1.67×10^{13}	57.7 - 60.1	4.8 - 7.0	$7.0 \times 10^4 - 6.3 \times 10^5$	23.7 - 23.9	0
CH_4 - extra loose	1.7×10^{14}	59.6	3.0	1.6×10^{14}	51.6	6.0	$1.67 \times 10^{12*}$	53.5 - 58.0		1.7×10^5	31.0							
CH_4 - loose	1.5×10^{13}	59.6	4.0	4.7×10^{14}	69.4	6.0	$1.67 \times 10^{12*}$	62.0		2.8×10^4	31.0							
CH_4 - tight	3.4×10^{11}	59.6	4.0				$1.67 \times 10^{12*}$	69.0		3.0×10^4	35.3							
H_2	1.6×10^7	45.7	4.6	1.6×10^{18}	88.8	0				20	22.3		1.67×10^{13}	72.8 - 76.9	8.0 - 9.8			
Methanol	1	59.6	0															
Aldehyde	1	59.6	0															
Tar	4.5×10^{13}	52.5	3.0	7.6×10^{11}	37.4													
Tar (2)				2.0×10^{17}	75.3													
Cracking Reactions																		
Paraffin	1.5×10^{11}	54.8	0													4.9×10^8	48.8++	0
Olefin	2.1×10^7	43.7	0													2.2×10^9	53.1°	0

All rate constants in sec-kcal-mol-K system of units
* Small variation in A allowed for
+ Data for propane
° Data for ethylene
++ Data for methane

Table 4

Kinetic Constants for Combustion of Various Species

Species	Single Step				Two Step		Quasi-Global	
	A	E_A	a	b	A	E_A	A	E
CH_4	1.3×10^8	48.4	-0.3	1.3	2.8×10^9	48.4	4.0×10^9	48.4
CH_4	8.3×10^5	30	-0.3	1.3	1.5×10^7	30	2.3×10^7	30
C_2H_6	1.1×10^{12}	30	0.1	1.65	1.3×10^{12}	30	2.0×10^{12}	30
C_3H_8	8.6×10^{11}	30	0.1	1.65	1.0×10^{12}	30	1.5×10^{12}	30
C_4H_{10}	7.4×10^{11}	30	0.15	1.6	8.8×10^{11}	30	1.3×10^{12}	30
C_5H_{12}	6.4×10^{11}	30	0.25	1.5	7.8×10^{11}	30	1.2×10^{12}	30
C_6H_{14}	5.7×10^{11}	30	0.25	1.5	7.0×10^{11}	30	1.0×10^{12}	30
C_7H_{16}	5.1×10^{11}	30	0.25	1.5	6.3×10^{11}	30	1.0×10^{12}	30
C_8H_{18}	4.6×10^{11}	30	0.25	1.5	5.7×10^{11}	30	9.4×10^{11}	30
C_8H_{18}	7.2×10^{12}	40	0.25	1.5	9.6×10^{12}	40	1.5×10^{13}	40
C_9H_{20}	4.2×10^{11}	30	0.25	1.5	5.2×10^{11}	30	8.8×10^{11}	30
$C_{10}H_{22}$	3.8×10^{11}	30	0.25	1.5	4.7×10^{11}	30	8.0×10^{11}	30
CH_3OH	3.2×10^{12}	30	0.25	1.5	3.7×10^{12}	30	7.3×10^{12}	30
C_2H_5OH	1.5×10^{12}	30	0.15	1.6	1.8×10^{12}	30	3.6×10^{12}	30
C_6H_6	2.0×10^{11}	30	-0.1	1.85	2.4×10^{11}	30	4.3×10^{11}	30
C_7H_8	1.6×10^{11}	30	-0.1	1.85	1.9×10^{11}	30	3.4×10^{11}	30
C_2H_4	2.0×10^{12}	30	0.1	1.65	2.4×10^{12}	30	4.3×10^{12}	30
C_3H_6	4.2×10^{11}	30	-0.1	1.85	5.0×10^{11}	30	8.0×10^{11}	30
C_2H_2	6.5×10^{12}	30	0.5	1.25	7.8×10^{12}	30	1.2×10^{13}	30
$CO^{(1)}$	4.0×10^{14}	40	1*	0.25	-	-	-	-
$CO^{(2)}$	1.3×10^{14}	30	1*	0.50	-	-	-	-
$H_2{}^{(3)}$	3.2×10^{15}	40	1*	0.25	-	-	-	-
$NH_3{}^{(4)}$	4.9×10^{14}	39	0.86	1.04	-	-	-	-

Table entries in units of cm-sec-mol-kcal-K
*Rate expression also includes $[H_2O]^{1/2}$ dependence.

(1) Dryer and Glassman, 1973   (2) Howard et al., 1973
(3) Levy et al., 1983   (4) Branch and Sawyer, 1973

Table 5

Reaction mechanism used in quasi-global mechanism for CO-H_2-O_2 system. Reverse rates computed from relevant equilibrium constants.

Reaction	A	n	E_a
$H + O_2 = O + OH$	2.2×10^{14}	0.0	16.8
$H_2 + O = H + OH$	1.8×10^{10}	1.0	8.9
$O + H_2O = OH + OH$	6.8×10^{13}	0.0	18.4
$OH + H_2 = H + H_2O$	2.2×10^{13}	0.0	5.1
$H + O_2 + M = HO_2 + M$	1.5×10^{15}	0.0	-1.0
$O + HO_2 = O_2 + OH$	5.0×10^{13}	0.0	1.0
$H + HO_2 = OH + OH$	2.5×10^{14}	0.0	1.9
$H + HO_2 = H_2 + O_2$	2.5×10^{13}	0.0	0.7
$OH + HO_2 = H_2O + O_2$	5.0×10^{13}	0.0	1.0
$HO_2 + HO_2 = H_2O_2 + O_2$	1.0×10^{13}	0.0	1.0
$H_2O_2 + M = OH + OH + M$	1.2×10^{17}	0.0	45.5
$HO_2 + H_2 = H_2O_2 + H$	7.3×10^{11}	0.0	18.7
$H_2O_2 + OH = H_2O + HO_2$	1.0×10^{13}	0.0	1.8
$CO + OH = CO_2 + H$	1.5×10^{7}	1.3	-0.8
$CO + O_2 = CO_2 + O$	3.1×10^{11}	0.0	37.6
$CO + O + M = CO_2 + M$	5.9×10^{15}	0.0	4.1
$CO + HO_2 = CO_2 + OH$	1.5×10^{14}	0.0	23.7
$OH + M = O + H + M$	8.0×10^{19}	-1.0	103.7
$O_2 + M = O + O + M$	5.1×10^{15}	0.0	115.0
$H_2 + M = H + H + M$	2.2×10^{14}	0.0	96.0
$H_2O + M = H + OH + M$	2.2×10^{16}	0.0	105.0

Table entries in units of cm-sec-mol-kcal-K

Rate Constant $= AT^n \exp(-E_A/RT)$.

Table 6

Comparative Time Scales of Some Processes

Temperature (K)

	1000	1500	2000
Tar release	10^{-3}	10^{-6}	10^{-8}
H_2 release	10^{2}	10^{-1}	10^{-3}
Tar cracking	10^{1}	10^{-3}	10^{-5}
Tar combustion[1]	10^{0}	10^{-3}	10^{-4}
Tar combustion[2]	10^{-1}	10^{-3}	10^{-4}

1. Assuming "cyclic hydrocarbon" global rate.
2. Assuming benzene global rate.

All time scales in seconds.

208

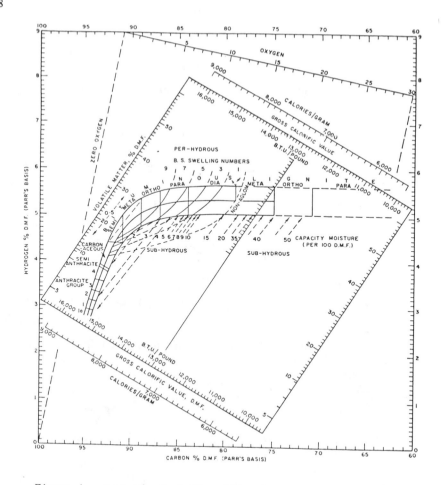

Figure 1. Seyler's Coal Classification Chart

International classification, class number	0	1.	2	3	4	5	6	7	8	9
		5	10	15 Volatile-matter parameter a/ 20	25	30 b/	14,000	13,000 12,000 Calorific-value parameter a/ 11,000		10,000
ASTM classification, group name	Meta an- thra cite	Anthracite	Semianthracite	Low-volatile bituminous coal	Medium-volatile bituminous coal	High-volatile A bituminous coal	High-volatile B bituminous coal	High-volatile C bituminous coal and subbituminous A coal		Subbituminous B coal

a/ Parameters in International system are on ash-free basis; in ASTM system, they are on mineral-matter-free basis.
b/ No upper limit of calorific value for class 6 and high-volatile A bituminous coals.

—Comparison of Class Numbers and Boundary Lines of International System With Group Names and Boundary Lines of ASTM System.

Figure 2. Coal Classification Criteria

209

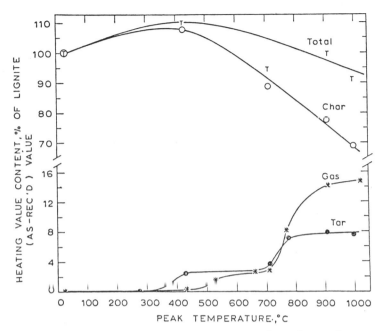

Figure 3: Distribution of heating value content of products from
lignite pyrolysis to different peak temperatures under
inert gas. Heating Rate - 1000°C/s. 100% on ordinate
corresponds to 17.8 kcal/g (9900 BTU/lb), lower heating
value (as received). Suuberg (1977)

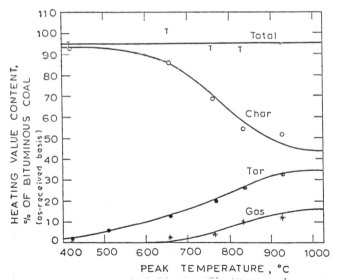

Figure 4. Distribution of heating value content of products from Pittsburgh
high volatile bituminous coal pyrolysis to different peak temperatures under inert gas. Heating Rate - 1000°C/s followed by 4-10
second hold at 1000°C. 100% on ordinate represents 21.6 kcal/g
(12000 BTU/lb) lower heating value. Suuberg et al. (1979)

210

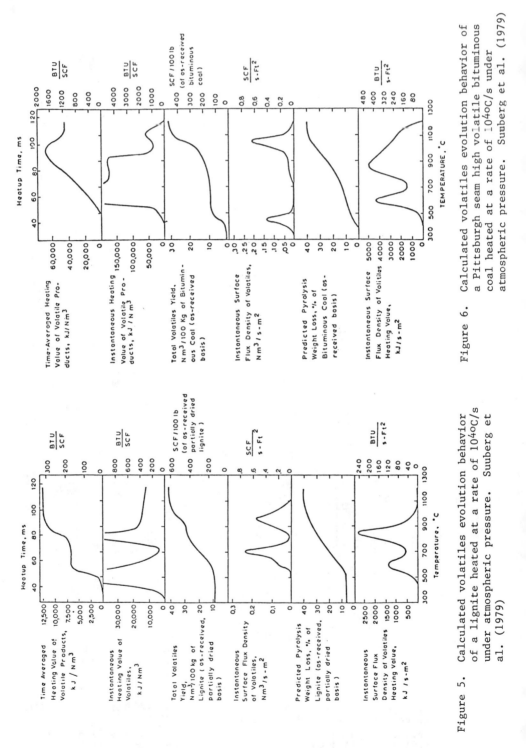

Figure 5. Calculated volatiles evolution behavior of a lignite heated at a rate of 10⁴ºC/s under atmospheric pressure. Suuberg et al. (1979)

Figure 6. Calculated volatiles evolution behavior of a Pittsburgh seam high volatile bituminous coal heated at a rate of 10⁴ºC/s under atmospheric pressure. Suuberg et al. (1979)

DISCUSSION

J.H. Pohl

Most properties of coal can be correlated using test at standard conditions, maceral content and C(d.a.f.). Have you investigated correlations of results from standard conditions and these two parameters ? We have found that flame stability in laboratory pilot scale combustors, and boilers can be correlated with the heating value of the volatile matter based on coal (as received). In addition, the heating value of the volatile matter can be correlated with carbon (daf). These correlations have been tested with a large number of coals and several boiler trials. Have you investigated such correlations ? If the kinetics of devolatilization are accepted, the results show that tar formation and destruction at PC flame conditions is rapid and has little influence of flame performance. Is this true, and if so why has so much effect been expended on defining tar formation for applications to PC flames ?

R.H. Essenhigh

This is a complex of several questions that need to be given much more careful attention with :

1. Maceral content. There are increasing numbers of claims for the influence of maceral content and, intuitively, one would expect some influence. I have not looked at this parameter so I cannot comment from my own experience. However, correlation of some properties with maceral content may be only that : a correlation ; and the maceral content could be only an index of some other factor. Nevertheless, establishing char correlations is still an important and valuable point step.

2. Heating value of VM content. Again, I have not been in a position to establish correlations between flame stability and the VM content heating value, but the information referred to is most interesting and it will be a great service to the combustion field if this can be fully published. Again, however, the question comes up : is the correlation between flame stability and combustible volatiles an actual property of the system or is it only a correlation ?

3. Tars. Tars can be correlated with the pyridine extract which has also been identified as the easily-evolved (combustible factor) of the volatiles. I believe Wheelers (1913) point is still valid that the volatiles do not emerge together, but over a period of time (though short), particularly the hydrogen. It is certainly the case that tars contain a very high proportion of the heating value present in the volatiles released from many coals. Then it might not be unreasonable to expect that the true correlation would be between the flame stability and the tars (but the total combustible volatile is an adequate index of the tars if tar data

212

are lacking). The amount of tar release may also be important
in determining sooting tendency.

K.H. Van Heek

How have been established the heating values of the
tars ? I see difficulties, as their qualities change with coal
type and reaction conditions, on the one hand, and as the
amounts received from laboratory experiments are small for
proper analysis, on the other hand.

E.M. Suuberg

In the absence of direct measurements, we assumed a heating value of tar 8.9 Kal/g (lower heating value). This value
is "typical" of bituminous coal tars, but we agree completely
that there will likely be variations with the rank of parent
coal. We simply did not have the data.

J.B. Howard

Regarding the correlation, or lack thereof, between tar
yield and elemental composition of the coal, some recent
progress have been made at MIT (Ko, G., Peters, W.A., and
Howard, J.B., FUEL, in press) using not tar yield in general,
but the "ultimate" tar yield observed under low pressure where
the effect of secondary reactions are approximately negligible. The correlation uses the contents of C, H., O and S in a
simple functional form approximately representative of the
chemical bonding, or functional group characteristics, believed to be important in tar formation. The correlation exhibits
an encouraging reduction in scatter, relative to that of
previous tar correlations, using data for a range of coal
types and from several countries.

E.M. Suuberg

This is indeed an important development. Of course it
still leaves the problem of correcting vacuum yields to
atmospheric pressure yields but as indicated in the reference
by Suuberg in the Chemistry of Coal Conversion, this kind of
correlation may also be possible. Further work is clearly
required on this latter point, though the only concern I would
reiterate is that one must make sure that enough samples are
used in order to provide reliable statistics. In connection
with this, I'll point out that a fairly simple correlation
does seem to exist between tar yield and H/C ratio at low H/C
ratios (below about 0.75, if one leaves out experiments in
which tar cracking is likely, such as fluid beds, shock tubes,
and entrained flow reactors). But whether a particular
correlation is satisfactory (an uncertainty of + 5 % absolute
is seen in a simple correlation with H/C, below about 0.75),
depends upon how demanding the correlation's user is.

H. Jüntgen

You have found that tar yield is not a clear function of H/C ratio. Reasons may be, that tar formation is very much complex and consists on several different mechanisms.

1) Role of chemical structure of coal : tar formation is not only dependent on cleavage of weak bonds of coal macromolecule, but also on hydrogen donor effects due to hydroaromatic structures in coal. So it must be expected, that not only H/C ratio, but the CH distribution is relevant to tar formation.

2) The theory of metaplast formation as a first step of pyrolysis is not relevant to distillation of compounds trapped in the holes of macromolecules (= mobile phase). This follows from the fact, that tar formation begins before gas formation. Metaplast formation is also not relevant, if transport processes in the particle are fast (e.g. in the case of very small particles).

E.M. Suuberg

We agree entirely that better correlations may be possible by including more structural features of the coal, or by performing a more elaborate correlation, such as that suggested by Professor Howard's Group. We wanted here, however, to emphasize that there is no simple correlation with rank of H/C ratio. As you point out, it may be desirable to include other chemical characteristics such as donatable hydrogen content (sometimes substituted for by aliphatic hydrogen content), oxygen functional content, and maybe some characteristic of the macromolecular structure. In addition to these, we have tried to indicate that other variables must also be considered.

These include : - cation and mineral content
- particle size
- external gas pressure.

Inclusion of all these additional variables will make such correlations unattractive to end users whose main interest is in pulverized coal combustion (particularly if the determination of values for the variables involves elaborate chemical characterizations, such as spectroscopic or wet chemical determination of donatable hydrogen). Having established a case for the lack of simple correlations, now we are in a position to begin to ask what the additional key variables are, and how much accuracy we want in exchange for how much information we must provide. With respect to this effort, it should be noted that there is still a paucity of reliable, information on tar yields from pyrolysis ; additional experimental data will be needed, particularly from heated grid type experiments.

2) We would be a bit cautions about agreeing with these

conclusions True, a great deal of tar formation precedes most gas formation, but there is some gas formation at very low temperatures of pyrolysis, so some bond breaking is certainly occurring at the temperatures of tar formation. Indeed, the bonds that break to form the extractables (metaplast) need not be the same as those that break to form gas. Also a large amount of bond breakage may not be necessary to form large amounts of extractables, if the macromolecular structure is not too highly crosslinked.

As regards metaplast transport in the particle, little is known about time scales, because little is known about mechanisms. For example, we have shown (Suuberg & Sezen, Proc. 1985 Int. Conf. on Coal Sci. Pergamon 1985) that even a diffusivity of metaplast species within the particle of order 10^{-6} cm^2/s can be consistent with observed tar formation behavior. For a 100 m particle, the characteristic diffusion time from the center of the particle is far longer than the characteristic time for tar release shown in table 6. This does not "prove" that liquid phase diffusion is the mechanism for tar escape (bubble transport may be quite important), but it does suggest that one cannot immediately conclude that transport is "fast". Again, this area requires further work.

P.T. Roberts

Volatile matter energy content : Estimates of volatile matter energy content, expressed as the difference between coal and rapidly pyrolysed char, show that it amounts to 12 - 14 mJ/kg for coals in the approximate volatile matter range 20-40 % w (db). This is a small (50 %) proportion of the energy release needed to raise all combustion air to realistic flame temperatures in p.f. boiler suggesting that char particle combustion is important in early flame development, i.e. on a timescale comparable with burner mixing times.

R.H. Essenhigh

The contribution of the char in the so-called "primary flame ball" has clearly been underestimated in the past. The value you gave verbally for the flame temperature with no char combustion of about 800°C makes the point even more clearly. Smoot's flame temperatures (for his type I flames) generally did not exceed 1000°C, which is why there was little char combustion, and only the volatiles were burning. Goldberg temperatures at a whole-coal equivalence ratio of 1.5 were as high as 1410°C, but the air quantity was substantially reduced.

At 10% excess air his flame temperature dropped to about 1000°C, and this was only achieved with up to 20% or 25% of the char also burning.

This makes flame stability appreciably dependent on the char combustion efficiency in the "flame ball". This was investigated by Waibel and myself and reported in a 1970 paper

at one of the periodic Penn State Coal Conferences. The
results were summarised in ch. 3 in the book on Coal Processing Technology, edited by Wen and Lee, published by Addison-
Wesley in 1979. Those tentative results showed that flames for
many conditions would extinguish with only a small drop in
char combustion efficiency.

I.W. Smith

In order to correlate e.g. tar yields with coal properties (C, H/C, etc...) it may well be possible to measure
yields at equilibrium for successive well specified reaction
conditions. How far do we have to go to allow for various
heating rate, final temperature, system fluid dynamics, etc. ?

E.M. Suuberg

We agree that the smaller the range of conditions for
which such correlations are to be performed, the better will
be the correlations. For example, to eliminate from consideration particle size, pressure, and vapor phase secondary
reactions will eliminate all these difficult to quantify
variables. Nevertheless note that most data shown in table 3
are for heated grids operated at comparable heating rates and
temperatures. Even with this limited subset of data, no simple
correlations are yet apparent.

HETEROGENEOUS COMBUSTION

HETEROGENEOUS COMBUSTION OF RESIDUAL COKE PARTICLE

G. PRADO, D. FROELICH and J. LAHAYE

LABORATOIRE ENERGETIQUE ET COMBUSTION
ECOLE NATIONALE SUPERIEURE DE CHIMIE DE MULHOUSE
3 RUE A. WERNER 68093 MULHOUSE CEDEX (FRANCE)
and

CENTRE DE RECHERCHES SUR LA PHYSICO-CHIMIE DES SURFACES SOLIDES
24 AVENUE DU PRESIDENT KENNEDY 68200 MULHOUSE (FRANCE)

The combustion of pulverized coal involves two major steps : (1) The devolatilization, occuring during the initial heating, (2) the subsequent combustion of the porous, solid residue resulting from the first step (char). These two steps are strongly inter-dependent and, for some conditions, might occur simultaneously, also, char combustion is generally much slower (by a factor of about 10) than particle devolatilization.

Modelling of the heterogeneous char combustion requires the coupling of chemical reactivity and mass transfer, and is complicated, as outlined by SMOOT and SMITH (1) by many factors :

- Coal structural variations,
- Diffusion of reactants,
- Reaction by various reactants (O_2, H_2O, CO_2, H_2),
- Particle size effects,
- Pore diffusion,
- Char mineral content,
- Changes in surface area,
- Fracturing of the char,
- Variations with temperature and pressure,
- Moisture content of the raw coal.

Some of these factors have been taken into account in experimental work, and several reviews (1-15) summarize the results, with in common the theoretical approach of reactivity and mass transfer outlined below.

1 - Theoretical calculation of burning rates :

It is now established that pulverized coal combustion is controlled by chemical kinetics at low temperature, oxygen pore diffusion at moderate temperature, and oxygen bulk surface diffusion at high temperature. The extent of each of these three zones depends on the nature and size of coal particles.

Zone I : (T ≃ 600°C) :

In this zone, chemical reactions are slow enough to have negligible effects on oxygen transport inside pores. Oxygen concentration profile is constant at the surface and inside the char particle (Fig. 1a), which burns internally, at constant diameter.

Zone II : (T ≈ 600 - 800°C) :

Chemical reactions are fast enough to affect oxygen concentration. Oxygen pore diffusion is the limiting factor, and its concentration decreases to zero at the center of the particle (Fig. 1b), which burns internally and externally.

Zone III : (T > 800°C) :

Chemical reactions are so fast that oxygen is entirely at the outer surface of the particle (Fig. 1c). Oxygen diffusion in the boundary layer controls the burning rate. The particle burns at constant density, with no effect of chemical reactivity and porosity.

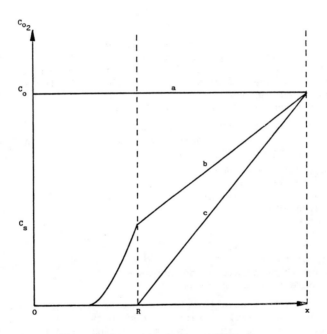

Figure 1 : Oxygen concentration profiles in the vicinity and inside the particle for the three regimes

The modelling of burning rates necessitates the computation of char reactivity and of oxygen diffusion.

1.1 Char reactivity :

The intrinsic reaction rate, R, at the internal surface of a porous solid, may be written as :

$$R = k\ c^m\ \ kg/m^2.s$$

with :   R :   intrinsic reaction rate $kg/m^2.s$

            k :   rate constant $kg/m .s . (kg/m^3)^{-m}$

            c :   local oxygen concentration kg/m^3

            m :   true order of reaction

For char surface, it is usual to consider active sites, related to the intrinsic reactivity :

$$R = k_t [S] m_c C^m$$

with : k_t : intrinsic rate constant $(kg/m^3)^{-m} s^{-1}$

            $[S]$: active sites concentration m^{-2}

            m_c : carbon atom mass (kg)

Very often, the intrinsic reactivity is reported relative to the external surface reactivity R_s :

$$R_s = k C_s^m$$

C_s : oxygen concentration on particle outer surface

and a global reactivity R_m is defined, expressed as a rate of mass loss per unit mass of the particle :

$$R_m = \eta A_g R_s \quad kg/kg.s$$

with : A_g : specific surface of the sample, m^2/kg

            η : effectiveness factor

η is the ratio of the actual combustion rate to the rate in the absence of pore diffusion resistance existed. It corresponds to the fraction of specific surface involved in the reaction, assuming the local intrinsic reactivity equal to R_s.

For zone I : $\eta = 1$

    zone II : $A_e/A_g < \eta < 1$

    zone III: $\eta = A_n/A_g \ll 1$

with : $R_s = m_c k_t [S] C_s^m$

            $R_m = \eta A_g [S] m_c k_t C_s^m$

R_m is proportional to the active surface $A_g[S]$.

Often, R_m is plotted against oxygen concentration on particle outer surface :

$$R_m = k_s C_s^m$$

with : k_s : apparent rate constant $(kg/m^3)^{-n}$ kg/kg.s

            n : apparent order of reaction

The apparent and intrinsic rate constants are usually written as Arrhenius expressions :

$$k_s = A_a \exp(-E_a/RT)$$
$$k_t = A_t \exp(-E_t/RT)$$

E_a : apparent activation energy (kJ/mole)

E_t : true activation energy (kJ/mole)

A_a : apparent pre-exponential factor $(kg/m^3)^{-m}\ kg/m^2.s$

A_t : true pre-exponential factor $(kg/m^3)^{-m}\ s^{-1}$

Finally the global reactivity per unit mass R_m is easily converted in global reactivity per unit of external surface R_{ext} :

$$R_{ext} = \gamma\ \sigma_a\ R_m\ (kg/m^2.s)$$

with : γ : characteristic size of particle (ratio of volume to external surface). (m).

σ_a : particle density (kg/m^3).

1.2 Oxygen Diffusion :

The oxygen diffusion to the outer surface is described by the FICK law :

$$J_s = h_D (C_o - C_s)$$

with : J_s : oxygen mass flux at the outer surface $(kg/m^2.s)$

h_D : oxygen diffusion mass transfer coefficient (m/s)

C_o : oxygen concentration in bulk gas (kg/m^3)

The oxygen diffusion mass transfer coefficient h_D can be computed from SHERWOOD's number Sh for a sphere :

$$Sh = \frac{h_D\ d}{D_a} = 2 + 0,6\ Re^{1/2}\ Sc^{1/3}$$

with : Re : Reynold's number

S_c : Schmidt's number

d : particle size (m)

D_a : bulk gas diffusion coefficient (m^2/s)

For small particles, the slipping velocity is usually very small, and the particle can be considered as stationary relative to the gas. Consequently :

$$Re = 0 \quad \text{and} \quad h_D = 2\ \frac{D_a}{d}$$

From mass balance consideration, the oxygen flux to the surface must be equal to the oxygen leaving the surface and contained in CO or CO_2 molecules. This provides a relation between the oxygen flux the global reactivity :

$$\gamma\ \sigma_a\ R_m = \Lambda\ 2\ \frac{D_a}{d}\ (C_o - C_s)$$

with : Λ : stoichiometric factor

$$\Lambda = M_c/(M_g . \nu_g)$$

M_c/M_g : atomic mass ratio of C and reactant

$$v\,g\ =\ 1\quad \text{and}\quad \Lambda\ =\ 3/4 \qquad \text{for} \qquad C\ +\ \tfrac{1}{2}\,O_2\ \longrightarrow\ CO$$
$$v\,g\ =\ 2\quad \text{and}\quad \Lambda\ =\ 3/8 \qquad \text{for} \qquad C\ +\ O_2\ \longrightarrow\ CO_2$$

For a spherical particle, $\gamma\ =\ d/6$ and :

$$R_m\ =\ \frac{12\,\Lambda\,d_a}{\sigma\,d^2}\ (C_o\ -\ C_s)$$

1.3 Discussion :

a) The dependence of D_a with temperature T and pressure P is written :

$$D_a\ (T,\ P)\ =\ D_a\ (T_o,\ P_o)\ \left(\frac{T}{T_o}\right)^{7/4}\ \left(\frac{P_o}{P}\right)$$

Replacing mass concentration c, by molar fraction x :

$$C_o\ -\ C_s\ =\ \frac{M_g\,P}{RT}\ (X_o\ -\ X_s)$$

M_g : Molar mass of gas

One obtains :

$$R_m\ \alpha\ T^{3/4}\ d^{-2}$$

It follows that small particles will be controlled by chemical kinetics in a wider temperature range than large particles. At a given temperature, small particles might burn in the zone II regime, while large particles burn in the zone III regime.

b) In zone III, $C_s = 0$,   and the maximum combustion rate is :

$$R_{m,D}\ =\ \frac{12\,\Lambda\,D_a}{\sigma\,d^2}\ C_o$$

As long as $R_m\ <\ R_{m,D}$, chemical kinetics control partially the combustion rate.

c) In zone III, it is easy to compute the burn-out time t of a particle, in the pure diffusion regime :

$$R_{m,D}\ =\ -\ \frac{1}{m_p}\ \frac{d\,m_p}{dt}$$

with :

$$m_p\ =\ \frac{\pi}{6}\ \sigma_a\ d^3$$

m_p : particle mass

The integration gives :

$$d_o^2\ -\ d^2\ =\ \frac{8\,\Lambda\,D_a\,C_o\,t}{\sigma_a}$$

and the particle burn-out time (d = 0) is :

$$t\ =\ \frac{\sigma_a}{8\,\Lambda\,D_a\,C_o}\ d_o^2$$

Experimentally this expression is valid for $d_o\ <\ 200$ μm.

d) Very often, the reaction order is assumed to be 1 . In that case :

$$R_{ext}\ =\ k_{e,o}\ C_o\ =\ K_{e,s}\ C_s\ =\ K_D\ (C_o\ -\ C_s)$$

with : $k_{e,o}$: chemical rate constant relative to oxygen concentration C_o

        $k_{e,s}$: chemical rate constant rapported to oxygen concentration C_s

        k_D : diffusion rate constant

one obtains :

$$C_s = \frac{k_D}{k_D + k_{e,s}} C_o$$

and
$$1/k_{e,o} = 1/k_{e,s} + 1/k_D$$

This is a well known relation, indicating the dependence of carbon burn-out with chemical kinetics and diffusion. Diffusion is controlling the process for small values of $k_D/k_{e,s}$.

From above : $k_D \ \alpha \ T^{3/4} /Pd$

$k_{e,s} \ \alpha \ \exp(-E/RT)$

$k_D / k_{e,s} \ \alpha \ T^{3/4} \exp(E/RT) / Pd$

Diffusion regime is favoured by high temperature, high pressure and large particles.

For example, at atmospheric pressure, T = 1773 K, the regime is purely diffusional for particles > 100 μm, and purely kinetically controlled for particles < 1 μm (6).

e) The diffusion of oxygen through the pores, assumed to be cylindrical and of constant diameter has been computed using THIELE modulus (7). Details are available in ref. 2 and 4. With these assumptions, it is possible to show that the global reactivity per unit of mass R_m is :

– in zone I, independent of d_o, and consequently R_{ext} is proportional to d_o.

– in zone II, R_m is inversely proportional to d_o, and consequently R_{ext} is independent of d_o. For the expression $R_m = k_s C_s^n$ the apparent reaction order n is related to the true reaction m by n = (m + 1) / 2 and the apparent activation energy is :

$$E_a = E_t / 2$$

1.4 Conclusion :

Three regimes corresponding to three zones govern the particle combustion (Fig. 2).

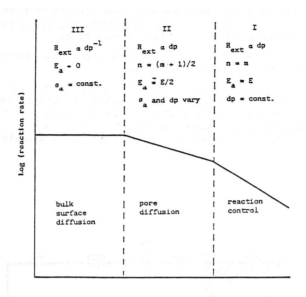

Figure content

Figure 2 : Rate controlling regimes for heterogeneous
char oxidation.
(figure used with permission from Smith, 1979)

Zone I :

There is no diffusion limit, internally or externally. The effectiveness factor $\eta = 1$, and the apparent activation energy is equal to the
true activation energy. The apparent reaction order is also equal to the
true reaction order. The particle burns internally, at constant diameter,
with a global reactivity computed per unit of external surface (R_{ext})
proportional to the initial diameter d_o.

Zone II :

There is some diffusion resistance through the pores, but not through
the external boundary layer. The effectiveness factor is between A_e/A_g
and 1. The apparent activation energy is half of the true activation
energy, and the apparent reaction order n is related to the true reaction
order m by $n = (m + 1) / 2$. The particle burns internally and externally,
with decreasing diameter and density, and the global reactivity R_{ext} is
independent of diameter.

Zone III :

The reaction rate is only limited by oxygen diffusion in the external
boundary layer. The apparent activation energy is close to 0, and the
apparent reaction order is 1. The effectiveness factor is equal to A_e/A_g,
i.e. very small. The particle burns at constant density, and the global
reactivity R_{ext} is inversely proportional to the diameter.

2 - Experimental results :

In the last 20 years, many authors have studied carbonaceous material combustion in a variety of equipments and conditions, including : drop tube furnace, thermobalance, choc tube, flat flames of gas...

The three zones of combustion have been experimentally observed (8-11), and previous hypothesis validated.

Most of the results are summarized in the review of SMITH (4) who plotted the reaction rate per unit of external surface (R_{ext}) and the intrinsic reaction rate (R_i), against temperature in Arrhenius form, for coke, chars, and various carbons. The curves of SMITH are reported on Fig.3 and 4. Details on coal types and conditions are available in ref. (4) and (12).

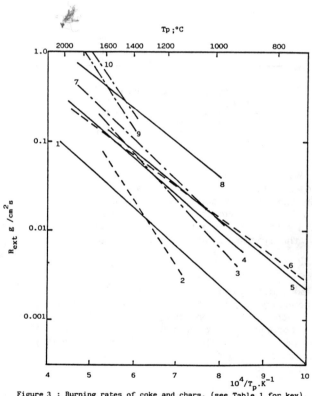

Figure 3 : Burning rates of coke and chars. (see Table 1 for key).
(Figure used with permission from Smith, 1982)

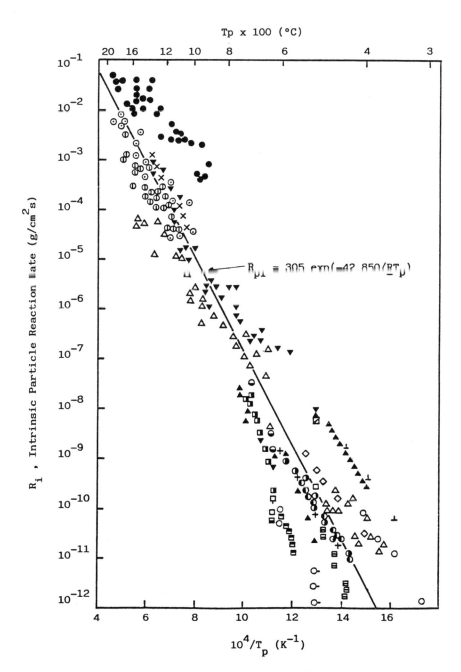

Figure 4 : Intrinsic reactivity of several porous carbonaceous solids in oxygen
(at an oxygen pressure of 0.1 mPa (1 atm)). (Figure used with permission from Smith, 1982). (See ref. 4 for key).

In summary, it is apparent that values of R_{ext} differ for different types of coal by a factor of up to 100. Apparent activation energy and apparent order of reaction differ also widely. (Table 1).

Table 1

Data for coal chars in Fig. 3

Line no.	Parent coal	Pre-exponential factor $(g/cm^2 s \ (atm)^n)$	Activation energy (kcal/mol)	Apparent order of reaction	Particle size (µm)
1	Petroleum coke	7.0	19.7	0.5	18,77,85,88
2	East Hetton, swelling bituminous coal, UK	635.8	34.0	1.0	72
3	Brodsworth, swelling bituminous coal, UK	111.3	24.1	1.0	31
4	Anthracites and semi-anthracites, UK and western Europe	20.4	19.0	1.0	78,49,22 72,42
5	Millmerran, non-swelling sub-bituminous coal, Australia	15.6	17.5	0.5	85
6	Yallourn brown coal, Australia	9.3	16.2	0.5	89.49
7	Ferrymoor, non-swelling bituminous coal, UK	70.3	21.5	1.0	34
8	Whitwick, non-swelling bituminous coal, UK	50.4	17.7	1.0	27
9	Pittsburgh seam, swelling bituminous coal, USA	4187.0	34.0	0.17	16
10	Illinois No. 6, swelling bituminous coal, USA	6337.0	34.1	0.17	13

R_{ext} combines the separate effects of the intrinsic reactivity of the carbon and the extent and accessibility of the internal pore surface of the char. It is therefore very important to derive intrinsic rate expressions (Fig.4). This eliminates effects of different pore sizes and surface areas. However the dispersion of experimental points illustrates the wide range of reactivity, leading SMITH to conclude that, pragmatically, there is a need to determine R_{ext} for engineering purposes, for each char considered at relevant temperature and oxygen concentrations, and for a range of particle sizes for a given char.

Also, there is still a strong need to gain unifying understanding of carbon reactivity.

To meet these two objectives, it is necessary to design new experimental techniques allowing to separate as clearly as possible the different parameters involved : temperature, heating rate, oxidant concentration, properties of the char... This implies to measure combustion rate on isolated particles, in a well controlled environment.

2.1 Measurement of combustion rate on isolated particles :

Several laboratories have recently designed experiments allowing the measurement of combustion rates of isolated particles.

TIMOTHY (13) measured temperature evolution of the coal particle surface during combustion in a drop tube furnace, and proposed a model of particle gasification.

MITCHELL (14-15) derived the combustion rate of a particle in a flame from measurement of particle temperature during a short time, and proposed a new technique applicable to a drop tube furnace (16).

FROELICH (17) and EYNAUD (18) designed a reactor allowing the continuous observation of a single coal particle burning in a controlled environment. To illustrate the capabilities of this technique, a description of the reactor follows.

2.2 Drop tube furnace for the combustion of single particle :

A schematic of the equipment is on Fig. 5. Coal particles are placed in a rotating disc, and fall individually in the furnace, in a controlled quasi-stagnant atmosphere. For each particle, the following parameters are measured :

- ignition delay,
- volatile matter combustion time,
- char combustion time and temperature.

Fig. 6 illustrates the measured emittance of the burning particle at two wavelengths and the corresponding temperature, computed from PLANCK law, with a grey body assumption. This assumption is probably valid for char combustion but not for volatile matter flame. A third wavelenght has been added recently to check this hypothesis, and derives more precise informations.

Detailed heat transfer balance, taking into account the radiative and convective flux to the particle, allows the determination of the heat flux due to the chemical reaction :

$$C + \tfrac{1}{2} O_2 \longrightarrow CO$$

at the surface of the particle.

From this chemical heat flux, the reactivity R_{ext} and the mass loss are computed (Fig. 7). These results are for a bituminous coal (FREYMING type), diameter 80-100 μm. The average value of R_{ext} is 8×10^{-2} kg/m^2.s, at a temperature of 1750 K. On the Arrhenius plot of Fig. 4, this corresponds to $\frac{1}{T} = 5,7$ and $R = 8 \times 10^{-3}$ g/cm^2.s, in excellent agreement with the Arrhenius line drawn.

230

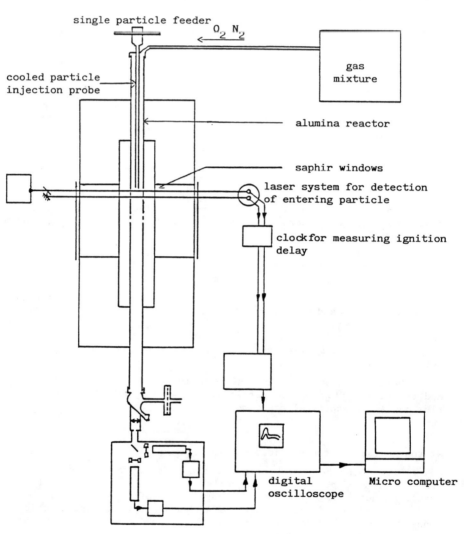

single particle feeder

O_2 N_2

gas mixture

cooled particle injection probe

alumina reactor

saphir windows

laser system for detection of entering particle

clock for measuring ignition delay

digital oscilloscope

Micro computer

Two (or three) colors pyrometer

Figure 5 : Drop tube furnace and diagnostics for single particle combustion measurement

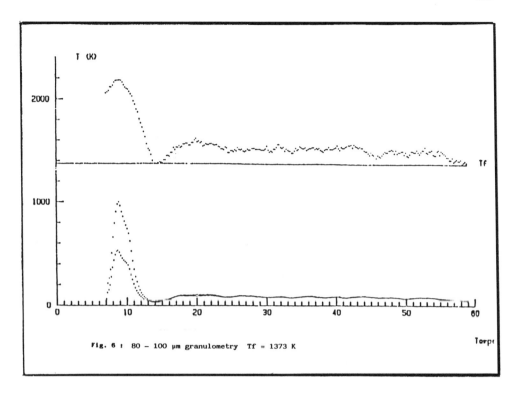

Fig. 6 : 80 - 100 μm granulometry   Tf = 1373 K

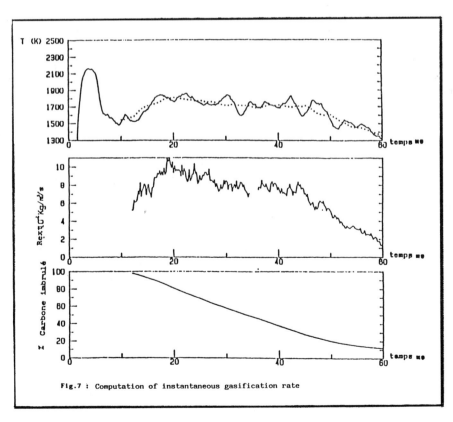

Fig.7 : Computation of instantaneous gasification rate

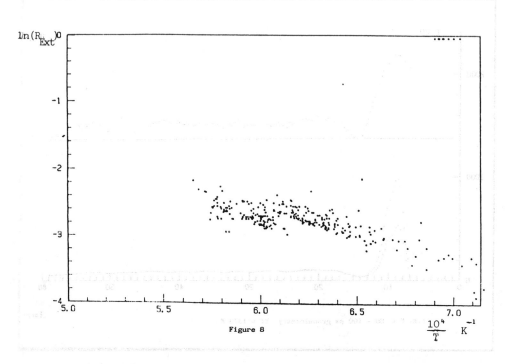

Figure 8

Coal being a very heterogenous material, it is important to study a large number of particles for a given sample. Fig. 8 illustrates the wide dispersion of measured reactivity. This figure corresponds to the combustion of a bituminous coal 100-120 µm, Freyming type, for a furnace temperature of 1400 K to 1700 K, burning in air. Approximately 100 particles were measured at each furnace temperature. It appears that for a given particle temperature, R_{ext} lies within a factor of 10. This might be due to experimental uncertainties, but probably also reflects the maceral composition of the coal, and the variation of specific surface according to the considered maceral.

The least square line drawn on Fig. 8 corresponds to :

$$R_{ext} = 2,43 \exp \left(-\frac{55771}{T}\right) \text{ kg/m}^2.\text{s}$$

The apparent activation energy is 48.0 kJ/mole, which would correspond to a combustion in zone II, and a true activation energy of 96.0 kJ/mole, in good agreement with other results on bituminous coals.

3 - Other important considerations for char reactivity :

3.1.   We have considered so far only reactivity of char relative to oxygen. In industrial applications, char burns in a mixture O_2-CO_2-H_2O-CO-H_2-N_2. Very few studies of char reactivity in this type of environment have been published yet (19-20). In a recent study related to blast furnace tuyere applications, we have prepared, in a drop tube furnace, coal-char particles at different temperatures, for a heating rate of approximately 10^5 K/s, under N_2 or N_2/O_2 atmospheres. These chars have been submitted to gasification in a $CO_2/CO/N_2$ atmosphere, heated up to 1500°C at 10°C/mn in a Mettler thermobalance. The CO_2 reactivity tests were conducted by C. OFFROY and J.L ROTH, at I.R.S.I.D., Maizières-les-Metz (France). The observed mass losses for a South African low volatile coal are reported in figure 9. A strong increase of reactivity of the char in this CO_2 atmosphere is observed with increasing devolatilization temperature. The same experiment with a bituminous coal results in smaller difference of reactivity. In all cases, the reference coke particles are much less reactive than the coal-char particles. This is coherent with results of WELL (19), and further data are needed.

3.2.   Many authors have stressed the importance of extent of internal surface to correlate reactivities, specially at moderate temperature. Another phenomenon recently observed in the drop tube furnace at Mulhouse and at M.I.T., should be taken into account. High speed cinematography of particles burning in pure oxygen indicates clearly that some particles explode in several fragments during devolatilization (18). The conditions (mainly heating rate and temperature) of this phenomenon have to be clarified, the practical implications being important.

3.3.   Finally, the possible swelling or shrinking of the particle during devolatilization has to be taken into account in the model of char reactivity. We have systematically measured the particle size distribution of the coal before and after devolatilization, under nitrogen or air, for temperatures ranging from 1100 K to 1700 K. For all conditions, the size distribution remains constant, indicating that no swelling occurs, although the swelling index of this coal in standard test is 3 - 5.

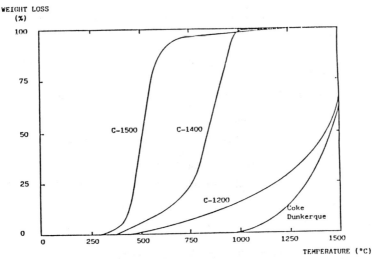

Figure 9 : Char and coke weight losses in the thermobalance

Conclusion :

The concluding remarks made by SMITH, at the 18th Symposium of combustion, four years ago, are certainly pertinent to open the discussion at this workshop. In summary :

- Where required for engineering purposes, it is necessary to determine the R_{ext} for chars made from coals under conditions appropriate to the type of combustion system of interest. The response of R_{ext} to temperature, oxygen concentration and particle size needs to be known for each type of char.

- There is a strong need to gain a unifying understanding of carbon reactivity, and to clarify why the intrinsic reactivity R_i, varies so widely for different carbons.

- The roles of impureties and the atomic structures of carbons need further investigations.

- To satisfy these needs, new experiments are required, to give information on particle temperature and velocity, coupled with accurate measurements of combustion rates in atmospheres and heating environments appropriate to the combustors considered.

The new generation of drop tube furnace experiments on single particles, coupled with modern diagnostics techniques, should bring some answers to these questions.

Bibliography :

1) SMOOT L.D. and SMITH P.J.
Heterogenous Char Reaction Processes, Coal Combustion and Gasification The Plenum Chemical Engineering Series.
Series Editor : Dan LUSS - New-York and London p. 77-110 (1985).

2) ESSENHIGH R.H. - Fundamentals of coal combustion, second supplementary volume, Chemistry of coal utilization (H.A HELLIOTT, ed.), Wiley, New-York, 1153 (1981).

3) LAURENDEAU N.M - Heterogenous Kinetics of coal char gasification and combustion, Progr. Energy Combust. Sci. 4, 221-270 (1978).

4) SMITH I.W - The Combustion Rates of coal chars : A review, 19th Symposium (International) on Combustion. The Combustion Institute, Pittsburgh, PA, 1045 (1982).

5) SKINNER F.D and SMOOT L.D - Heterogenous reactions of char and carbon, Pulverized coal Combustion and Gasification (L.D SMOOT and PRATT Eds.), Plenum, New-York, 149 (1979).

6) BEER J.M and ESSENHIGH R.H, Nature, 187, 1106 (1960).

7) THIELE E.W : Ind. Eng. Chem. 31, 916 (1939).

8) FIELD M.A : Comb. and Flame, 13, 237 (1969).
14, 237 (1970).

9) FROBERG R.W and ESSENHIGH R.H. 17th Symposium (International) on Combustion. The Combustion Institute, Pittsburgh, PA, 179 (1979).

10) TYLER R.J, WOUTERLOOD H.J and MULCAHY M.F.R. - Carbon, 14, 271 (1976).

11) SMITH I.W : Combustion and Flame, <u>17</u>, 421 (1971).

12) SMITH I.W : Fuel, 57, 409 (1978).

13) THIMOTHY L.D, SAROFIM A.F and BEER J.M - 19th Symposium (International) on Combustion (The Combustion Institute). p. 1123 (1982).

14) MITCHELL R.E and Mc LEAN W.J - 19th Symposium (International) on Combustion (The Combustion Institute). p. 1113 (1982).

15) MITCHELL R.E and Mc LEAN W.J - SANDIA Report, 163 rd Meeting of Electronical Society, SAN FRANCISCO (1983).

16) TICHENOR D.A, MITCHELL R.E, HENCKEN K.R and NIKSA S. - 20th Symposium (International) on Combustion (The Combustion Institute) p. 1213 (1984).

17) FROELICH D. - Experimental study and modelling of coal particle combustion. PH.D. dissertation thesis, University of "Haute-Alsace", february 20, 1985.

18) EYNAUD B. PH.D. dissertation, University of "Haute-Alsace", to be published (end of 1986).

19) WELLS W.F, KRAMER S.K, SMOOT L;D and BLACKHAM A.V - 20th Symposium (International) on Combustion (The Combustion Institute) p. 1539 (1984).

20) T. ADSCHIRI and T. FURUSAWA - Fuel, p. 927 Vol. 65, July (1986).

21) SAROFIM A. personnal communication.

NOMENCLATURE

A_a : apparent pre-exponential factor $((kg/m^3)^{-m} kg/m^2 s)$

A_e : external (geometrical) area (m^2/kg)

A_g : specific pore surface area (m^2/kg)

A_t : true pre-exponential factor $((kg/m^3)^{-m} s^{-1})$

C : local oxygen concentration at the vicinity of the surface (kg/m^3) $0 < C < C_s$

C_o : oxygen concentration in bulk gas (kg/m^3)

C_s : oxygen concentration on particle outer surface (kg/m^3)

d : sphere diameter (m)

D_a : bulk gas diffusion coefficient (m^2/s)

E_a : apparent activation energy (kJ/mole)

E_t : true activation energy (kJ/mole)

h_D : oxygen diffusion mass transfer coefficient (m/s)

J_s : oxygen mass flux at the outer surface $(kg/m^2 s)$

k : rate constant $((kg/m^3)^{-m} \times kg/m^2 s)$

$k_{e,o}$: chemical rate constant rapported to oxygen concentration C_o (m/s)

$k_{e,s}$: chemical rate constant rapported to oxygen concentration C_s (m/s)

k_s : apparent rate constant $((kg/m^3)^{-m} kg/m^2 s)$

k_t : intrinsic rate constant $((kg/m^3)^{-m} s^{-1})$

k_D : diffusion rate constant (m/s)

m   : true order of reaction

m_c : carbon atom mass (kg)

m_p : particle mass (kg)

n   : apparent order of reaction

P   : pressure (Pa)

R   : reaction rate $(kg/m^2 s)$

Re   : Reynold's number

R_{ext} : global reactivity per unit of external surface $(kg/m^2 s)$

R_m : global reactivity per unit of mass (kg/kg s)

$R_{m,d}$: maximum global reactivity per unit of mass (kg/kg s)

[S]   : active sites concentration (m^{-2})

Sc   : Schmidt's number

Sh   : Sherwood's number

t   : particle burn-out time (s)

T   : temperature (K)

X   : molar fraction

η : effectiveness factor

γ : characteristic size of particle (ratio of volume to external surface) (m)

σ_a : particle density

DISCUSSION

E. Suuberg

In between the volatiles flame and the char ignition, both of which are obviously high emission periods, it is possible that you have a non luminous H_2 flame around the particle, which shields the surface from O_2, which is why the char does not begin to burn immediately.

G. Prado

It is certainly a possibility, but we have no experimental evidence of it.

P.R. Solomon

You showed the intensity of radiation going to zero between the volatile ignition and the char ignition. How general is this phenomenon ?

G. Prado

This phenomenon is dominant at temperature above 1500 K. Its occurence decreases with temperature. At 1100 K, it becomes a rare event.

E. Saatdjian

During devolatilization, a soot cloud forms around the particle. Your temperature (measured) curves exhibit 2 peaks. Should the initial part of the curve be corrected ? and how ? Higher or lower ? What do you suggest ?

G. Prado

As soot emission is dominating during devolatilization, the grey body asumption is no longer valid. A 1/T dependance of the emissivity may be assumed, resulting in a significant decrease of the computed temperature. We are presently developing a three wavelength pyrometer to experimentally clarify this point.

P.J. Jackson

We have regularly measured CO concentration at 10 m - 15 m downstream of burner in large pulverized coal flames. We observed high CO concentration in char burn-out region. CO_2 develops later e.g. 20-25 m downstream.

D.R. Hardesty

1) Comment : Our work at Sandia and the works of Seeker et al at EER have shown that particle bursting, jetting or rapid spinning (1 re/msec) are relatively rare phenomena. These studies were done at oxygen concentrations of 3 to 10 %, with

a range of coals (from lignites to bituminous coals), for particles less than 150 m at gas temperatures less than 1700 K We certainly do occasionally see such events; recently, experiments with a Canadian subbituminous coal (20% ASTR volatile content) at Sandia by Mitchell have shown a higher incidence of (presumably) particle fracturing during devolatilization. We suspect that such phenomena during devolatilization at very hight oxygen concentrations (hence, high particle temperature and event high heating rates) is somewhat pathological in nature.

2) Several questions on the optical technique :
  a) I am suspicious of your measurements which show temperatures only a 1000 K higher than Tgas. I am concerned that with such "hot-walled" drop tube reactions, your collected radiation is dominated by wall-emitted (and particle scattered or reflected) radiation. Have you estimated the error due to such factors ?

  b) What can you say about particle emissivity as a function of particle burnout ? We (and others) have extensive SEM micrographies which clearly show the accumulation, growth and coalescence of mineral matter particles on the char surface during burnout. With increasing burnout the surface becomes literally covered with such material.
  How does the grey-body assumption hold-up with burnout?

  c) How do you discriminate against radiations from the coalesced, soot-like structures which invariably form (presumably) due to cracking and condensation of heavy tars in the vapor phase following pyrolysis or devolatilization of bituminous coals. We and Seeker have observed that these structures oxidize during the same period during which the char is consumed.

G. Prado

1) We found that for the bituminous coal studied (Freyming) particle explosions is a very frequent event in pure oxygen above 1100 K, and a rare event in air for the same conditions. Clearly, more experiments are needed.

2) (a) The equipment has been very carefully designed to eliminate interferences of theses radiations from the furnace, so only the scattered and/or reflected radiations interfere with our measurements. These interferences increase with wall temperature. At 1200 K, the error is estimated to be 40 K, and increases at 100 K around 1500 K. This limits the operating conditions, mainly for lower oxygen concentration operations.

  b) We do not observe any peculiar trend in computed temperature during char burnout, which could be due to the non-validity of the grey body assumption. However, we are developing a three wavelength pyrometer to check this hypothesis.

c) With the rotating disc technique, we have extremely dilute coal particle concentration, and the soot separates well spatially with the char. Furthermore, from fast movie experiments, we have no evidence of soot burnout during char combustion in our reactor. We do not believe that soot and char combustion interfere significantly for our conditions.

T.F. Wall

Your result indicate a range of reactivity for chars from a single coal. The final burn-out of interest in practical systems will be determined by the least reactive material. Would you comment on the possible analysis of your data in terms of a normal distribution of (say) the pre-exponential constant with a fixed activation energy for the reactivity equation.

G. Prado

We have not modeled yet our results in terms of a gaussian distribution of kinetic parameters, but we certainly retain the suggestion for the near future.

J.B. Howard

It would be of interest to perform the measurement of the different types of ignition at smaller particle sizes than those used in your work so far. One would predict the hetero- geneous ignition to become more prevalent as particle size decreases, and the suggested experiments would allow this prediction to be tested. Could you go down to say, 25 or 30 μm with your equipment ?

G. Prado

With the rotating disc technique, we are limited to parti- cle larger than 100 μm. We can use a fluidized bed technique for smaller particles, but with no control of the exact number of particles introduced. This does not prevent some statistics to be made on the different types of ignition, with a reduced precisions. We will try to test the hypothesis in future ex- periments.

G. de Soete

1) You reminded us that the coke oxidation may be chemical- ly controlled, or diffusion controlled (pore diffusivity, bulk diffusivity), which leads to these three "zones". It should he added that, on top of that, principly in each of these "zones", the reaction can be either desorption controlled or adsorption controlled. Since the apparent reaction order (a) with respect to (O_2) concentration gives interesting infor- mation on the type of control (a=0 for chemical desorption control; a=1 for chemical adsorption control; a=0,5 for dif- fusional desorption control, etc...), this apparent order should always be determined in a kinetic study.

2) Even above 1000°C, dual site desorption yielding CO_2 may occur when the covered sites fraction is close to unity (thus in <u>desorption control</u> case); another case where the desorption rate of CO_2 may be important, is the case of catalytically assisted oxidation, as may occur if the ash content is important.

R.H. Essenhigh

1) Past literature on swelling shows that it appears to be suppressed partially or completely by high heating rates, with a best estimate a few years ago of about 10 % swelling or less. This is consistent with your results of zero swelling.

2) It is always condescending to congratulate very competent workers for very competent work but it is particularly pleasing to see the recognition of the importance of large number measurements for proper statistical accuracy.

3) The choice of rate equation can sometimes have a significant influence on predicted combustion efficiency in final burn-out, particularly if the temperature is falling sharply. It may be that in testing different models, they can be most sensitive to the final burn-out stages.

4) Do you know the activation energy of the plot you gave? With the precision of measurement you are arriving at, and I hope you will achieve, it should justify the use of a better equation than the semi-empirical "nth" order rate equation. Do you plan to use a more complete rate equation in the future ?

G. Prado

The true activation energy for this plot is about 100 kJ/mole. Thanks for your comments, which should help us to improve our analysis of burning rates.

J.H. Pohl

Kobayashi, Sarofim, and I observed both swelling and non-swelling behaviour for seam = 8; The swelling behavior depended on temperature and heating rate (there could not be separated). The coal swelled at lower temperature, but not at and above a calculated temperature of 1150 K. In response to Professor Essenhigh, measurements routinely made for a wide variety of coals at EER in a pulverized coal flame show 99.9 percent removal of hydrogen within 0.5 seconds. Do you think the observed volatile flame is burning soot.

G. Prado

The yellow, intense emission of the volatile flame evidence in color movies suggest strongly that the volatile flame contains burning soot.

P.R. Solomon

I would like to comment further on the question of swelling
and its dependence on heating rate. We find results similar to
those reported by Pohl and Sarofin. That is, for swelling
coals, the amount of swelling can be observed to go through a
maximum as a function of increasing heating rate. These
results are in inert atmospheres and dont require a flame. We
followed up these observations by examing the chars in a
scanning electron microscope.

For maximum swelling the char consists of smooth cenospheres without blow holes. At higher heating rates (achieved
using helium transfer gas), the cenospheres are smaller and
have one or two blow holes. At even higher heating rates, the
char particles consist of even smaller particles consisting of
many bubble cells with a blow hole in each cell.

K.H. Van Heek

1) Is the fact, that you find no swelling possibly due to
the maceral composition of your coal, as inertinite does not
swell and some particle shrinking can compensate the swelling
of vitrinite particles ?

2) You showed an increasing reactivity (CO_2) of the chars
with increasing temperature of preparation. However, the main
factor must be the higher rate of heating at higher final
temperature as normally temperature treatment reduces reactivity even at relatively short residence times (some secs).

G. Prado

1) As the entire size distribution of the particle, and not
only the average values, remain constant during heat up, we do
not believe in the possibility of compensating effects.

2) We totally agree with your comment.

POLLUTANTS IN COAL COMBUSTION

POLLUTANT FORMATION AND DESTRUCTION

A.F. SAROFIM

Department of Chemical Engineering

MASSACHUSETTS INSTITUTE OF TECHNOLOGY
CAMBRIDGE, MA   02139
(U.S.A.)

1. INTRODUCTION

One of the principal deterrents to the increased use of coal is the concern with noxious combustion-generated emissions. The pollutants of greatest concern have changed over the centuries (Table 1) and can be seen to be related to either the combustion process itself, for example polycyclic aromatic hydrocarbons, soot, and the oxidation of atmospheric nitrogen, or to contaminants in coal such as mineral constituents, sulfur, and organically bound nitrogen. In this presentation, an overview will be presented of the mechanisms governing the formation and, sometimes destruction, of the different pollutants.

TABLE 1. Coal Consumption: Past, Present, and Future Environmental Issues

Century	Dominant Problem	Effect
12th	Smoke (Soot and PCAH)	Inconvenience
	Ash + SO_x	Impairment of respiratory function
	Acid Rain (SO_x, NO_x)	Damage to aquatic life, vegetation, and main-made objects
21st	Greenhouse Effect (CO_2, N_2O)	Climatic

Although the broad coverage of all emissions has the disadvantage of a superficial coverage on each, it has the merit of showing the interdependence of the processes governing the formation of different pollutants and the potential danger of reducing the emission of one pollutant at the expense of increased emission of another. The sequence of presentation will first cover the mechanisms of formation followed by a discussion of their interdependence.

2. POLYCYCLIC AROMATIC HYDROCARBONS AND SOOT

Polycyclic aromatic hydrocarbons (PAH) and soot are the pollutants generated by coal that were of major concern in the early days of coal burning. The emission of both are very strongly dependent on combustion conditions. They are produced in high yields by the small poorly designed combustors that were originally used for coal, typically consisting of grates supporting the burning and pyrolysing coal lumps placed under the surface of the charge being heated. The pyrolysis products emerging from the grate were rapidly quenched and carried into the ambient environment. The emissions of PAH and soot from modern large scale pulverized coal fired boiler,by contrast, are found to be low. Interest in the PAH is mainly for purposes of evaluating the emission associated with coal use relative to that of other sources such as diesel engines and wood stoves and in determining its role as a precursor to soot. The interest in soot in coal combustors is increasingly because of the dominant role it plays in determining the radiative flux from the flame zone. This section will provide a short summary of the effects of coal type and operating conditions on the concentration and composition of PAH and soot.

A schematic of the processes governing the emission of soot and PAH is provided in Fig. 1. The coal, on heating, releases volatiles including PAH. Secondary pyrolysis of the lighter volatiles provides an alternative route to PAH. The products of pyrolysis may undergo further decomposition under fuel-rich conditions to give compounds with larger ring structures and ultimately soot. On mixing with air at high temperature they will be oxidized.

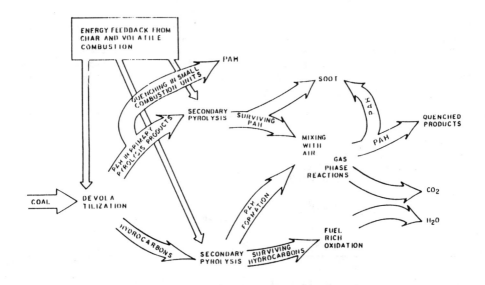

FIGURE 1. Schematic of processes governing the emission of PAH and soot[3]

The primary products of pyrolysis will have compositions that will be closely related to that of the parent coal[1]. The PAH from coal can be distinguished from that produced by pyrosynthesis reactions from lower hydrocarbons by the presence of a higher degree of alkyl side-chain substitution and heteroatoms such as S and N[2]. The yield of the PAH is a strong function of coal rank, increasing with rank from low values for low rank coals such as lignites, passing through a maximum for high-volatile bituminous coals, and decreasing to low values for anthracites[3]. This trend is shown in Fig. 2, in which are provided the PAH yields obtained on the pyrolysis of coals for a residence time of about 0.3 seconds at

temperatures of 900 to 1700 K. In the order of increasing carbon content
the five coals studied were a Montana lignite, a Montana Rosebud
subbituminous coal, a Pittsburgh hvA bituminous coal (PSOC 997), a
Pocahontas #3(PSOC 130), and a Primrose anthracite (PSOC 869).

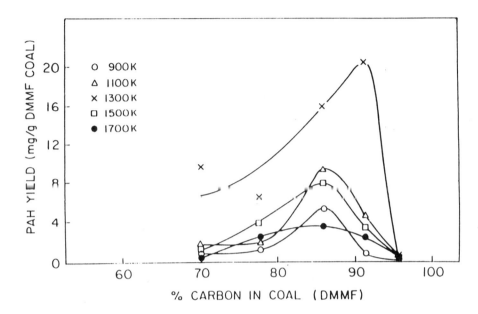

FIGURE 2. PAH yield during the pyrolysis of coals of different rank (%
carbon content) at temperatures of 900 to 1700 K (Residence time
≅ 0.35)[3].

The pyrolysis products generated by coal will undergo further
decomposition in the absence of oxygen. Detailed study of the
transformation products show a steady decline in the total concentration
of PAH, with the side-chain substituted compounds decreasing at a faster
rate than the non-substituted compounds, and with an increase in the
relative amounts of the higher ring compounds. The increase in soot yield
at the expense of the PAH yield is evident in Fig. 3. From Fig. 3 it is
apparent that the sum of the soot and PAH yields are approximately
constant so that the ultimate soot yields at high temperatures will show
the same dependence on coal type as that shown in Fig. 2 for PAH[3,4]. The
decrease in alkyl substitution with increasing secondary pyrolysis[6] is
shown in Fig. 4, and is of importance in determining the mutagenicity of
the PAH[5].

In practice the soot and PAH concentrations will be determined by a
balance between the rates of formation and oxidation. Studies of the
oxidation of coal particles dispersed in laminar oxidant streams show the
formation of a luminous spherical flame around individual coal
particles[7-9]. From the magnitude of the intensity of the radiation

estimates of the maximum soot yields of up to five percent of the coal weight have been obtained[9]. When the oxygen concentration falls below a

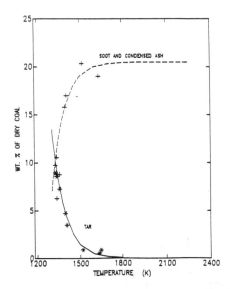

FIGURE 3.
Effect of pyrolysis temperature on the yields of tar and soot for a high-volatile bituminous coal ($\tau \cong 0.35$)[4].

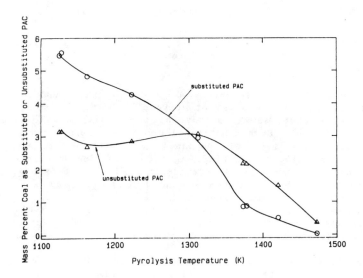

FIGURE 4. Effect of pyrolysis temperature on the yields of alkyl substituted and unsubstituted PAH[6].

critical value, the rate of oxidation cannot keep up with that of burnout,
and the soot forms a contrail several micrometers in diameter and tens of
micrometers long[7],[8]. In such laminar flames, the rate of soot oxidation
is expected to be described by the Nagle Strickland-Constable relations
under fuel-lean conditions and by OH radical attack under fuel rich
conditions[10]. For turbulent flames the rates of PAH and soot oxidation
have been found to be controlled by the rate of mixing of the fuel-rich
eddies containing the soot and the oxygen-rich ambient gases[11]. Analyses
based on the turbulent mixing model of Magnussen have been found to
adequately correlate the rate of decay of soot concentrations in turbulent
flames[11],[12].

The size of soot particles in flames are generally found to follow the
predictions of coagulation theory, with the major source of uncertainty
being the present lack of understanding of when the particles will stop
growing individually and form aggregates[13],[14]. The mean particle size for
spherical particles can be related to soot volume fraction f_v and
residence time by

$$d = (\pi/6)^{-1/3}/(k_{theory} f_v t)^{2/5} \qquad (1)$$

where k_{theory} is the rate coefficient for collision between particles with
allowance for a self-preserving size distribution[15]. Similar relations
have been found to apply to the soot produced by coal in a laminar flow
pyrolysis experiment, in which the growth of the particles was by both
surface growth and coagulation[16].

In summary, the concentrations of soot and PAH are determined by the
difference between the processes governing their formation--strongly
influenced by coal type and pyrolysis conditions--and destruction--
dependent on the mixing of fuel- and oxygen-rich eddies and temperature.
The amounts of soot and PAH correlate with the tar yields of coals and,
under comparable combustion conditions, follow the progression
anthracites<lignites<bituminous coals. The emissions from combustion
systems are determined by the time available for mixing with oxygen of the
fuel-rich PAH-containing eddies before they contact heat transfer surfaces
and are quenched. This is demonstrated graphically in Table 2 which
presents the concentrations of selected PAH in the effluent from small
coal-fired stokers and a large power-plant together with representative
concentrations in pyrolysis products; the results suggest that the
pyrolysis products are rapidly quenched in the small units but there is
adequate time for mixing and burnout in the larger units. Although the
soot and PAH concentrations in the effluents of large boilers are small
the localized concentrations of soot in the flame zone may be large enough
to significantly augment flame radiation. Soot yields in the flame zone
of utility boilers of the order of one percent of the total coal carbon[17]
have been reported and a highly sooting bituminous coal has been shown to
have much higher emissivity than an anthracite of comparable particle
size[18].

TABLE 2. Comparison of PAH, evolved (ng/J) during the pyrolysis of Montana Lignite and Pittsburgh High Volatile-A Bituminous coals with that emitted by various sources

	Methyl Anthracenes/ Phenanthrenes	Fluoranthene	Pyrene	Phenan- threne/ Anthra- cene
Pyrolysis				
1. Montana Lignite	–	1.72	2.71	10.95
2. Pittsburgh High- volatile Bituminous	5.89	6.32	11.68	19.04
Residential Stoker Coal- fired Boiler, 200 KW				
1. High-volatile Bitumi- nous (Elkhorn #3)	3.16	3.12	2.12	5.16
2. Western Subbituminous (Colorado)	0.37	0.096	0.132	0.72
Coal-fired Power Plant	0.50×10^{-5}	0.07×10^{-5}	0.4×10^{-5}	7.6×10^{-5}

3. NITROGEN OXIDES

Coals have a high nitrogen content, of the order of 1 to 2 percent by weight, and therefore the oxidation of fuel nitrogen is an important contributor to the NO_x emissions. In order to determine the relative contributions of fuel nitrogen and atmospheric nitrogen to coal-generated NO_x, Pershing and Wendt conducted pilot-scale combustion studies using both air and argon/oxygen/carbon dioxide mixtures as oxidants with the carbon dioxide being added in such concentrations as to give the same flame temperatures for the two oxidants[19]. Their results showed that the fuel nitrogen contribution was dominant under most conditions of current interest. However, prior to the modification of furnace and boiler design for purposes of reducing NO_x emissions, the fixation of atmospheric nitrogen had been found to account for significant production of nitrogen oxides for high temperature operations such as those encountered in cyclone burners and in furnaces with high heat release rates per unit area of wall surface[20].

The oxidation of atmospheric nitrogen is easily suppressed by the reduction in peak combustion temperature by staging the air supply, by the reduction in combustion intensity, or by internal recirculation of cooler combustion products by aerodynamic means. The emphasis here will be on the formation of nitrogen oxides by the oxidation of the organically or fuel bound nitrogen in coal. Most of the nitrogen is believed to be bound in the heterocyclic polyaromatic ring moieties that constitute the organic matter in coal[21]. The processes that control the fate of this coal bound nitrogen is shown schematically in Fig. 5. Part of the nitrogen in coal is released during heating. The volatiles undergo a sequence of pyrolysis and oxidation reactions which determine the distribution of the nitrogen between the ultimate products of molecular nitrogen and nitric oxide. (Recent findings of significant amounts of nitrous oxide will be discussed later). In parallel with the gas phase reactions, the nitrogen that is retained by the char will also be partially converted to nitric oxide.

Char can reduce nitric oxide, providing another route from the coal nitrogen to N_2. It is clear from a consideration of the pathways in Fig. 5 that the overall conversion of the coal nitrogen to nitric oxide will be determined by the oxidation and temperature histories to which the coal and its volatile products are subjected. The examination of the individual steps will be discussed first followed by the presentation of data on the overall process.

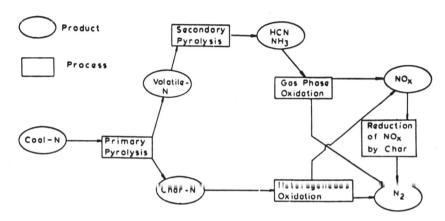

FIGURE 5. Schematic of processes occuring during the formation of nitric oxide from coal nitrogen.

The heating of coal until a constant weight is approached shows that the organically bound nitrogen is more refractory than the other volatile elements with the exception of sulfur (Fig. 6). At the ASTM temperature for devolatilization of 1223 K (750 °C), the char may retain 30 to 80 percent of the original coal nitrogen. It takes temperature of 2100 K to devolatilize most of the coal nitrogen. The nitrogen is evolved mostly with coal tars, as determined from heated grid or pan experiments in which the volatile products are rapidly quenched after release. The nitrogen released with the volatiles is found to correlate well with the tar released (Fig. 7). The high molecular weight products are expected to rapidly decompose to lower molecular weight compounds and HCN is expected to be the dominant intermediate for the nitrogeneous compounds, both as a primary product of decomposition and a secondary product generated by the reaction with hydrocarbons of any amines produced during the primary decomposition step.

The kinetics of devolatilization of the nitrogen content of coal is of importance to the design of strategies for NO_x reduction. In a staged combustor it is important that all the nitrogen be released in the fuel rich first zone since any nitrogen retained by the char will be oxidized downstream under lean condition in the second stage, conditions which favor the formation of NO. This view point is supported by the observation that the reduction in coal particle size will reduce the production of NO under staged conditions, since size reduction favors early release of the coal-bound nitrogen. Despite the importance of nitrogen devolatilization kinetics suggested by the preceding arguments, models of coal nitrogen oxidation by Smoot and coworkers[22] have been

successful in matching experimental NO values using the simple assumption
that the fractional nitrogen release by coal equalled the fractional
weight loss of the carbon. Clearly, the importance of nitrogen
devolatilization kinetics on NO_x formation is still uncertain.

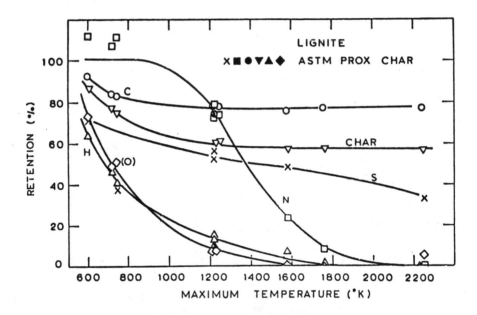

FIGURE 6. Effect of temperature on the retention of selected elements by
the char produced on the crucible heating of a char (time varied
to obtain asymptotic yield).

The conversion of the nitrogen compounds to NO and N_2 was identified by
Fenimore[23] as the reaction of a fuel nitrogen intermediate, NX, with
either an oxidant, primarily OH, to form NO or with NO to form N_2. The
kinetics of the reaction are either treated using global rate expressions,
drawn mainly from the studies of De Soete[24], or using a reaction set
involving the major primary reactions[25]. The major component of the
detailed mechanism is based on the seminal research of Haynes[27]
identifying the pathway leading from cyano compounds, which appear to be
dominant in the hydrocarbon rich flames, to the NH_i radicals which are the
key intermediate, NX, postulated by Fenimore. Recent observations[28] of
significant concentrations of N_2O in the effluent of a number of
combustors suggests that the gas phase reactions may be more complicated
than was first thought and reactions such as $NCO + NO = N_2O + CO$, studied
by Perry[29], may be important. Other reactions of importance that are not
yet adequately understood are the hydrocarbon—NO reactions. These are of
importance because they provide a means of recycling the NO to HCN
providing further opportunities for conversion to N_2. This set of
reactions provides the basis for the process of "reburning" NO. It is
believed that CH and CH_2 are hydrocarbon radicals participating in the

reduction of the NO in reburning, but uncertainty persists because of the difficulties of calculating the concentrations of these radicals in flames.

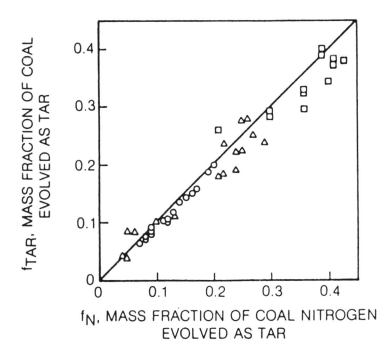

FIGURE 7. Correlation of evolution on pyrolysis of coal nitrogen with that of tar: For varied heating rates, and a Rocky Mountain Bituminous (triangles, Appalachian Bituminous (squares), and a subbituminous (circles) coal[26].

The reactions of the nitrogen in the char are incompletely understood. The char nitrogen is believed to be oxidized to NO which is then partially reduced to N_2 while diffusing out of a particle. Limited information is available on the reduction of the NO by char and the augmentation of the reduction by $CO[30]$. These reactions are catalyzed by inorganic constituents in the char and are also influenced by the heteroatom content of the char[31], effects that need to be studied in greater detail.

The complexity of the reactions involving nitrogenous compounds can be appreciated by examining part of the data obtained in a comprehnsive study of the effects of fuel equivalence ratio and coal type on the composition of the emission of the first stage of a staged combustor. The data in Fig. 8 show the effect of air/fuel ratio on the composition, prior to the introduction of secondary air, of the nitrogenous species for a bituminous coal and a lignite. As expected as the stoichiometric ratio is decreased below one, the nitric oxide is reduced but one sees increasing amounts of ammonia and hydrogen cyanide, with the ammonia being dominant for low rank coals. A study of a large number of coals of different rank has led to correlations for the fuel nitrogen conversion to NO under fuel-rich and fuel-lean conditions. These correlations[32] can serve as bases for determining the effect of coal type on NO_x emissions until more fundamental models are developed.

4. MINERAL MATTER

Understanding of the transformation of mineral matter is important because of its influence on fouling, slagging, and heat transfer in boilers, on the performance of particulate control equipment, and on the health and ecological effect of prticles escaping to the atmosphere. In order to be able to improve the performance of existing coal-combustors it is desirable to understand the processes that govern the size distribution and chemical composition of the ash within and at the exit of a combustor or furnace.

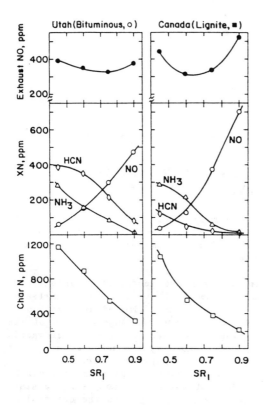

FIGURE 8.
Effect of air/fuel stoichiometric ratio in the first stage on the concentrations of char N, NH_3, HCN, and NO from the first stage and the exhaust NO for a staged combustor burning a Utah bituminous coal and a Canadian lignite[32].

The processes governing the transformation and deposition of ash are complex and are only partially understood. The mineral matter in pulverized coal is distributed in various forms. Some occurs in an essentially carbon-free form, and is designated as extraneous. Some occurs as mineral inclusions, typically 2-5 μm in size, dispersed in the carbonaceous coal matrix. Some is atomically dispersed in the coal either as cations on carboxylic acid side chains or in porphyrrin-type structures. The behavior of the mineral matter will depend strongly on the chemical and physical state of the mineral inclusions.

During combustion the mineral inclusions will decompose and fuse. Most of the mineral matter will adhere to the char surface but some is released as micron-size particles[33-35]. As the char surface recedes the ash inclusions are drawn together and coalesce to form larger ash particles. Char fragments are released and take with them ash inclusions and ash adhered to the surface. As each fragment burns out an ash particle is produced the size and composition of which is determined by the ash carried with the particular fragment. The residual fly ash size and composition distribution is therefore determined by the fragmentation patterns of the ash.

The exothermic oxidation of coal chars results in high particle temperatures which result in a partial vaporization of the mineral or ash inclusions. Both volatile compounds, such as those of the alkali metals (Na, K), and less volatile compounds, such as the salts of Si, Mg, Ca, and Fe are found to evolve during char combustion. Early evidence for ash vaporization was provided by the elemental analysis of the fly-ash produced in cal-fired utility boilers [36-39]. Volatile salts such as those of sodium, zinc, arsenic and antimony were found to preferentially concentrate in the smaller ash particles, with the concentration often correlated with the inverse of particle diameter (d^{-1}) or of the square of the particle diameter (d^{-2}). Surface analysis of the residual ash particles has revealed that the volatile trace species were also concentrated at the surface of the residual ash particles[38-40]. These phenomena have been explained in terms of vaporization of the volatile elements in the high temperature combustion zone followed by their condensation on the surface of the residual fly-ash particles in cooler regions of a furnace[32,38,41]. The volatilization of the oxides of the refractory oxides of Fe, Si, Mg, Ca can be greatly enhanced by reduction in the locally reducing zone within a coal particle to either the more volatile suboxide (SiO) or element (Fe, Mg, Ca). The volatilized suboxides or metals are reoxidized in the particle boundary layer and form a fine submicron aerosol.

The submicron particles in fly ash have been studied only recently[41-47]. Laboratory studies have shown that these particles are produced by the homogeneous condensation and subsequent coagulation of a portion of the vaporized fly ash. Measurements on full-scale utility boilrs[48] also showed the presence of a submicron particle mode, which accounted for 0.2 to 2.2 percent of the fly ash at the inlet to the electrostatic precipitator of the six boilers studied, but as much as 20 percent of the ash at the outlet.

The progress made to date in understanding the vaporization and condensation of the mineral constituents of fly ash can be described in the context of the schematic shown in Fig. 9 of the fate during coal combustion of the mineral constituents of coal. The vaporization of the mineral matter is important in the fouling of tubes, by providing bonding agents for the attachment of fly ash particles to surfaces. This can occur either by the direct condensation of vaporized species on the tube to form the glue for particles hitting the tube surface or by reaction of vaporized species with the surface of ash particles to form a sticky coating on the particles. Both mechanisms occur in practice. Sectioning of tube deposits shows an enrichment of potassium sulfate near the tube surface as would be expected from direct condensation on the tubes. Surface reaction of silica with sodium salts to form a low viscosity molten surface layer has been shown to occur at combustion conditions by Wibberley and Wall[49,50].

The vaporization process is also important because of its role in the formation of the submicron aerosol which is difficult to collect. The vaporization rate is a function of the vapor pressure of the individual constituents and is a function of composition and temperature[51], as shown in Fig. 10 by the fractional retention of selected elements of a lignite which has been pyrolyzed for a second at various temperatures. The vaporization rate of elements has been modeled including allowance for the reduction of the refractory oxides[52]. The condensation and growth of the mineral constituents generate an aerosol; the size distribution of the particles are described[43,44] by relations very similar to Eq. 1.

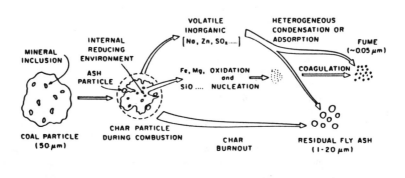

FIGURE 9. Schematic of transformation of mineral matter in coal during coal combustion[45].

There is a general understanding that the size of ash produced during coal combustion decreases with increasing comminution and with decreasing mineral content of the parent coal particles. There are, however, no predictive models which will permit the estimation of the reductionachievable in the size of ash particles when coal is fine ground or beneficiated and furthermore how the size is affected by combustion conditions. The major gap is in determining the number of size of the fragments produced during char burnout. Recent developments in the application of percolation theory show promise for providing this information. One model, due to Kerstein[53], is the treatment of char as a number of sites or bonds. The simulation of the combustion of a char particle using such a model is shown in Fig. 11.

5. SULFUR OXIDES

The sulfur contained in coals will generally be converted to SO_2 in a combustor, with small amounts in the form of SO_3. For low rank coals some of the sulfur may be retained by the alkaline ash. This section will describe briefly the strategies for the in-furnace capture of sulfur by the injection of limestone and the conversion of SO_2 to SO_3.

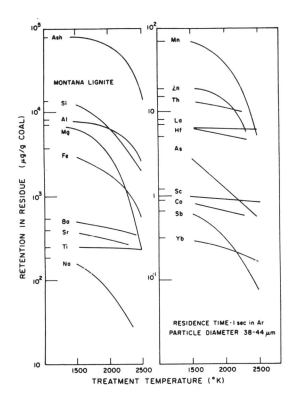

FIGURE 10.
Retention of different elements by the char produced by pyrolysis of coal for 1 second at different temperatures[51].

An appreciation for the strategy of sulfur capture by limestone may be obtained from consideration of that attainable at equilibrium. In Fig. 12 the fractional sulfur retention by the solid is shown as a function of temperature and the fuel equivalence ratio (the reciprocal of the air/fuel ratio used earlier). Under fuel rich conditions, the sulfur will be retained as CaS. Any CaS formed, however, will be rapidly converted to CaO on encountering oxygen at combustion temperature so that the CaS route does not appear to be attractive; but it has been proposed that the sulfur could be rejected as CaS in the slag of a cyclone combustor operated fuel rich and this option is under active consideration. The alternative is to capture the sulfur as $CaSO_4$ by injecting the limestone upstream of the temperature at which the sulfate is stable (i.e., at a position at which the temperature is higher than about 1450 K). Care must be taken, however, not to inject the limestone at a much higher temperature since that may result in the loss of surface area of the lime produced through sintering[54].

The injection of limestone in furnaces for sulfur capture has, however, only met with modest success. Data by Pershing and coworkers[55] shown in Fig. 13 is illustrative of the sulfur capture by calcined limestone as a function of residence time. After an early rapid rise in the sulfur

258

content of the lime an asymptotic conversion is approached. The
explanation for the slowing down of the reaction is that the calcium
sulfate product, because of its high specific volume, plugs up the pores
in the limestone preventing further access by the SO_2 to the unreacted
lime. Reducing the size of the limestone, either by crushing or by rapid
heating to obtain particle fragmentation, will result in higher sulfur
capture[54]. At present, in the U.S., limestone hydrates with high specific
surface areas are being utilized to obtain a higher sulfur capture, albeit
at the expense of a higher sorbent cost.

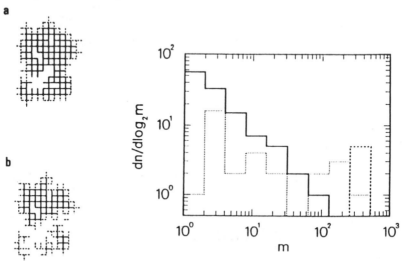

FIGURE 11.   Percolation model of char burnout[53]. Inset (a) Initial
              representation of unignited (solid) and ignited (dashed)
              bonds (b) Later stages of burnout showing four fragments.
              Right-hand graph shows fragment size distribution initially
              (dashed), at 38% (dotted) and 78% (solid) burnout; n =
              number of fragments, m = mass of fragment.

The second topic to be discussed under sulfur is that of SO_3 formation,
both because of its role in decreasing the resistivity of fly ash for
purposes of improving the performance of electrostatic precipitators and
because of the potential adverse health effects of sulfuric acid produced
at lower temperatures from the SO_3. Equilibrium favors SO_2 at flame
temperatures. Some SO_3 is formed by the reactions of O radicals present
in supra-equilibrium concentrations in the flame zone but it is
shortlived[56]. Most of the SO_3 produced in a furnace is from the catalyzed
oxidation of SO_2. It appears from well-defined laboratory experiments
that the aerosol produced by the vaporization and condensation of mineral
constituents in coal provide the high surface areas needed for appreciable
conversion of the SO_2 to SO_3. Some results of the temperature dependence
of the SO_3 yields are shown in Fig. 14 for a model system in which
Sherocarb particles doped with an iron salt were used to generate the
aerosol[57]. A maximum in conversion is observed because of the trade-offs
between kinetics which govern the reaction at low temperatures to

equilibrium which is controlling at high temperatures. A study of different coals showed a high yields of SO$_3$ for an Illinois No. 6 coal which produced a high iron content aerosol and low yields for coals, such as a Montana lignite, which produced an aerosol with a high alkali content[58].

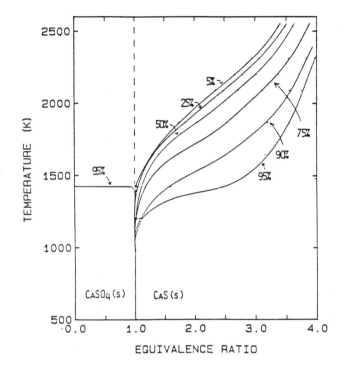

FIGURE 12.
Sulfur retention at equilibrium for an Illinois bituminous coal as a function of fuel/air equivalence ratio and temperature.

% SULFUR RETENTION FOR WET ILLINOIS

6. INTERACTIONS

The formation mechanisms of the different pollutants are dependent on temperature and air/fuel ratio in such a way that the adjustment of a parameter to reduce one problem may create another. Some of these interactions are:

  * Reduction of NO$_x$ formation by staging air may result in increased emissions of soot and PAHs. A further example of interactions of hydrocarbon and NO chemistry is provided by returning.

  * Staging of air to reduce NO$_x$ has been found[59] to increase, under some circumstances, the emissions of fine particles produced from the mineral constituents in coal, the vaporization of which is augmented under reducing conditions.

  * The emissions of NO$_x$ have been correlated with those of fine inorganic particulate matter[60].

  * It has been observed that sulfur in a fuel may affect the emissions of NO because of the interdependence of the free radical chemistry

260

influencing both[61].

* NO_x may be reduced by soot and char.

The above examples show that the nitrogen chemistry may be influenced by the sulfur, carbon, and hydrocarbons in a flame, that the sulfur chemistry is influenced by the fate of the mineral matter, and that changes in temperature and fuel/air ratio impacts a number of pollutants but that the changes may be of the same or opposite sign. Such interactions can only be anticipated by a good mechanistic understanding of the processes governing each of the pollutants. There remains many fruitful opportunities for research to guide the development of cleaner coal-combustion systems.

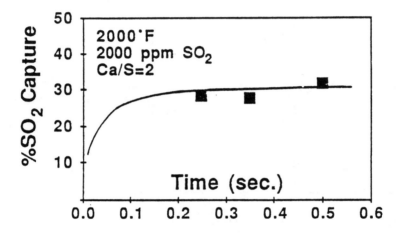

FIGURE 13.   Percent SO_3 Capture as a function of residence time. Comparison of experiments (data points) with predictions of pore plugging model (solid line).

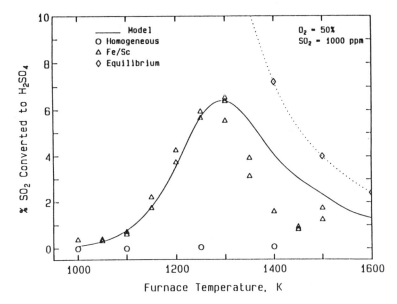

Model, Homogeneous, Iron Catalyzed
and Equilibrium Conversions

FIGURE 14.   Fraction of SO_2 converted to sulfuric acid as a function of
furnace   temperature   at   equilibrium   (diamonds),   for
homogeneous oxidation (circles), catalyzed by rion aerosol
produced by combustion of spherocarb particles impregnated
with an iron salt (triangles)[57].

7. ACKNOWLEDGEMENT

The author has drawn on material developed under NIEHS Center Grants
5 P30 E SO2109 and 2 PO2-ESO2429-05, in addition to the literature in
preparing this manuscript. The author wishes to acknowledge assistance
provided by Judy Wornat.

8. REFERENCES

1. Solomon, P.R.,   and   Colket,   M.B.,   Seventeenth   Symposium
(International) on Combustion, 131-143, The Combustion Institute,
Pittsburgh, PA, 1979.
2. Lee, M.L., Prado, G.P., Howard, J.B., and Hites, R.A., Biomedical
Mass Spectrometry, 4, 182-6 (1977).
3. Mitra, A., Sarofim, A.F., and Bar-Ziv, E., "The Influence of Coal
Type on the Evolution of Polycyclic Aromatic Hydrocarbons During
Coal Devolatilization", Aerosol Science and Technology. (In Press).
4. Nenniger, R.D., Howard, J.B., and Sarofim, A.F., "The Sooting
Potential of Coals", Proceedings 1983 International Conference on
Coal Science, pp 521-524, Pittsburgh, International Energy Agency,

1983.

5. Silverman, B.D., and Lowe, J.P., Cancer Biochem Biophys, 6 89-94 (1982).

6. Wornat, M.J., Ph.D. Thesis in Chemical Engineering, Cambridge, MA. (In Progress).

7. Seeker, W.R. Samuelsen, G.S., Heap, M.P., and Trolinger, J.D., Eighteenth Symposium (International) on Combustion, 1213-1224, The Combustion Institute, Pittsburgh, PA 1981.

8. McLean, W.J., Hardesty, D.R., and Pohl, J.H., Eighteenth Symposium (International) on Combustion, pp. 1239-1247, The Combustion Institute, Pittsburgh, PA, 1981.

9. Timothy, L.D., Froelich, D., Sarofim, A.F., and Beer, J.M., "Soot Formation and Burnout During the Combustion of Dispersed Pulverized Coal Particles", Twenty-First Symposium (International) on Combustion. (In Press).

10. Neoh, K.G., Howard, J,B., and Sarofim, A.F., "Soot Oxidation in Flames", in "Particulate Carbon Formation During Combustion", Eds. Siegla, D.C., and Smith, G.W., Plenum Press, N.Y., 1981.

11. Magnussen, B.F., Hjertager, B.F., Olsen, J.G., and Bhadwi, D., Seventeenth Symposium (International) on Combustion, pp. 1383-1393, The Combustion Institute, Pittsburgh, PA, 1979.

12. Abbas, A.S., and Lockwood, F.C., J. Inst. Energy, 58, 112-115 (1985).

13. Prado, G., and Lahaye, J., "Physical Aspects of Nucleation and Growth of Soot Particles" in "Particulate Carbon Formation During Combustion, Eds. Siegla, D.C., and Smith, G.W., Plenum Press, N.Y., 1981.

14. Haynes, B.S., and Wagner, H. Gg., Prog. Energy Combust. Sci., 7, 229-273 (1981).

15. Graham, S.C., and Homer, J.B., Faraday, Symp., 7, 85 (1973).

16. Nenniger, R.D., Ph.D. Thesis in Chemical Engineering, M.I.T., Cambridge, MA, 1986.

17. Wall, T.F., and Stewart, I., McC., Fourteenth Symposium (International) on Combustion, pp. 689-697, The Combustion Institute, Pittsburgh, PA, 1973.

18. Riviere, M., J. Inst. Fuel, 29, 9 (1956).

19. Pershing, D.W., and Wendt, J.O.L., Ind. and Eng. Proc. Des. and Dev., 18, 60-66 (1979).

20. Crawford, A.R., Manny, E.W., Gregory, M.W., and Bartok, W., "The Effect of Combustion Modification on Pollutants and Equipment Performance of Power Generation Equipment", In the Proceedings of the Stationary Source Combustion Symposium, Vol III, EPA-600/2-76-1520 (NTIS PB 257146) June, 1976.

21. Jungten, H., and Klein, J., Org. Geochem., 33, 647 (1972).

22. Smith, P.J., Hill, S.C., and Smoot, L.D., Nineteenth Symposium (International) on Combustion, pp 1263-1270, The Combustion Institute, Pittsburgh, PA, 1982.

23. Fenimore, C.P., Combustion and Flame, 26, 249 (1976).

24. De Soete, G.G., Fifteenth Symposium (International) on Combustion, pp 1093-1102, The Combustion Institute, Pittsburgh, PA, 1975.

25. Miller, J.A., Branch, M.C., McLean, W.G., Chandler, D.W., Smooke, M.D., and Lee, R.J., Twentieth Symposium (International) on Combustion, pp 673-684, The Combustion Institute, Pittsburgh, PA, 1985.

26. Freihaut, J.D., Zabielski, M.F., and Seery, D.J., Nineteenth Symposium (International) on Combustion, pp 1159-1167, The Combustion Institute, PA, 1983.

27. Haynes, B.S., "Kinetics of Nitric Oxide Formation in Combustion" in "Alternative Hydrocarbon Fuels: Combustion and Chemical Kinetics", Eds. Bowman, C.T., and Birkeland, J., Progress in Astronautics and Aeronautics, Vol. 62, AIAA, 1978.

28. Ho, W.M., Wofsy, S.C., McElroy, M.B., and Beer, J.M., "Sources of Atmospheric Nitrous Oxide from Combustion" (Accepted for publication in J. Geophys. Res., 1986.

29. Perry, R.A., J. Chem. Phys., 1985.

30. Levy, J.M., Chan, L.K., Sarofim, A.F. and Beer, J.M., Eighteenth Symposium (International) on Combustion, p 111, The Combustion Institute, Pittsburgh, PA, 1980.

31. De Soete, G., "Reduction of Nitric Oxide by Solid Particles" in "Pulverized Coal Combustion: Pollutant Formation and Control" MSS submitted to EPA, 1980, to be published 1987.

32. Chen, S.L., Heap, M.P., Pershing, D.W., and Martin, G.B., Nineteenth Symposium (International) on Combustion, pp 1271-1280, The Combustion Institute, Pittsburgh, PA, 1982.

33. Raask, E., J. Inst. Fuel, 44, 339-344 (1968).

34. Ramsden, A.R., J. Inst. Fuel, 41, 451-454 (1968).

35. Sarofim, A.F., Howard, J.B., and Padia, A.S., Combust. Sci. Technol, 16, 187-204 (1977).

36. Natusch, D.F.S., Wallace, J.R., and Evans, C.A., Jr. Science, 183, 202-204 (1974).

37. Ragaini, R.C., and Ondov, J.M., J. Radioanal Chem., 37, 679-691 (1975).

38. Keyser, T.R., Natusch, D.F.S., Evans, C.A., Jr., and Linton, R.W., Environ. Sci. Technol., 12, 769 (1978).

39. Linton, R.W., Williams, P., Evans, C.A., and Natusch, D.F.S., Anal. Chem., 49, 1514 (1977).

40. Linton, R.W., Loh, A., and Natusch, D.F.S., Science, 191, 852-854 (1975).

41. Desrosiers, R.E., Riehl, J.W., Ulrich, G.D., and Chiv, A.S., Seventeenth Symposium (International) on Combustion, p 1395, The Combustion Institute, Pittsburgh, PA 1979.

42. McElroy, M.W., Carr, R.C., Ensor, D.S., and Markowski, G.R., Science, 215, 13-19 (1982).

43. Taylor, D.D., and Flagan, R.C., Aerosol Science and Technology, 1, 103-117. (1982).

44. Neville, M., Quann, R.J., Haynes, B.S., and Sarofim, A.F., Eighteenth Symposium (International) on Combustion, pp 1267-1282, The Combustion Institute, Pittsburgh, PA, 1981.

45. Haynes, B.S., Neville, M., Quann, R., and Sarofim, A.F., J. Coll. Interface Sci., 87, 266-278 (1982).

46. Schultz, E.J., Engdahl, R.B., and Graukenberg, F.T., Atmos. Environ. 9, 111-119 (1975).

47. Schmidt, E.W., Giescke, J.A., and Allen, J.M., Atmos. Environ., 10, 1065-1069 (1976).

48. Flagan, R.C., Seventeenth Symposium (International) on Combustion, pp 97-104, The Combustion Institute, Pittsburgh, PA, 1979.

49. Wibberley, L.J., and Wall, T.F., Fuel, 61, 87-92 (1982).

50. Wibberley, L.J., and Wall, T.F., Fuel, 61, 93-99 (1982).

51. Mims, C.A., Unpublished Research Results, M.I.T., (Personal Communication).

52. Quann, R.J., and Sarofim, A.F., Nineteenth Symposium (International) on Combustion, pp 1429-1440, The Combustion Institute, Pittsburgh, PA, (1982).

53. Kerstein, A.R., and Edwards, B.F., "Percolation Model for Simulation of Char Oxidation and Fragmentation Time-Histories", accepted for publication in Chemical Engineering Science, 1987.

54. Cole, J.A., Kramlich, J.C., Seeker, W.R., and Heap, M.P., Environ. Sci. Technol., 19, 1065-1071 (1985).

55. Pershing, D.W., and Newton, J., DOE Contractor's Workshop, EER, Irvine, CA, May, 1986.

56. Squires, R.T., J. Inst. Energy, 55, 41-46 (1982).

57. Atherton, C., S.M. Thesis in Chemical Engineering, M.I.T., Cambridge, MA, 1985.

58. Neville, M., e-lab, pp 1-2, The Energy Laboratory, M.I.T., Cambridge, MA, July-Sept., 1983

59. Linak, W.P., and Peterson, T.W., Aerosol Science and Technology, 3, 77-96 (1984).

60. McElroy, M.W., and Carr, R.C., "Relationship Between NO_x and Fine Particle Emissions", Proceedings of the Joint Symposium on Stationary Combustion NO_x Control, Vol. V, U.S.A. EPA and EPRI, IERL-RTP-1087, EPA, R.T.P., N.C., Oct., 1980.

61. Wendt, J.O.L., and Eckmann, J.M., Combustion and Flame, 25, 355-360 (1975).

62. Torres-Ordonez, R., Sc.D. Thesis in Chemical Engineering, M.I.T., Cambridge, MA, 1986.

DISCUSSION

John H. Pohl (Energy Systems Associates)
I have three questions relating to particulate formation and destruction in flames.
1) Is there a quantitative relationship between tar and soot yields?
2) Is the time constant required to produce a self-preserving particle size distribution long compared to the residence time in a boiler?
3) Should the burning time for soot agglomeration be calculated using the diameter of the individual soot particles, the diameter of the agglomerate or something in between?

A.F. Sarofim
In response to the three questions
1) The soot in pyrolysis experiments was found to be formed at the expense of the tar. The sum of the yields of tar and soot was approximately constant under conditions at which the asymptotic volatile yield was attained.
2) The time constant required to approach a self-preserving particle size distribution depends on the soot volume fraction but is typically of the order of milliseconds, much shorter than the residence time in a boiler.
3) The answer depends upon the Thiele modulus or the relative rate of diffusion into a soot aggregate and the rate of reaction at the surface of the primary particles. At the low temperatures at which soot contrails persist, the primary soot particle size should be used in calculating the burning time from the Nagle Strickland-Constable equations.

265

P.R. Solomon (Advanced Fuel Research)
What is the current thinking on the predominant primary nitrogen gas from pyrolysis, HCN or NH₃? If HCN is the predominant species as many pyrolysis experiments suggest, then what is the mechanism for NH₃ production in flames?

E. Suuberg (Brown University)
1) Has there been a study to demonstrate directly the applicability of Nagle Strickland-Constable constants to burnout of the vapor-phase pyrolyzed tars (soot)? The concern is that these materials may be quite different than the carbons, blacks and soots studied previously (then may be more amorphous).
2) Has there not been evidence of direct formation of significant amounts of HCN and NH₃ during primary pyrolysis (e.g. Blair et al., 16th Symposium)?

A.F. Sarofim
HCN and NH₃ in coal flames may be formed by primary pyrolysis of amine and cyano substituents on the aromatic rings within the coal structure, by the decomposition of tars produced from the coal into lighter gases, and by gas phase reactions involving the gas-phase primary and secondary nitrogen-containing species. It is to be expected, therefore, that the HCN and NH₃ content of pyrolysis and combustion gases will be strongly dependent on coal composition and the temperature/oxidation history of the coal. Only fragmentary information is available on these processes. The contributions of side chain substitutions are believed to be of secondary importance on the basis of models proposed for the coal structure and the similarity in the trends of evolution of tars and nitrogen compounds during pyrolysis. The observations of HCN and NH₃ by Blair et al. could well be due to secondary pyrolysis reactions which are difficult to suppress completely. The secondary pyrolysis of tars can yield varying amounts of HCN and NH₃ depending on the composition of the tar, as explained by Dr. Jungten (v. infra). Another factor is that gas phase reactions with hydrocarbons have consistently been found to yield HCN in fuel-rich hydrocarbon flames, conditions which would be more likely to be encountered in the flames of bituminous coals rather than lignites. There are conditions, however, when gas phase reactions favor the formation of NH₃ (Roby, R.J. and Bowman, C.T., presentation at 21st Symposium on Combustion). The situation is far from being settled...
The Nagle Strickland-Constable constants have been obtained for electrode carbons and have been applied with success to carbon blacks. They can be considered to provide an order of magnitude estimate for the burnout times of the soot produced in the vicinity of a burning coal particle. The issues raised by Dr. Suuberg on the effect on burning rates of soots of their tar content, particle morphology, and porosity are ones that need to be addressed before reliable estimates of burnout times can be obtained.
In laminar flame systems these effects may be important. In turbulent flames the rate of micromixing of oxidant with fuel rich eddies is often controlling and the kinetic constant used for soot burnout then become of secondary importance.

H. Juntgen (Bergbau-Forschung GmbH)
1) Can you give more information on the side chain substitution of coal

generated PAH? Methods used for analysis? Carcinogenic effect of substitution?

2) Have you investigated chlorine compounds in gas phase? Kinds of compounds? Mechanism of formation?

As to the formation of different N containing species during pyrolysis and combustion: At low temperatures formation of NH_3 and HCN is also a function of substances burnt: e.g. pyrrolidine tends to form higher concentration of NH_3, pyrrol tends to form higher concentrations of HCN and NO. Therefore coals behave differently due to the different occurrence of N compounds in their tars. At higher temperatures HC radicals are formed which can lead to a partly intermediate reduction of N compounds to NH_3 (remarque to the comment and question of P.R. Solomon with reference to the paper presented at the workshop.)

A.F. Sarofim

The differentiation of PAH emitted by coal from that generated by other fuels on the basis of the greater extent of alkyl substitutions on the PAH rings was reported first by Hites and coworkers. They drew this conclusion from analyses by GC/MS of products emitted during the combustion of coal in a simple flame system. More recent studies in our laboratory of coal pyrolysis products, obtained at times and temperatures representative of those encountered in pulverized coal flames, show that the extent of side chain substitution decreases with the severity of secondary pyrolysis, achieved by increasing either the residence time or temperature. A novel steric exclusion chromatographic technique developed by Lafleur et al. (Lafleur,A.L., and Wornat, M.J., "Multimode Retention in High-Performance Steric Exclusion Chromatography with Poly-divinylbenzene", submitted to Analytical Chemistry) was used for the analysis. We have yet to complete the mutagenicity studies on the effect of side chain substitution but Silverman and Lowe (Ref. 5) have shown that the position of methyl substitution on the aromatic rings had a significant effect on the biological reactivity of methylated PAHs. Referral is made to Dr. Jackson's comments for the fate of chlorine in coal.

P.J. Jackson (Central Electricity Generating Board)

Firstly, in response to Dr. Juntgen's question, our research has shown that early in the combustion, virtually all the chlorine in coal is converted to hydrogen chloride in the gas phse. I have been working on the subsequent behavior of chloride with respect to its deposition, and there is appearing a strong correlation of this with the calcium concentration in the gas. (Unpublished work). With regard to calcium, may I take this opportunity to thank Professor Sarofim for his information about the behavior of organically-bound calcium during char burnout--it confirms what we had suspected.

Further, our deposition studies have included minor elements in coal, such as lead, zinc, phosphorous and boron, the former three of which featured in Professor Sarofim's presentation. Incidentally, we have recently shown that even in high temperature staged p.f. combustion in a large boiler, phosphorous was not apparently reduced to elemental form, and later enriched in the depositing ash, as is the case with fixed bed combustion including high temperature reducing conditions. Since we are interested in the behavior of boron, I ask Professor Sarofim he has any information on its release during combustion.

G. de Soete (Institut Francais de Petrole)
At the time being, how good (or how bad?) is our understandig of the heterogeneous NO formation from char bound nitrogen, especially taking into account the competition of CO enhanced NO reduction on the char? Is our knowledge on that particular point (of competition between heterogeneous oxidation and reduction) speculative? Or do there exist results from fine fundamental studies to support it?

A.F. Sarofim
The factors governing the formation of nitric oxide from char bound nitrogen are understood qualitatively. Oxygen penetrating a char particle will oxidise the nitrogen and carbon non-selectively (the NO/CO ratio of the primary products wil be the same as the N/C ratio in the char). As the NO and CO diffuse out of a char particle the NO will be partially reduced to N_2 as a consequence of both the direct and the CO enhanced reduction of NO by C. The rates of reduction will be catalysed by inorganic constituents such as ion-exchanged Ca in coal and will be influenced by the concentration gradients in a particle which are functions of the pore-size distribution, the particle size, and the temperature. In principle, models utilizing the kinetic parameters of the individual reactions developed by Dr. de Soete could be used for this purpose. Calculations have been carried out of the reduction of NO in pores by a homogeneous reaction with CO (Wendt, J.O.L., and Schulze, O.E., A.I.Ch.E. Jl., 22, 102-110, 1976). There is a need to extend such models to incorporate the influence of heterogeneous reactions.

H. Kremer (Ruhr-Universitat Bochum)
Having seen your photograph of burning coal particles with a soot containing cloud I would like to ask first whether there is an influence of particle density on soot formation. My second question is concerning the importance of soot radiation as compared to that of char and ash particles in connection with radiation modeling.

A.F. Sarofim
The amount of soot produced by dilute suspensions of coal particles increased when the interparticle distance decreased to the point where the volatile flames surrounding different particles overlapped. This would be expected from the decreased access of oxygen to the volatiles when the particles are closely spaced, which results in locally fuel-rich regimes in which soot formation is favored. The major interest in soot formation in flames is for the modelling of radiation. The emission from soot will dominate that from char and ash in the flames of bituminous coals in which the peak soot amount is expected to exceed one percent of the coal feed. Early evidence for the dominant contribution of the soot radiation was provided by the comparison of the radiation from a high volatile bituminous flames with that from an anthracite in an International Flame Research Foundation study.

K.H. van Heek (Bergbau-Forschung GmbH)
You mentioned an interesting chemistry, the reduction of MgO by CO.
1) What is the experimental evidence of these reactions and are they confirmed e.g. by using the pure oxides as model substances?
2) I assume, that not all of the minerals mentioned are in the state of

oxides. Are there similar reduction reactions for other compounds?

A.F. Sarofim
    The reduction of oxides present in minerals to the more volatile
suboxides or elements is supported by a number of independent
observations. The enhancement of the rate of vaporization of SiO_2 by such
a mechanism had been first reported in the coal literature by Raask and
Wilkinson to explain the fume produced in an oxygen-blown gasifier, by a
number of investigators in the aerospace industry who found similar
reactions occur in the glass-fibre reinforced phenolic heat shields during
reentry of space vehicules and by model studies in our laboratory on the
vaporization of coal ash in the presence and absence of char. Most of the
studies to date have focussed on the reduction of oxides. Reduction of
other species may lead to a similar enhancement of the vaporization of
mineral constituents. For example, the reduction of pyrites present in
the coal may undergo the following sequential reactions which are
thermodynamically favored at combustion conditions:

$$FeS_2 = FeS + 1/2\ S_2\ \ ;\ \ FeS + CO = Fe + COS$$

At this time, very few of the possible reactions have been established.

J. Lahaye (Centre de Recherches sur la Physico-Chimie des Surfaces Solides
          C.N.R.S.)
    You have mentioned that the concentrations of some minerals (silica in
particular) in fly ash aerosol is higher than the one expected from their
vapor pressure. Did you take into account that the vapor pressure in
equilibrium with droplets is higher than the vapor pressure in equilibrium
with a bulk liquid (Kelvin equation).

A.F. Sarofim
    The Kelvin equation giving the increase in vapor pressure over the
surface of a droplet was not found to be of importance in the vaporization
studies. However, it had to be taken into account in estimating the
condensation of the vapors of inorganic constituents at low cooling rates
under which the vapor pressure driving force for condensation was low and
the radius (10 nm) of the aerosol onto which the vapors were considering
was small.

MINERAL MATTER IN PULVERIZED COAL COMBUSTION.

P.J.JACKSON

CENTRAL ELECTRICITY GENERATING BOARD, MARCHWOOD ENGINEERING LABORATORIES, MARCHWOOD, SOUTHAMPTON, SO4 4ZB, UNITED KINGDOM.

1. INTRODUCTION

A mineral is a substance having a definite chemical composition and atomic structure, formed by natural inorganic processes (1,2). The term "mineral matter" as applied to coal is most often used to represent the inorganic impurities as they exist before the coal is heated significantly. Here, is considered the behaviour of any coal constituent which yields an inorganic residue during combustion of the pulverised fuel.

Detailed reviews of the mineralogical species identified in coals and their modes of association have been published (3-22).

The most popular and probably the most specific method used for identifying minerals by their structures is X-ray diffraction analysis (18,23) and differential thermal analysis has been used most effectively for following changes with thermal treatment (24-27). These techniques are often supplemented by optical or electron microscopy (3) and by rationalised chemical analysis such as has been applied to deposits derived from p.f.-fired boilers (28).

In this paper, a fully detailed consideration is not possible: it reproduces largely the treatment of the subject presented at a University of Newcastle, N.S.W. lecture course (29) which is relevant to pulverised coal combustion in large water-tube boilers, with the addition of the author's more recent views. Descriptions of many details of the subject matter are given in Raask's recent book (30).

2. THE NATURE OF MINERAL MATTER IN COAL.

Being a sedimentary deposit, coal is widely variable in its overall composition, and in its content of inorganic matter and its mode of association with the chemical entities not strictly part of the organic matter of the coal substance. As mined, most coals contain very significant proportions of the adjoining strata: shales, sandstones, etc., either interbanded or just mixed. All coals contain also some elements pertinent to this study, particularly the alkali- and alkaline-earth metals and sulphur, which are associated with the organic material itself on an atomic scale: either chemically combined, or as ions adsorbed from ground waters. Yet other minerals, notably carbonates and sulphides have been formed in voids in the coal seam, such as between fracture surfaces of hard coals, by being deposited from the percolating water.

Table 1 represents a consensus of the published information on minerals identified as being present in hard coals. For all coals, the proportions of each of these minerals is widely variable, depending on the sedimentation facies at the time the organic material was deposited, later cycles of inundation, metamorphic processes, infiltration and deposition by ground waters and the selection of strata during mining. Most mechanically-mined hard coals of Permian and Carboniferous (Westphalian, Pennsylvanian) origin include mineral matter composed of over 80% w/w silicates and quartz, less than 15% w/w carbonates and less than 10% w/w sulphides. Brown coals and lignites mostly have a lower content of clay minerals and substantially more calcium, magnesium and sodium, combined either as carbonates, sulphates or organically with the coal substance; most of their sulphur is also present in organic combination (8,31,32). Relevant to the subject of this paper is the variability of composition of the clay mineral illite and of the carbonates.

TABLE 1. Major Minerals Naturally Occurring in Coals.

Silicates

| Quartz | SiO_2 |
| Biotite | $K(Mg,Fe)_3(AlSi_3O_{10})(OH)_2$ |

Major silicates with clay minerals. Quartz present as sediment, round to angular grains – 1mm to <1μm.

Clays

Montmorillonite	$Al_2Si_4O_{10}(OH)_2 \cdot x\ H_2O$
Illite-sericite	$KAl_4(AlSi)_8O_{20}(OH)_4$
Kaolinite	$Al_4Si_4O_{10}(OH)_8$
Halloysite	$Al_4Si_4O_{10}(OH)_8$
Chlorite	$(Fe,Mg,Mn)_4(Al,Fe)_8Si_4O_{20} \cdot (OH)_8$
Smectites	$(Ca,Na)(Al,Mg,Fe)_4(Si,Al)_8O_{20}(OH)_4 \cdot nH_2O$

Finely-dispersed in coal substance; principal constituents of shales; Crystals < 1μm. Impregnation of joints in brown coal.

Sulfides

Pyrite	FeS_2
Marcasite	FeS_2
Sphalerite	ZnS
Galena	PbS
Chalcopyrite	$CuFeS_2$
Arsenopyrite	$FeAsS$
Millerite	NiS

In nodules and crystals 10cm to <1μm in coal substance and in joints and cleats. Pyrite and marcasite are principal sulphides. Principal form of iron in most hard coals.

Carbonates

Calcite	$CaCO_3$
Dolomite	$(Ca,Mg)CO_3$
Siderite	$FeCO_3$
Ankerite	$(Ca,Fe,Mg)CO_3$
Witherite	$BaCO_3$

In hard coals present mainly in joints and cleats. Principal form of calcium and magnesium, secondary form of iron.

TABLE 1. Continued.

Sulfates
 Barite $BaSO_4$
 Gypsum $CaSO_4 . 2H_2O$
 Anhydrite $CaSO_4$
 Bassanite $CaSO_4 . \frac{1}{2}H_2O$

Deposited from ground waters.

Chlorides
 Halite $NaCl$
 Sylvite KCl
 Bischofite $MgCl_2.6H_2O$

Occasionally present as discrete crystals, often as ions adsorbed on coal substance (vitrinite): can be recrystallised from solutions after wetting.

Oxides and Phosphate
 Hematite Fe_2O_3
 Magnetite Fe_3O_4
 Rutile TiO_2
 Apatite $Ca_5(PO_4)_3(F,Cl,OH)$

Combined with or Adsorbed on Organic Matter
 Calcium   Ca
 Magnesium   Mg
 Sodium   Na
 Sulphur   S
 Vanadium   V
 Boron   B
 Nickel   Ni
 Iron   Fe

---

3. CHANGES DURING PREPARATION FOR COMBUSTION.
 The act of mining releases pressure on most underground coals, permitting loss of gases and water; strata are broken and this process is continued in the pulverizing of the coal. Some of the coarser-grained mineral matter is separated by these processes from coal substance but the more finely-disseminated material is still "inherent" in at least the larger coal particles: this includes clay minerals, pyrites and the elements combined with or adsorbed on the internal surface of the organic matter. If the coal is treated by a gravity-separation process the proportion of the mixed inorganic material, especially shales and pyrites, which are of higher density, is substantially reduced. Wet washing may also redistribute soluble material, such as salts of calcium, magnesium or sodium. With the common practice of recirculating washery water, washing leaves significant films of the solution on the superficial surface of the coal: the solutes might later be leached out during storage (33,34) or crystallised on evaporation of the water either naturally or during the drying of the coal when it is ground. Some of the fine magnetite used to increase the bulk density of

the washery water is left on the surface of the coal: this can increase the iron content of low-ash coals.

The products of coal mills are classified by aerodynamic means and since the density of most of the inorganic minerals is significantly greater than that of the coal, the particle size of separated mineral matter in the p.f. fed to the burners is generally lower than that of the relatively clean coal (10,11,21).

4. BEHAVIOUR DURING HEATING.

Firstly, there is a loss of "inherent" water, adsorbed on the coal substance or loosely combined with mineral matter, at temperatures below about 500 K. The next significant change, as the pulverized coal passes into the ignition zone of the flame, is the decomposition of the organic matter to yield complex hydrocarbon vapours ("volatile matter") at temperatures over 500 K, depending on the structure of the coal. Of the elements combined organically or physically adsorbed, chlorine and sulphur are volatilised and others such as calcium and sodium will tend to remain in the char, possibly to be volatilised at a later stage of the burning of the particle when its temperature has risen to over, say, 1000 K. Information on the details of these processes is at present sparse and bears significantly on the mechanism of the formation of species later responsible for the ash behaviour in depositing and reacting on boiler surfaces. Any sodium chloride present as such in the coal is volatilised rapidly as particle temperatures exceed 1300 K (35). In the early stages, an appreciable proportion of the organically-combined sulphur is released in the volatile organic fragments, later to be pyrolised to yield elemental sulphur, its hydrides and eventually oxides. Chlorine is most likely volatilised as hydrogen chloride at an early stage in the flame. At temperatures around 600 K, pyrite and marcasite decompose to pyrrhotite (FeS) and sulphur, these ultimately oxidising to iron and sulphur oxides.

The carbonates yield CO_2 and the respective metal oxides at temperatures between 900 and 1200 K, the oxides, depending on their cationic composition, are partially converted to sulphates as they cool below about 1200 K. Kaolinite and other clay minerals undergo a series of structural changes as combustion proceeds and the temperature of the residual ash rises. Micas and the related illites cleave easiest along the planes of potassium ions and since they are not mechanically strong, pulverizing exposes a considerable area of this surface. Potassium would be relatively easily released from this surface by heating (36) and there is at least circumstantial evidence that potassium can be exchanged for sodium (37-39) during combustion; the possibility of hydrogen chloride reacting to form volatile potassium chloride (40) also should not be ignored. The composition of the atmosphere surrounding the mineral matter during its heating, which varies greatly with time, also affects the direction and rate of many of the chemical reactions, for example, the decomposition and oxidation of sulphides and the decomposition of carbonates, the latter being stabilised by the generally high partial pressure of carbon dioxide.

Close proximity of different mineral components and their decomposition products bring about other interactions: two examples of relevance to the later behaviour of the products are as follows. Carbon and silicon dioxide can react to produce silicon monoxide vapour at temperatures of 2000 K (4): the SiO subsequently oxidises to SiO_2 as submicron spheres of very high specific surface, capable of reacting with other gaseous species or of depositing by diffusion processes on to other particles or boiler surfaces. Calcium and sodium oxides residual from the decomposition of coal substance (e.g. vitrinite which is comparatively rich in these elements) can react locally with the residues of clay minerals during the burning-out stages of coke combustion. This yields small silicate spheres possessing a surface glaze rich in calcium and/or sodium and therefore of melting point much lower than that of the parent silicate (41-44).

The size distribution of the mineral matter itself, as well as that of the overall p.f. containing it, is important in determining the composition and size-consist of the products, which in turn affect so strongly the pattern of deposition behaviour in the plant. As the devolatilised coke burns away, at temperatures up to 2000 K, some of the finer particles of mineral residues are separated at the burning surface and disperse as the carbon is converted to carbon monoxide. Others, if touching, tend to coalesce and this process is aided, when they are of sufficiently low viscosity, by the negative angle of contact between the silicate glass and the carbon surface of the supporting char particle (45). So, some silicate particles could be formed larger than the clay or other mineral crystals from which they were derived, but this seems to be of relatively rare occurrence. The principal result of the non-wetting of the silicates by the carbon is the production of many individual silicate particles, mostly glassy spheres, from each coal particle containing clay mineral crystals. The fusion of silicate minerals has been studied in some detail, relating the rate of rounding of the particle shape to temperature and viscosity; for a British coal, only a few percent remained angular when heated for up to 0.5 s at 1600 K (46). Quartz in particular requires temperatures over 1700 K to become rounded: the particles larger than about 50 μm, e.g. sand in the coal, remained unfused at 1800 K, a temperature typical of the hotter zone of a p.f. flame.

5. EFFECTS OF MINERAL MATTER ON COMBUSTION.

High speed photography of burning single particles of coal (47) has shown that during the combustion of most of a char residue, individual particles of inherent mineral matter are separated and lost to the surrounding gases. Towards the end of the burning process, the remaining ash largely disintegrates. Thus, there is a low probability of steric hindrance of oxygen access to and carbon monoxide escape from the reacting surface, which would occur if a layer of coherent ash accumulated over the carbon. The non-wetting of fusing silicates in the high temperature region near a combusting surface also inhibits such an ash cover and aids the removal of the inorganic material. So, it seems unlikely that char reactivity is significantly reduced by ash-forming constituents of the char particles unless ash content is extraordinarily high, or the ash is highly refractory.

Conversely, several of the metallic elements likely to be retained in the char after devolatilisation are known to enhance, or catalyse, the oxidation of carbon: these include manganese, sodium, lead and the alkaline-earths, as reported by Heap et al (Table 2). Although this shows negligible effect of vanadium, probably because of its low abundance in most coals, vanadium and iron in particular catalyse the combustion of residual oil char (48,49) and would undoubtedly act similarly if present at similar concentrations (up to 10^{-3} atomic) in devolatilised coal.

TABLE 2. Relative Potential Enhancement of Char Combustion Rate by
Organically Combined Metallic Elements in Coal (22).

Element	Min.	Max.	Element	Min.	Max.
Mn	2000	100,000	Ca	1	400
Na	150	10,000	Mg	8	150
Cs	100	4,000	Sr	1	60
Pb	30	2,000	Ag	1	50
Ba	10	800	Be, B, Al, Cd, Ni, Au, V		1

6. COMBUSTION PRODUCTS.

The flame thus contains a mixture of partly decomposed coal, with
mineral residues and gaseous species. The coal particles in the flame
consist of part evolving volatiles, part burning char, and part ash
residues of different sizes relative to the parent coal particles. The
residues of reacting mineral matter are surrounded by gas containing
volatilised inorganic material - hydrogen chloride, sulphur, sulphur
oxides, sodium and potassium compounds and trace-element radicals,
sulphides, chlorides and oxides. The major constituents of the gas are
water vapour, oxygen, carbon monoxide and dioxide with a background of
nitrogen and high concentrations of ions and radicals taking part in the
combustion reactions. The concentrations of any of these constituents
are extremely variable, as is the local temperature, and the distribution of residence times is likewise spread widely about a mean of 1
second. It is from this witches' brew that deposition can first take
place, on the walls of the combustion chamber.

These are the immediate products of heating the coal and there is
scope for such a wide variety of chemical reactions that calculations
modelling the history of specific components have so far been restricted
to relatively simple, apparently homogeneous, cases such as sulphur
oxides, carbon and nitrogen oxides and water (50,51). Inferences about
the course of chemical reactions in the pulverized coal flame have been
made from either the later products in suspension or after deposition, or
by analogy with chemically much simpler systems with very few potentially
interacting components.

We know that at the end of the main combustion period, designated the
"flame", with temperatures falling by radiation to the walls, we can
begin to describe products which are reproducible and recognisable; in
most plant this state of affairs is achieved at a gas temperature of
about 1650 K, before most of the products pass near to convectively-
cooling surfaces. Some of the mineral residues are solid - some iron
oxides and some silicates as spheres, larger quartz particles and smaller
particles of the oxides of calcium, magnesium and iron as angular
fragments. For most coals, these are accompanied by a much higher proportion of spherical particles of glassy silicates of relatively low
viscosity, many of sub-micron diameter and even smaller spheres of silica
condensed from the vapour.

The volatilised alkali-metals are especially important in bonding
deposits and forming corrosive layers in them. They are most probably
present as hydroxide vapour initially, this reacting with hydrogen
chloride or sulphur dioxide and oxygen in stages, to form species which

result in deposits of sodium and potassium carbonates, sulphites,
chlorides and sulphates (52).   The proportions of chloride and sulphate
depend on the temperature history and the concentrations of the principal
reactants, all of which are very variable with respect to time, the
mixing pattern downstream of the burner and the composition of the fuel.
Low residence times and low (or even "negative") oxygen concentrations
favour higher proportions of hydroxide, carbonate or chloride, but
ultimately the thermodynamics favour near-complete conversion to sulphate
(35).   At around 1400 K the rate of formation of sulphur trioxide is at
a maximum (51).

7. DEPOSIT FORMATION.

At the walls of the combustion space, or as the combustion products
are cooled by tubes over which they flow, deposition of some of the
solid, liquid or gaseous species takes place: most of this deposition is
irreversible, and is selective in terms of its particle size, density and
composition.   A classification of the nature of the depositing material,
the dominant dynamic processes of deposition and their effect on the
plant is given in Fig.1. (29).

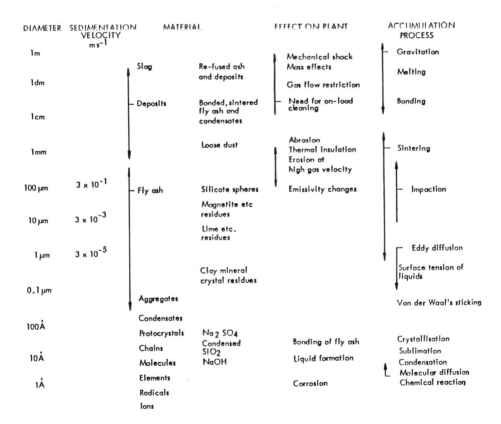

FIGURE 1.   FLUE GAS-BORNE MATERIAL IN P.F. FIRING
            ARRANGED ACCORDING TO ITS PARTICLE SIZE.

This, a summary of the behaviour of the principal participating elements
in Fig.2 and a diagram of the sequence of chemical reactions (Fig.3),
illustrate the following description of deposit growth on heat-exchange
surfaces.

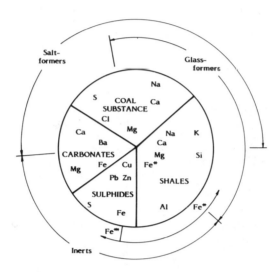

FIGURE 2. CLASSIFICATION OF BEHAVIOUR OF MINERAL MATTER IN COAL DURING
COMBUSTION. MAJOR ELEMENTS: OUTER. MINOR ELEMENTS: INNER.

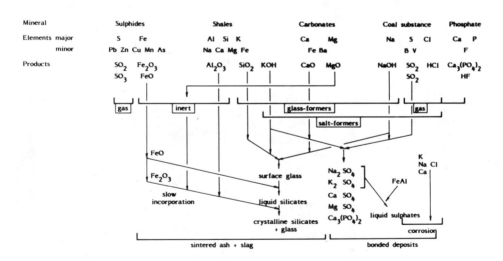

FIGURE 3. DEPOSIT GROWTH PROCESSES IN PULVERIZED COAL COMBUSTION.

7.1 Deposit Growth.

Most clean tubes exposed to p.f. combustion products exhibit first the
formation of a pale-coloured layer of softish fine-grained dust: this is
composed of fly ash particles mostly less than 10μm in diameter, enriched
in sodium and often in potassium. When this layer is several tens of
micrometres thick, the value depending inversely on the local temperature
and heat flux, coarser fly ash particles adhere to the surface, forming
irregularities; this material accumulates much faster than that forming
the initial layer. The rate of accretion of ash is higher at the apices
of the projections thus formed, which in the case of tubes over which gas
is flowing transversely, are usually larger at the upstream positions 30°
to 60° from the direction of the gas movement. Thus, two "ears" form,
which grow upstream preferentially at the tips of any unevenness in the
surface. Generally, in between these ears, a "dead" space fills with
more loosely-bonded fly ash, which becomes enclosed when the "ears" grow
together to form a roughly V-section deposit on the upstream face of the
tube. After several hours, the deposited ash in gas at temperatures
over about 1000 K turns from its original black, grey and colourless
state to various shades of brown, red or purple, and the deposit becomes
harder and tougher. Downstream, a deposit forms, similar in character
to the forward-facing one, but much thinner; on near-horizontal tubes,
loose dust can accumulate by gravity and if undisturbed this forms a
lightly bonded crust which preserves its shape as a triangular-section
deposit of included angle 20° or less. Wall tubes, or those in closely-
spaced elements or platens, show similar deposit characteristics,
modified according to the local gas-flow patterns, deposit growth
proceeding in an upstream direction.

Gas temperatures and heat fluxes determine whether the parts of the
deposits remote from the cooling effect of the tube achieve temperatures
at which fusion of the whole of the ash particles takes place. If this
happens, firstly small areas, then an extensive liquid surface of dark
coloured glass is presented to the gas, sometimes punctured by gas blow-
holes, on which fly ash particles stick and become incorporated in the
slag. As the deposit increases in thickness, its mass may exceed its
ability to retain a rigid structure, and it then falls or flows. In
areas of very high gas temperature, say, over 1600 K, and/or high heat
flux (over 150 kW m^{-2}), the temperature gradient can exceed 200 K mm^{-1},
resulting in a thin slag layer close to the metal; more usual values for
the high temperature gradient are 30 to 100 K mm^{-1}. Depending on the
viscosity of the molten ash and local conditions of atmospheric
composition and temperature, slag layers can be a few millimetres to
several centimetres thick - lumps forming in corners or as "eyebrows"
over burners may be 0.5 m or more thick and have a mass of over 100 kg.

7.2 Deposition-Erosion Processes.

A detailed analysis of the physical processes involved in forming
deposits was published in 1966 (53). The conclusions, confirmed by
other workers (54-56), were that the deposit constituents present as
vapour in the flue gas are deposited mainly by molecular diffusion, and
that the smaller particles, say around 1 μm in diameter, are deposited by
eddy diffusion and the larger ones, say over 10 μm in diameter, mostly
by inertial impaction. Transfer through the boundary layer is likely to
be affected by both thermal diffusion and electrostatic attraction.
Retention of material reaching the surface is, for discrete particles,by
van der Waals forces; if a liquid film is present, very strong surface

tension forces dominate. Bishop (57) demonstrated the presence of a liquid "primary" layer, corresponding to the sodium/potassium/chloride/ sulphate quaternary eutectic region (m.p. down to 787 K), in deposits derived from high chlorine coals, and the equimolecular complex sulphate $Na_3K_3Fe_2(SO_4)_6$ melts at 825 K: both of these temperatures are well below that of the surface of most high temperature superheater tubes.

The alkali metal compounds play a very important part in deposit formation, in subsequent reactions within the deposit and most probably in corrosion of the underlying metal. Sodium chloride and sulphate are deposited as condensates from vapour and dewpoints corresponding to the known vapour pressure are exhibited (38,40,54,57). The rate of deposition of sodium sulphate has been shown to accord with convective mass transfer theory (55,57).

As a corollary to the deposition of fly ash, consideration should be given to the behaviour of the solid particles which bounce off. Those larger than about 10 μm have sufficient momentum to penetrate the boundary layer and impact the surface; the kinetic energy of the particles may cause local fusion at their surfaces (58). For the more refractory particles moving at higher velocities, the energy released on impaction is transferred to other materials at the surface - deposited dust or even that of the substrate itself, normally iron oxides, and this can result in their removal to the passing gas stream. Bishop (57) concluded that this leads to less deposition on the middle of the upstream surface of a tube. In the extreme case, erosion of the protective oxide scale or metal takes place, often at an embarrassing rate. The approach velocity (V) of the particles is a dominant factor (metal loss $V^{3.5}$), as is the angle of incidence, and angular quartz grains abrade more effectively than glassy spheres (59). In boilers, significant erosion occurs where local gas velocities exceed 30 m s^{-1}: this can be caused by faults in boiler design, extensive fouling of a tube bank or misalignment of sootblowers which entrain coarse deposit debris.

7.3 Reactions in deposits.

The large temperature gradients within the porous deposits set up a high temperature fractionation system, in a temperature range within which the alkali metal compounds have significant vapour pressures, and this results in migration of sodium and potassium compounds towards the cooler inner layers. Most of the ash particles react to some extent with the alkali metal hydroxides, chlorides and sulphates to form lower-melting silicates, hematite from magnetite via iron chloride, and iron and aluminium complex sulphates, often molten at the local temperatures. This type of deposit, considerably enriched in sodium, potassium, calcium and magnesium, is bonded by the local growth of glass or crystals of new compounds within the deposit. Their strong lattice bonding confers high mechanical strength to accompany the significant elasticity afforded by the porous structure. It is these properties which make such "bonded" deposits difficult to remove on-load. Off-load removal is also hindered by the insolubility in water of silicate glasses and calcium sulphate; the progressive sulphation of lime particles after deposition (60) can result in the formation of very hard bonded deposits on primary superheaters and economisers (41). Development of a strong adhesive bond by the formation of an iron-rich silicate enamel immediately over the tube-metal oxide, has been demonstrated by Raask (61), who estimated that when its rupture strength exceeded 10 kN m^{-2}, sootblowing would not break it.

The sintering of silicate particles has long occupied the attention of

investigators in this field.   Generally, bonds are formed between the
fly ash spheres at temperatures in the range 1100 to 1700 K.   Development of sintering, usually measured as ultimate compressive strength
(62,63), progresses smoothly with temperature and time and is dependent
on the particle size and its distribution and on the composition of the
surface layers of the ash.   Reaction of sodium and calcium at the
surface of silicate glass particles has been shown to penetrate up to
about 40A (44), and to form a low-melting, low-viscosity glaze which
initiates sticking at temperatures as low as 1300 K.   The rate of coalescence of particles depends on the mobility of bonding in the glass at
their surfaces; this is controlled by temperature and the concentration
and nature of the cations in the structure.   Those known to be of
particular significance in industrial coal combustion are sodium,
potassium, calcium and ferrous iron; of potential activity but of
unproven participation are boron, lead and zinc.

Since ferrous ions in the silicate glass result in a much more labile
structure than the corresponding concentration of ferric ions, the state
of oxidation of iron in pulverised coal ash and its deposits is of great
significance to slagging propensity (64,65): a prime example of the
important interaction of coal mineral matter and its environment.

The sticking of particles results in the gradual accretion of ash onto
plant surfaces, and confers mechanical strength to the deposits.   The
relationship of these functions to the local temperature is consistently
of the form shown in Fig.4, which was derived empirically.   The position
of the inflections in this curve in relation to both axes and the
gradient of the steep rise are of vital significance to the successful
design and operation of pulverised coal-fired boilers.

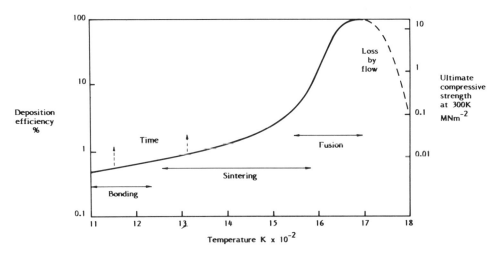

FIGURE 4.   DEPENDENCE OF ASH DEPOSITION AND DEPOSIT STRENGTHS
ON LOCAL GAS TEMPERATURE

The rate of deposition of ash depends on the mass concentration in the
gas approaching a target, and on its particle size distribution (see
Fig.1).   The use of deposition efficiency in recent research excludes
these factors: it is calculated as the ash which sticks on a target, as

a proportion of the ash expected to impact on it (66). Much recent research to relate features of this behaviour, simply, to the chemical composition of coal ash and derived functions (67) has yielded too wide a scatter for accurate forecasting of the behaviour of fuels. A more subtle and realistic approach is to take note of the mode of combination of the elements in question, via the mineralogical constitution, thus allowing for the appropriate variation in lattice and glass bonding energy: an increasing proportion of effort is now being devoted to this.

7.4 High Temperature Corrosion.

Enrichment of the inner layers of deposits in the more volatile components of the depositing material is unfortunate: many of the elements thus concentrated close to the metal and its normally protective oxide are capable of reacting with them at rates which constitute corrosion, that is, loss of metal sufficient to cause operational embarrassment. The rates of reaction are strongly dependent on local temperatures, as well as on the composition of the metal and its oxide layers - indeed, choice of the latter affords an important means of control of this problem. Two areas of the boiler are particularly susceptible to the high rates of corrosion (over 50 nm h^{-1}), as follows.

Combustion chamber wall tubes, usually of mild steel, have been corroded at rates up to 800 nm h^{-1}. The circumstances of the attack are the local impingement of an early part of the p.f. flame, a high heat flux, and in various boilers high chlorine content coals (68), and coarse p.f. (more than 40% larger than 80μm or more than 2% larger than 300μm). Corrosion products almost invariably consist of high proportions of iron sulphides (FeS and FeS$_2$) accompanied in the high chlorine cases by chlorides. It appears that severe corrosion is caused by a combination of the factors listed above and can be alleviated by control of almost any one of them. In short-term circumstances where this is not possible, the CEGB has found a solution in the installation of coextruded tubes, having outer layers of alloys of higher corrosion resistance, such as 50:50 nickel:chromium or AISI Type 310 (25Cr:20Ni) steel (69).

High temperature superheater and reheater tubes, particularly those in the higher gas temperatures (over 1300 K) are subjected to corrosion by the complex alkali-metal sulphates, liquid in the resulting temperature range at the tube surface. The mechanism of this corrosion has been described (69-71). When high chlorine coals are burned and especially at local sub-stoichiometric combustion conditions (72), it is possible for chloride to be included in the alkali-metal salts and this apparently penetrates between the metal grains before sulphur or oxygen and accelerates and wastage (Fig.5). High K:Na ratios in the deposit lower the melting point of the complex sulphates and result in higher rates of corrosion; this relates back to the behaviour of the mineral matter in the flame. Alleviative measures are more difficult to apply in this region of the boiler and the most economical relief has been afforded by installing coextruded tubing, in this case with an inner alloy of high creep strength.

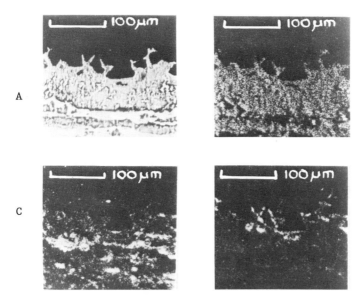

A

C

FIGURE 5.   ELECTRON PROBE IMAGES OF CORRODED SUPERHEATER TUBE METAL.

A: Electron Image
B: Oxygen
C: Suphur
D: Chlorine

(Courtesy of Electricité de France)

8. FURTHER RESEARCH.

Early investigations of the relationship between coal composition and problems in p.f. combustion due to ash misbehaviour resulted in simple classifications based on the concentrations of key elements, such as chlorine, sulphur or sodium.   More recent trends have been to examine the accuracy of forecasting using ratios of groups of elements, such as the "silica ratio" ($SiO_2 \times 100/SiO_2 + Fe_2O_3 + CaO + MgO$) or " slagging index" (Bases/Acids x %Sulphur) (66), but these, too, are proving insufficiently precise for industrial use without recourse to yet other fuel or operational factors.   These approaches hint at the realistic information required: the chemical and physical mobility of the complex mixture of raw materials at a range of energy contents, and in a variable environment, as exemplified in the broadest way by Fig.4.   A significant improvement in the accuracy of predicting behaviour during and after combustion should be yielded by taking note of the energetics of the mode of combination of those elements known to be relevant.

An increasing proportion of investigators in this field appear to be working on these lines.   The following is a check list of topics for future investigations.

1. Mineralogical constitution of the fuels (73).
2. Variations of the chemical composition of relevant minerals such as illites, carbonates, sulphides.
3. Transfer to and reactions in the vapour phase of key elements: sodium, potassium, sulphur, chlorine, silicon, iron.
4. Mobility of ions near the surface of mineral combustion residues.

282

5. Structural changes and viscous flow in glasses (74).
6. Vapour/condensed phase interactions.
7. Mass transfer phenomena for the wide variety of species in the combustion gases, in relation to their properties as elucidated above (56).
8. Chemical reactions and mass transfer by diffusion within deposited material, in particular the rate of equilibration of iron with the local oxidising potential.
9. Mechanical and thermal properties of deposits and their constituents at a realistic range of background physical and chemical conditions.
10. Last but not least, it is essential to ensure that any of these studies are closely related to the real industrial environment which is being modelled. Parallel work in operating plant and the conventional laboratory is most valuable (62,65) in this respect and it is strongly recommended to more workers.

9. ACKNOWLEDGEMENT.
This paper was written at the Marchwood Engineering Laboratories of the Technology Planning and Research Division and is published with the permission of the Central Electricity Generating Board.

10. REFERENCES.

1. Read H.H., (1947). Rutley's Elements of Mineralogy, Murby, London.
2. Hurlbut, C.S. (1959), Dana's Manual of Mineralogy, Wiley, London.
3. Mackowsky, M.T., (1955). Mitt. VGB. 38, 16.
4. Mackowsky, M.T., (1963). The Mechanism of Corrosion by Fuel Impurities, 80, Butterworth, London.
5. Mackowsky, M.T., (1968). Mineral Matter in Coal, Oliver & Boyd, Edinburgh.
6. Gumz, W., Kirsch, W. & Mackowsky, M.T., (1958). Schlakenkunde, Springer, Berlin.
7. Nelson, H.W. et al. (1959). A Review of Available Information on Corrosion and Deposits in Coal- and Oil-Fired Boilers and Gas Turbines. ASME, New York.
8. Grant, K. & Weymouth, J.H., (1962). J.Inst.Fuel, 35, 444.
9. Williams, F.A. & Cawley, C.M., (1963). The Mechanism of Corrosion by Fuel Impurities, 24, Butterworth, London.
10. Littlejohn, R.F. & Watt, J.D., (1963). The Mechanism of Corrosion by Fuel Impurities, 102, Butterworth, London.
11. Rayner, J.E. & Marskell, W.G., (1963). The Mechanism of Corrosion by Fuel Impurities, 113, Butterworth, London.
12. Brown, H.R. & Swaine, J.D., (1964). J.Inst.Fuel, 37, 422.
13. Kemezys, M. & Taylor, G.H., (1964). J.Inst.Fuel, 37, 389.
14. Marshall, C.E. & Tompkins, D.K., (1974). Paper 4 to Symposium "The Inorganic Constitutents of Fuel". Melbourne, Inst.of Fuel, London.
15. Dixon, K., Skipsey, E. & Watts, J.T., (1964). J.Inst.Fuel, 37, 485.
16. Dixon, K., Skipsey, E. & Watts, J.T., (1970a). J.Inst.Fuel, 43, 124.
17. Dixon, K., Skipsey, E. & Watts, J.T., (1970b). J.Inst.Fuel, 43, 229.
18. Gluskoter, H.J., (1977). Ash Deposits and Corrosion Due to Impurities in Combustion Gases, 3, Hemisphere, Washington DC.
19. Miller, R.N. & Given, P.H., (1977). Ash Deposits and Corrosion Due to Impurities in Combustion Gases, 39, Hemisphere, Washington DC.
20. Plogman, H. & Mackowsky, M.T., (1977). Ash Deposits and Corrosion Due to Impurities in Combustion Gases, Hemisphere, Washington DC.

21.Wall, T.F., Lowe, A., Wibberley, L.J. & Stewart, I.McC. (1979).
   Prog.Energy Combust.Sci., 5, 1.
22.Heap, M.P., Kramlich, J.C., Pershing, D.W., Pohl, J.H., Richter, W.F.
   & Seeker, W.R., (1986). Effects of Coal Quality on Power Plant
   Performance and Costs, Vol.4: Review of Coal Science Fundamentals.
   EPRI Document CS-4283, Vol.4. EPRI. Palo Alto, Calif. USA.
23.O'Gorman, J.V. & Walker, P.L., (1972). Mineral Matter and Trace
   Elements in U.S. Coals, U.S.Dept.of Interior R & D Report No.61,
   Interim Report No.2.
24.Warne, S. St.J., (1965). J.Inst.Fuel, 38, 297.
25.Warne, S. St.J., (1970). J.Inst.Fuel, 43, 240.
26.Warne, S. St.J., (1979). J.Inst.Fuel, 52, 21.
27.Bryers, R.W., (1978). Paper 78-WA/CD-4, ASME. New York.
28.Drummond, A.R., Boar, P.L. & Deed, R.G., (1977). Ash Deposits and
   Corrosion Due to Impurities in Combustion Gases, 243, Hemisphere,
   Washington DC.
29.Jackson, P.J., (1979). Pulverised Coal Firing – The Effect of
   Mineral Matter, L9, L10. Ed.I.McC.Stewart & T.F. Wall, University
   of Newcastle, N.S.W.
30.Raask, E., (1985). Mineral Impurities in Coal Combustion, Hemisphere,
   Washington, DC.
31.Gray, R.J. & Moore, G.F., (1974). Paper 74-WA/Fu-1, ASME, New York.
32.Sondreal, E.A., Gronhovd, G.H., Tufte, P.H. & Beckering, W., (1977).
   Ash Deposits and Corrosion Due to Impurities in Combustion Gases. 85,
   Hemisphere, Washington DC.
33.Daybell, G.N. & Gillham, E.W.F., (1959). J.Inst.Fuel, 32, 589.
34.Gillham, E.W.F., (1960). J.Inst.Fuel, 33, 193.
35.Halstead, W.D. & Raask, E., (1969). J.Inst.Fuel, 42, 344.
36.Jackson, P.J., (1963a). The Mechanism of Corrosion by Fuel
   Impurities, 190, Butterworth, London.
37.Jackson, P.J. & Ward, J.M., (1956b). J.Inst.Fuel, 29, 441.
38.Jackson, P.J. & Duffin, H.C., (1963). The Mechanism of Corrosion by
   Fuel Impurities, 427, Butterworth, London.
39.Raask, E., (1963). The Mechanism of Corrosion by Fuel Impurities,
   186, Butterworth, London.
40.Jackson, P.J., (1963b). The Mechanism of Corrosion by Fuel Impurities,
   484, Butterworth, London.
41.Jackson, P.J. & Ward, J.M., (1956a). J.Inst.Fuel, 29, 154.
42.Hosegood, E.A., (1963). The Mechanism of Corrosion by Fuel Impurities,
   71, Butterworth, London.
43.Boow, J., (1972). Fuel, 51, 170.
44.Natusch, D.F.S. et al. (1977). Anal.Chem., 49, 1514.
45.Raask, E., (1965). Paper No.65-WA/Fu-1, ASME, New York.
46.Raask, E., (1969a). Fuel, 48, 366.
47.Street, P.J., (1969). Ph.D.Thesis, University of Surrey.
48.Cunningham, A.T.S. & Jackson, P.J., (1978). J.Inst.Fuel, 51. 20.
49.Cunningham, A.T.S. & Jackson, P.J., (1979). ibid. 51, 231.
50.Gibb, J. & Joyner, P.L., (1979). CEGB Laboratory Note R/M/N1029,
   CEGB. London.
51.Squires, R.L., (1978). CEGB Laboratory Note R/M/N943, CEGB. London.
52.Dunderdale, J. & Durie, R.A., (1964). J.Inst.Fuel, 37, 493.
53.Samms, J.A.C. & Watt, J.D., (1966). BCURA Review No.254.
54.Brown, T.D., (1966). J.Inst.Fuel, 39, 378.
55.Jackson, P.J., (1977). Ash Deposits and Corrosion Due to Impurities
   in Combustion Gases, 147, Hemisphere, Washington DC.

56. Rosner, D.E., (1986). Transport Processes in Chemically Reacting Flow Systems, Butterworth, Stoneham, MA., USA.
57. Bishop, R.J., (1968). J.Inst.Fuel, 41, 51.
58. Duffin, H.C., (1960). NGTE Memo.No.341, HMSO, London.
59. Raask, E., (1969b). Wear, 13, 301.
60. Burdett, N.A., Hotchkiss, R.C. & Fieldes, R.B., (1979). The Control of Sulphur and Other Gaseous Emissions, Salford, I.Chem.E., London.
61. Raask, E., (1973). Mitt. VGB. (KWT), 4, 248.
62. Jackson, P.J., (1963). Mitt. VGB. 85, 220.
63. Gibb, W.H., (1977). Paper to VGB Conference, Essen, June, 1977, VGB. Essen.
64. Jackson, P.J., (1957). J.Appl.Chem., 7, 605.
65. Jackson, P.J. & Jones, A.R., (1984). Paper to Engineering Foundation Conference, Slagging and Fouling Due to Impurities in Combustion Gases. Engineering Foundation, New York, USA.
66. Jackson, P.J. & Jones, A.R., Paper to VGB. Conference, Slagging, Fouling and Corrosion in Thermal Power Plant, Essen. Feb-March, 1984.
67. Winegartner, E., (1974). Coal Fouling and Slagging Parameters, ASME. New York.
68. Jackson, P.J., (1964). Corrosion, I, 9.15, Newnes, London.
69. Cutler, A.J.B., Flatley, T. & Hay, K.A., (1978). CEGB Research, 8, 13.
70. Corey, R.C., Cross, B.J. & Reid, W.T., (1945). Trans.ASME, 67, 289.
71. Baker, D.W.C. et al, (1977). The Control of High Temperature Fireside Corrosion, CEGB.London.
72. Brook, S. & Meadowcroft, D.B., (1983). Corrosion Resistant Materials for Coal Conversion Systems, Applied Science Publishers, London.
73. Detaevernier, M.R., Platbrood, G., Derde, M.P. & Massart, D.L., (1985). J.Inst.Energy, 58, 24.
74. Kalmanovitch, D.P., Sanyal, A., & Williamson, J., (1985). J.Inst. Energy, 59, 20.

DISCUSSION

D.R. Hardesty

Would you please comment on the status of research and/or use of flame additives such as copper oxychloride which seem to be useful for influencing boiler deposits.
What do we know about the importance of the time/temperature history and local gas composition on the effectiveness of such additives ?

P.J. Jackson

I will restrict my reply to copper oxychloride : it is now reliably established that the mechanism of its action, in reducing the extent of ash particle aggregation and coalescence, is by inducing crystallisation at the surface of the silicate particles. This is an instance of my general remarks in sections 6,7.3 and 8 of my paper. It appears that the decomposition products of copper oxychloride in a pulverized coal flame-cuprous oxide and elemental copper-nucleate the crystallisation of spinels and possibly of mullite and since these compounds have melting points higher than the surface temperature of the ash particles, diffusion, and therefore bonding, between them is inhibited. Hence, deposits of greater porosity and lower mechanical strength are formed. Since the effects need be only at the surface of the particles (a very few atoms deep), very small mass proportions of the volatile copper compound are required - 1 or 2 ppm m/m more than this being redundant. The effectiveness of copper oxychloride was first established in boilers of the South-Western Region of the C.E.G.B. and the mechanism elucidated in laboratory studies between 1966 and 1982. Several well-controlled plant trials in other countries have confirmed the initial success. What still needs determining, as Dr Hardesty hints, is to quantify the effects in terms of deposition rate and mechanical strength of deposits under carefully-defined operational conditions such as local temperature and oxidising potential. It is probable that its effect will be less at very low oxygen partial pressure, when the majority of the ion will be in the ferrous form; plant experience to date indicates that temperature history of the ash and additive has little influence, in so far as they have been varied.

F.C. Lockwood

If the mathematical models could predict particle trajectories and temperature histories in the combustion chambers of a power station boiler, do you think that a mathematical submodel could currently be constructed which would allow one to predict the build up of slag deposits on the heat transfer surfaces ?

P.J. Jackson

If there were, equally closely defined, a relationship bet-

ween the response of the inorganic material in the fuel to its
temperature and chemical environmental history, the reply
would be - yes. However, such is <u>not</u> the case, and it is the
purpose of current studies of the behaviour of the inorganic
minerals and their products to derive such a relationship. In
this context, I should elaborate the point made in section 8
of my paper, that mathematical expressions have been constructed to represent simply the dependence of the sintering and
fusion of coal ashes on their chemical composition, but these
are limited in their application to narrow ranges of fuel
composition without incurring too great an error for satisfactory application in industry, and since high costs are at
stake, these errors dare not be accepted. In addition to the
variations in ash behaviour inherent in the heterogeneity of
the ash particle surface layers, there are those which are
dependant on only partially predictable factors of plant
operation, a scenario which is also being actively investigated.

A. William

Please can you give an indication of the behaviour of the
ash particles in the burning char particles.

P.J. Jackson

This subject is dealt with in some detail in section 4 of
my paper. Topics of current particles interest, on which quantitative information is rought, are :

(a) The proportion of silicon vaporised to SiO, its dependence on the mineralogical assemblage in the coal, and its
subsequent chemical history.
(b) Volatilisation, and reactions with silicon compounds,
of sodium and potassium.
(c) Reactions between calcium and magnesium compounds and
silicates during char burning.
(d) The ultimate particle size distribution (psd) of silicates, and its dependence on the psd of parent minerals and
factors such as (a), (b), (c) above.

H. Jüntgen

Relative to the catalytic behaviour of mineral components
in coal, I can give an example of the effect of the mode of
bonding of calcium in the coal substance and of its distribution in particles of coal. Calcium in German lignites (Rhei-
nische braun kohle) is combined with humic acids and therefore
evenly distributed in the coal particles, and these coals show
an extremely high reactivity with water vapour. Other lignites, which contain calcium in larger crystals of $CaCO_3$, show a
distinctly lower reactivity.

P.J. Jackson

I am grateful for Professor Jüntgen observations - it is

particularly timely, because I have recently investigated the effect of the degree of mixing of particulate chromium/iron oxide with graphite on the catalysis of its aerial oxidation. Similarly, dispersion of the metal atoms on the graphite surface by impregnating it with a nitrate solution yielded a much greater enhancement of reaction rate than the mixing of 40-to-60 μm particles.

INFLUENCE OF MINERAL MATTER ON BOILER PERFORMANCE

J. H. POHL

ENERGY SYSTEMS ASSOCIATES
TUSTIN, CALIFORNIA, USA

1.0 INTRODUCTION
    Mineral matter and ash increase pollution, reduce boiler performance, and increase the cost of power generation. Emissions of pollution are increased by compounds librated from the ash such as toxic metals and sulfur and by the particulate fly ash itself. Fly ash decreases the performance of the boilers by decreasing availability, capacity, and efficiency. Loss of one percent availability or capacity or a decrease in 100 Btu kW^{-1} hr^{-1} (105 kJ kW^{-1} hr^{-1}) increases the cost of generating power from a 1000 MWe plant by one million U.S. dollars per year (1).
    Decreased power plant performance and increased pollution is the result of abrasion and wear, slagging, decreased heat transfer, corrosion, fouling, erosion, electrostatic precipitator performance, and problems of waste disposal. Efforts to predict these problems in boilers are largely empirical and have only been partially successful. The ability to predict decreases in boiler performance caused by mineral matter and fly ash will reduce the cost of power generation by assuring that new boilers are properly designed to handle the proposed coal and that coals are properly selected for existing boilers.
    This paper reviews the current ability to predict problems in boilers caused by mineral matter and ash and suggests future areas of research which will improve the ability to predict the influence of ash properties on boiler performance.

2.0 ABRASION AND WEAR
    Mineral matter as well as coal abrades and wears pulverizers and feed pipes. Three empirical indices of abrasion have been used:

    1. The relative quartz value (2),

    2. The amount of free silica (3), and

    3. The total content of materials harder than the surfaces;
        i.e., SiO_2, Al_2O_3, Fe_2O_3.

The relative quartz value is the weight fraction of quartz particles which are greater than 44 microns in diameter. The relative quartz values correlated metal loss in a laboratory ball mill. It should be noted that the coals used had only small differences in the relative quartz value but that those small differences resulted in large differences in metal loss from the balls.
    The relative quartz value also predicted relative abrasion in operating plants. However, the relative quartz value over-predicted abrasion in all mills except one.

The percent free silica is calculated based on the ash composition and the ratio of SiO_2/Al_2O in kalonite,

$$FS = SiO_2 - 1.17\ Al_2O_3 \ . \tag{1}$$

The free silica has been shown to correlate with the wear on the blades of a laboratory blender for New Zealand coals, but has not been confirmed by field measurements.

The percentage of $SiO_2 + Al_2O_3 + Fe_2O_3$ in the coal has not been correlated with wear in laboratory, pilot scale, or power plant tests.

Field tests are insufficient to demonstrate the usefulness of any of the abrasion indices. Therefore, field tests on abrasion need to be conducted before more complicated methods to predict abrasion are developed.

Should more complicated methods to predict abrasion be desirable, then methods based more firmly on the mechanism of abrasion have a better chance of universal success. Such methods will need to be based on the minerals free from the coal (excluded mineral matter) during and after grinding as shown in Figure 1.

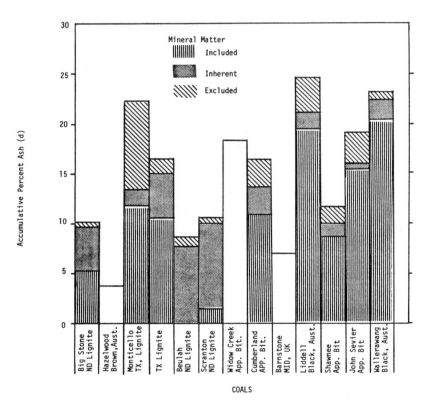

Figure 1. Mineral matter distribution in power plant coals (4).

New relations will also depend on the particle size and nature of the excluded mineral matter. The particle size distribution of the total mineral matter has been measured and shows particle sizes between 0.1 and 4 microns with mass mean particle sizes between 1 and 2 microns (5). Information is not available on the particle size distribution of excluded mineral matter nor the hardness and shape of this mineral matter.

3.0 SLAGGING

A number of empirical techniques exist to predict the slagging potential of coals. These techniques have been reviewed (1) and found to poorly predict the performance of operating boilers (6). Causes of the failure have been claimed to result from use of off design coals and operating conditions. However, the indices also poorly predict the performance of a wide range of coals burned under standard conditions in pilot scale furnaces.

There are many problems with the existing indices. First, the indices are based, for the most part, on bulk properties of the ash while slag deposits are known to form in layers (7) as shown in Figure 2.

Figure 2. Slagging deposit showing segregated development.

In addition, most slagging indices do not account for the total ash content of the coal, the operating conditions of the boiler, nor the flow patterns in the boiler.

Improving the ability to predict the slagging potential of coal ashes will require a better understanding of slag formation and deposition. This will require that flow patterns in boilers be observed and that deposits from boilers be extracted and analyzed in individual layers for several slagging and nonslagging coals and boilers. Hypotheses of the slagging mechanism can be formed from these results and then verified using controlled experiments in laboratory and pilot-scale furnaces.

4.0 HEAT TRANSFER

Heat transfer is currently estimated by both crude and sophisticated radiation models. These models require the emissivity of the gas and flame zone and the absorptivity and thermal conductance of the deposit on the boiler walls.

The emissivity of the gases has been well handled by methods developed by Hottel (8) and modified by Edwards (9).

The optical properties of soot are well known (10). However, the amount of soot formed is not well known and appears to vary with coal type (11).

The contribution of fly ash to the emissivity of the combustion cloud was uncertain in the past. However, recent work (12) has confirmed as suggested by Sarofim and Hottel (13), that fly ash is largely transparent and makes only a small contribution to gas cloud emissivity.

Many measurements have been made of deposit emissivity or absorptivity. Measured deposit emissivity decreases with increased temperature and varied between 0.4 and 0.95 (14). A highly reflective ash with an emissivity of 0.4 may upset the heat balance of some boilers, particularly during start up. However, heat transfer to the walls of most utility boilers is controlled by the thermal conductance of the deposits.

Limited measurements of the thermal conductivity (thermal conductance times the deposit thickness) have been made. The measured thermal conductivities varied from 0.02 to 2.0 kW m^{-1} k^{-1} (0.015 to 1.0 Btu hr^{-1} ft^{-1} F^{-1}) (15). Use of these values in heat transfer models with observed deposit thickness have been universally unsuccessful. Inspection of the deposits of Figure 2 and review of the techniques used to measure the thermal conductivity of deposits indicates the measured values are too low and that treating the deposits as solids results in unrealistic heat transfer.

Recent work by Viskanta (16) has included the structure of the deposits in determining thermal conductance. Heat can be transferred through the deposits by conduction, transmission, re-radiation, and possibly convection. These modes of heat transfer need to be taken into account in new measurements of thermal conductance in boiler deposits.

5.0 CORROSION

Corrosion is extremely complicated and involves many reactions as shown in Figure 3 (17).

One index of corrosion has been developed (18). This index empirically reproduces some of the mechanism shown in Figure 3. The index predicts an increase in corrosion with an increase in soluble alkalis, an increase in iron (a surrogate for sulfur in eastern U.S. coals), a reduction in calcium or magnesium (which will form stable compounds with sulfur). Comparison of predictions using this inddex with performance of operating boilers has been disappointing (19).

292

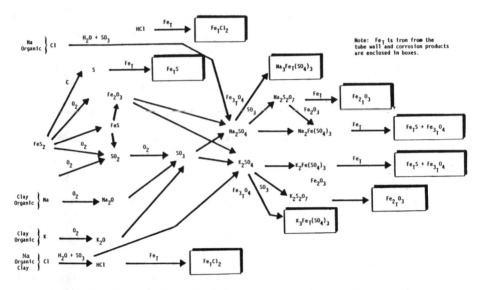

Figure 3.   Mechanism of corrosion (17).

6.0 FOULING

Again, many fouling indices have been proposed. These have been reviewed and found to be unreliable (1, 6) for the same reasons as the slagging indices.

A fouling index has been recently developed which satisfies many objections, and has correlated the performance of over fifty world coals (20). This index, Figure 4, relates the difficulty of removing a fouling deposit to the pounds of Na_2O per million Btu of coal (a parameter which accounts for the sodium content of the coal, the ash content of the coal, and the heating value of the coal), and the ratio of exit gas temperature (a measure of boiler design), and the initial deformation temperature of the ash under oxidizing conditions.

Figure 4 shows that coals with low sodium contents can be fired successfully near the initial deformation temperature (oxidizing) without forming difficult deposits, but coals with high sodium contents must be fired at conditions which produce temperatures considerably below the initial deformation temperature (oxidizing) of the ash to avoid deposits.

Only limited confirmation from boilers is available for this simple empirical relationship. Efforts should be directed to determining if such a simple relationship can adequately correlate boiler performance.

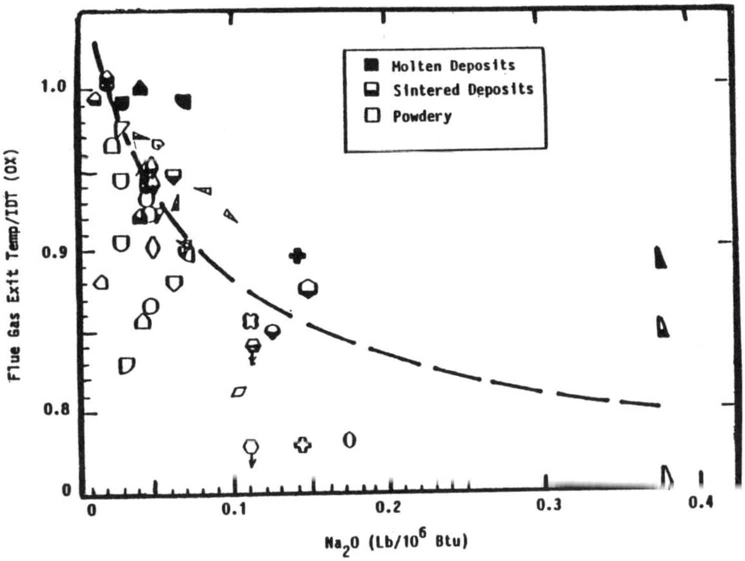

Figure 4. Correlation of fouling.

7.0 EROSION

An adequate empirical relationship of erosion has been available for some time (21). This relationship predicts that the rate of erosion is proportional to the ash loading, velocity raised to the 3.5 power, and a relative erosion index (I).

The relative erosion index must be measured. In the past, no satisfactory means of measuring the relative erosion index was available. Recently, Australian Coal Industry Research Laboratories, Ltd. (ACIRL) in Ipswich, Australia, and Combustion Engineering in Windsor, Connecticut, USA, have measured erosion on irradiated tubes. The results have been encouraging but need to be compared with operating experience.

8.0 ESP PERFORMANCE

The properties of the ash affect the efficiency of the electrostatic precipitator (ESP) and hence the emission of particulate matter. Precise mechanisms of ESP operations remain uncertain and arguments continue about the properties of fly ash which most affect ESP collection efficiency.

The properties most acknowledged to affect ESP performance is the particle size distribution of the fly ash. Figure 5 shows a correlation of calculated ideal efficiency (no sneakage or rapping re-entrainment) with measured particle size distribution for a large variety of coals.

The influence of fly ash resistivity, particle shape, and agglomerating properties are less certain. Operating the ESP at a voltage gradient so high as to cause back corna will reduce the collection efficiency of the ESP and higher voltage gradients will be present in ash layers with high resistivity. However, short of breakdown, the fly ash resistivity may have a small influence on collection efficiency than previously thought.

Two experimental techniques have been developed to estimate ESP collection efficiency. In the first, the particle size distribution, in-situ fly ash resistivity, voltage-amperage curves, and breakdown voltage are measured in a pilot-scale combustor. These are used to calculate the collection efficiency of the ESP using a model developed by Southern Research Institute in the United States. This model uses empirical factors for sneakage and rapping re-enhancement derived from measurements on operating ESP's in the United States. Figure 6 shows that the collection efficiencies calculated in this manner agree well with limited measurements on operating boilers.

The second method uses pilot scale precipitators to determine the voltage-amperage curves, and the collection efficiency versus specific collection area. The results are plotted on specially modified Deutch Equation paper. The result is straight, parallel lines of constant drift velocity on a plot of collection efficiency versus specific collection area. ACIRL in Ipswich, Australia has successfully compared measurements of this type with the collection efficiency measured in operating ESP's

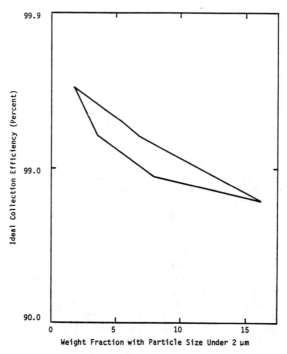

Figure 5.   Influence of fly ash particle distribution
on collection efficiency.

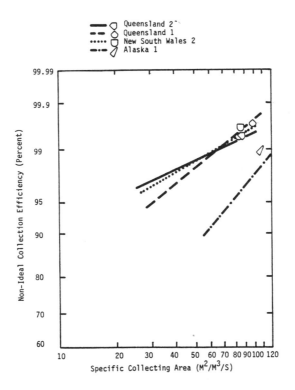

Figure 6.   Comparison of calculated and measured
efficiencies.

9.0 EMISSIONS
    Mineral matter also contributes to emission of pollutants other than
particulate matter. Volatile metals (in some cases reduced) are vaporized
in the flame zone (22). Many of these elements are toxic and condense on
or to form fine respirable particles.
    Experimental techniques have been developed to measure the rate of
evaporation of elements (22). Modeling techniques have been developed at
the University of Arizona which can predict condensation and coagulation of
the aerosol vapors.
    Mineral matter both emits and captures sulfur oxides. Sulfur from
pyrite in the mineral matter is released and oxidized to sulfur oxides.
Some of this sulfur oxide is captured in the alkalic metals of the fly
ash. The simple correlation shown in Figure 7 has been developed based on
power plant and pilot scale data. This relation predicts the capture
efficiency of fly ashes based on the Ca/S ratio. It should be noted, that

the correlation can be violated by drastically changing the time-temperature history of the process. This was done for the pilot scale data which falls off the correlation.

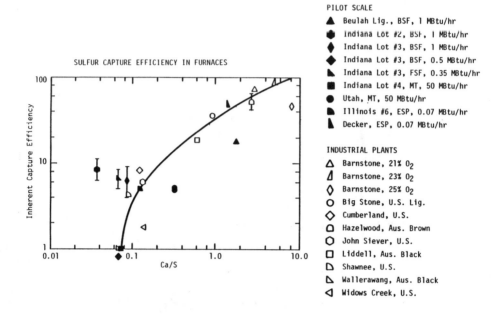

SULFUR CAPTURE EFFICIENCY IN FURNACES

PILOT SCALE

▲ Beulah Lig., BSF, 1 MBtu/hr
⬤ Indiana Lot #2, BSF, 1 MBtu/hr
◆ Indiana Lot #3, BSF, 1 MBtu/hr
◆ Indiana Lot #3, BSF, 0.5 MBtu/hr
◣ Indiana Lot #3, FSF, 0.35 MBtu/hr
■ Indiana Lot #4, MT, 50 MBtu/hr
● Utah, MT, 50 MBtu/hr
◣ Illinois #6, ESP, 0.07 MBtu/hr
◣ Decker, ESP, 0.07 MBtu/hr

INDUSTRIAL PLANTS

△ Barnstone, 21% O_2
◿ Barnstone, 23% O_2
◇ Barnstone, 25% O_2
○ Big Stone, U.S. Lig.
◇ Cumberland, U.S.
◖ Hazelwood, Aus. Brown
○ John Siever, U.S.
□ Liddell, Aus. Black
◺ Shawnee, U.S.
◺ Wallerawang, Aus. Black
◁ Widows Creek, U.S.

Figure 7.   Comparison of inherent capture efficiency between power plants and pilot scale furnaces.

10.0   WASTE DISPOSAL

Disposal of fly ash can present problems when certain elements are leached from the ash and enter the water supply. A laboratory technique has been developed to estimate the extent of fly ash leaching. Results from leaching performed on fly ash produced in a pilot-scale combustor are shown in Figure 8. Results show that some elements from the test coals were leached more extensively than common for U.S. power plants and some less. The accuracy of this test is unknown. Leaching of elements from the fly ash produced in pilot-scale combustors must be compared with that from fly ash produced in power plants and both must be compared with element leached into the ground water from disposal of power plant fly ash.

11.0   CONCLUSIONS

A great many empirical indices have been developed to predict the influence of mineral matter and ash on power plant performance. Some have been recently shown to predict plant performance poorly, others have not been adequately tested, and a few appear to be useful to predict plant performance, but need further power plant verification.

297

An urgent need exists to determine to what extent the unverified
indices can predict plant performance. Research is then required to
develop new indices which can predict performance to replace indices that
have failed. The new indices should be based on examination of the problem
in the power plant, formation of mechanistic hypotheses of the problem,
testing of these hypotheses in controlled laboratory and pilot-scale
equipment, and verifying the results with power plant data.

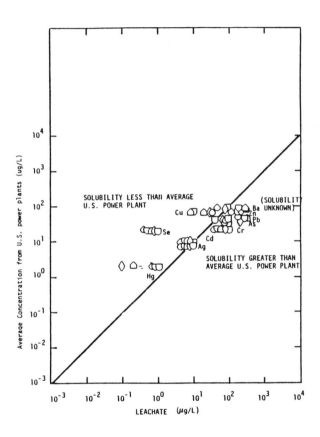

Figure 8.   Comparison of solubility of coal ashes with
            average solubility of U.S. power plant fly ashes.

REFERENCES

1. Heap, M.P., Folsom, B.A., Pohl, J.H.: Effects of Coal Quality on Power Plant Performance and Costs. Volume 1: State-of-the-Art Review, Summary, and Program Plan. Volume 1, EPRI CS-4283, 1986.

2. Goddard, C.W., Duzy, A.F.: Measuring Abrasiveness of Solid Fuels and other Materials by a Radiochemical Method. ASME Paper No. 67-WA/Fu-1, P. 7, 1967.

3. Douglas, L.S., Ditchburn, R.G.: Radiometric Assessment of Abrasiveness of Coal in Pulverizers. Fuel, pp. 532-534, 1980.

4. Heap, M.P., Pohl, J.H., Folsom, B.A., Mehta, A.K.: The Influence of Mineral Matter on Boiler Performance: A review of Status. Presented at the Third Engineering Foundation Conference on Slagging and Fouling due to Impurities in Combustion Gases. Copper Mountain, Colorado, 1984.

5. Sarofim, A.F., Howard, J.B., Padia, A.S.: The Physical Transformation of the Mineral Matter in Pulverized Coal under Simulated Combustion Conditions. Combustion Science and Technology, Volume 16, pp. 187-204, 1977.

6. Barrett, R.E., Murin, J.M., Mack, G.A., Dimmer, J.P.: Evaluating Slagging and Fouling Parameters. Presented at the Fall Meeting of the American Flame Research Committee, 1983.

7. Raask, E.: Deposit Constituent Phase Separation and Adhesion in Mineral Matter and Ash in Coal, K.S. Vores, ed., ACS Symposium Series 301, p. 303, Washington, D.C. American Chemical Society, 1986.

8. Hottel, H.C. Sarofim, A.F.: Radiative Transfer. New York: McGaw-Hill Book Company, 1967.

9. Edwards, D.K.: Radiation Heat Transfer Notes. Washington, D.C.: Hemisphere Publishing Corporation, 1981.

10. Lee, S.C., Tien, C.L.: Optical Constants of Soot in Hydrocarbon Flames. Eighteenth Symposium (International) on Combustion, pp. 1159-1166. Pittsburgh, Pennsylvania: The Combustion Institute, 1981.

11. McLean, W.J., Pohl, J.H., Hardesty, Dr.: Experimental Observation of the Early Stages of Combustion of Individual Coal Particles. Presented at the International Conference on Coal Science, Diisseldorf, Federal Republic of Germany, September, 1981.

12. Gupta, R.P.: Radiative Transfer due to Fly Ash in Coal-Fired Furnaces, Ph.D. Thesis, The University of Newcastle, Newcastle, Australia, 1983.

13. Sarotim, A.F., Hottel, H.C.: Radiative Transfer in Combustion Chambers: Influence of Alternate Fuels: Proceedings of the Sixth International Heat Transfer Conference, Volume 6, pp. 199-217. Washington, D.C.: Hemisphere Publishing Company, 1978.

14. Becker, H.B.: Spectral Band Emissivities of Ash Deposits and Radiative Heat Transfer in Pulverized-Coal-Fired Furnaces. Ph.D. Thesis. The University of Newcastle, Newcastle, Australia, 1982.

15. Wall, W.F., Lowe, A., Wibberley, L.J., Stewart, I.McC: Mineral Matter in Coal and the Thermal Performance of Large Boilers. Progress in Energy and Combustion Science, Volume 5, pp. 1-29, 1979.

16. Zunbrunner, D.A., Viskanta, R., Incropera, F.: Heat Transfer through Granular at High Temperature, Warme and Stoffubertragung, Volume 18, pp. 221-226, 1984.

17. Pohl, J.H. Heap, M.P., Mehta, A.K.: The Influence of Coal Quality on Boiler Availability due to Fireside Tube Wastage. EPRI Seminar/Workshop: Fossil-Fuel-jFired Boiler Tube Failures and Inspections, Miami, Florida, April, 1983.

18. Borio, R.W., Hensel, R.P., Ulmer, R.C., Grabowski, H.A., Wilson, E.B., Leonard, J.W.: The Control of High-Temperature Fire-Side Corrosion in Utility Coal-Fired Boilers. NTIS PB 183716, 1969.

19. Plumley. A.L.: Predicting and Assessing Tube Metal Wastage in Boilers Fired with Low-Rank Coals. Presented at the Low-Rank Coal Technology Development Workshop, San Antonio, Texas, June, 1981.

20. Pohl, J.H.: Pilot Scale Evaluation of the Influence of Mineral Matter on Boiler Performance. Second Annual Pittsburgh Coal Conference Proceedings, pp. 670-689, Pittsburgh, Pennsylvania, September, 1985.

21. Raask, E.: Tube Erosion by Ash Impact. Wear, Volume 13, pp. 301-315, 1969.

22. Mims, A.C., Neville, M., Quann, R.J., Sarofim, A.F.: Laboratory Studies of Trace Element Transformation During Coal Combustion. Presented at the National 87th AIChE Meeting, Boston, Massachusetts, August, 1979.

300

DISCUSSION

D.R. Hardesty

From your perspective, what fundamental research would you recommend on mineral matter transformation and/or deposition ?

J.H. Pohl

The problem areas of mineral matter in boilers are outlined in the paper. Those areas which need verification are : abrasion indices, fouling index, leachability and disposal of fly ash. Those which require formulation of new hypotheses are : slagging, heat transfer, and corrosion. The heterogeneous chemical and physical nature of slagging deposits needs to be determined. The deposits should be collected as functions of time and compared with deposits obtained by probes and in pilot scale combustors. These deposits need to be collected for a range of slagging and non-slagging coals, cross sectioned and analyzed. This will lead to hypotheses of slagging which can be verified in small scale experiments.

The thermal conductance (k/Δs) of the wall deposits often controls heat transfer in boilers. Measurement and interpretation of the thermal conductance is required. Such measurements can be made on deposits collected from boilers on heat flux probe. The surface temperature, deposit thickness, and heat flux must be measured in the presence of a realistic incident heat flux. The first two of these measurements are difficult. Further, the results need to be interpretated in terms of a porous, semi-transparent, re-radiating surface.

Corrosion indices are also poorly known. Some indication of the extent of corrosion can be obtained by depositing ash on partially protected corrosion samples, placing these samples in a controlled atmosphere for a series of times up to hundreds of hours. The depth and type of corrosion can be determined by comparing the corroded and non-corroded areas. The results could then be correlated with coal composition and boiler design.

J.B. Howard

You presented some fly-ash particle size distributions. Could you comment on how these distributions may relate to the particle size distribution of the coals burned ?

J.H. Pohl

I would like to but can't at this time. I suggest that the mineral matter and particle size distributions of pulverized coal be determined by density separatation, low-temperature ashing, and particle counting. This will yield the particle size distribution of excluded and included mineral matter. The amount of inherent mineral matter can be determined by

leaching. The amount and particle size distributions of the
various forms of mineral matter can then be correlated with
the particle size distribution of fly ash samples taken along
the axis for a number of coal and flame conditions. The
results need to be interpreted using models developped at MIT
and elsewhere for vaporization, condensation, and agglomeration of fly ash.

I.F. Wall

The question was raised regarding the number of ash particles formed from the combustion of each char particle. The
attached figure gives size distributions of ash from narrow
fractions of four coals. Given are size distributions :
- of the coal
- of low-temperature-ash (representing the original mineral
  matter)
- of ash from the combustion of the coal in a drop-tube furnace
- the size distribution of the ash expected if each coal particle generated one ash particle (1:1).

The four coals are of similar rank (VM:30-35% db) with
the following initial deformation and flow temperatures :
Eraring (1400/+1560°C), Bayswater (1250/1300°C), Greta (1400/
+1560°C), Muswellbrook (1270/1380°C).

Clearly, the fly ash is of a different size distribution
than that from the 1:1 model; with more fines and more coarse
fly ash particles being generated.

From the experiments the particle size distribution of
flyash was determined predominantly by the size of the p.f.,
the distribution of inorganic material between the coal
particles (inherent) and the extraneous minerals, and, the
swelling or fragmentation behaviour of the p.f. The particle
size of the low temperature ash was significant only where the
p.f. contained large amounts of extraneous minerals, for
example, the Bayswater coal. With the exception of the Eraring
coal, combustion temperature had little effect on the size
distribution of the flyashes.

P.f. size appeared to influence flyash size in two ways;
by determining the extent of agglomeration of mineral inclusions, with larger p.f. producing a larger mass mean size of
flyash, and, by affecting the extent of char fragmentation,
with larger p.f. tending to produce more fine ash. Char fragmentation in this case involves simple degradation due to
large size of the char.

Reference : L.J. Wibberley, Institute of Coal Research,
Report 86/1, 1986.

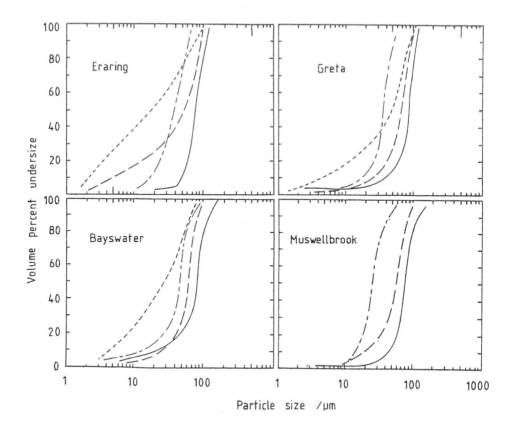

Figure :   Summary of particle size distributions for 63-90 µm
            bituminous p.f., showing the effect of coal type.
            Shown are particle size distributions of p.f., low'
            temperature ash, flyash and that calculated for
            flyash formed from shrinking core combustion.
            Combustion temperature 1580°C.
            ————— p.f – – – LTA —— — flyash ——-—— 1:1

R.H. Essenhigh

      Would you agree that a deposit cannot grow until the
first layer has been formed, in which case the initial deposit
is what matters, and analysis of aged deposits can be misleading.

J.H. Pohl

      I definitely agree. Samples of boiler deposits should be
obtained at different exposure times and analyzed as suggested
in the answer to Dr. Hardesty's question.

H. Jüntgen

In one of your slides you showed reactions between SO_3 and HCl resulting in $FeCl_2$. Can you give more comments on the significance of this or other reactions ? What is the source of SO_3 (decomposition of sulfates on sulfur oxidation) ? Has this reaction influence on emissions of chloride compounds ?

J.H. Pohl

I am sorry but the graph was confusing. Chlorides enter as NaCl, organic Cl and in clays. These chlorides then reacte with H to form HCl which can react to form $FeCl_2$ with iron. Sulfur trioxide usually is derived from the sulfur dioxide. The reactions of $SO_3 \rightarrow SO_2$ (not reverse) appears to be rapid. Results show that the amount of SO_3 retained follows equilibrium at the injection temperature. In a boiler, the SO may be increased toward the equilibrium concentrations by catalysis.

T.F. Wall

Can you please outline the evidence for the relative importance of conductivity (for thick deposits) and emissivity in limiting heat transfer in power station furnaces as well as the values of thermal conductivity for thick deposits one would back out from measured heat transfer and deposit thickness.

J.H. Pohl

Sensitivity studies using the Richter model have shown that uncertainties in the thermal conductance of wall deposits controls may be twice as important as uncertainties in wall emissivity and flame emissivity.

Crudely, thermal conductance $k/\Delta s$ is the important heat transfer component. Measured values of k at 1000 K are about 1 Kw/mK. Realistic deposits are 2-100 mm thick in operating boilers. However, deposit thickness of 1-5 mm usually must be specified to obtain agreement with measurements. This would results in a k of about 4 Kw/mK which may be considerably too high.

K.H. Van Heek

Could you please comment on the situation in the U.S.A. concerning either disposal or utilisation of the ashes and on the needs for relevant research ?

J.H. Pohl

Ash disposal is not yet a problem in the United States. Some fly ash in the United States is economicably sold for low density aggregates (cenospheres) and cement kilns. However, disposal of fly ash in many european and some asian countries is restricted.

STAGED COMBUSTION OF PULVERIZED COAL

H. Kremer [1], R. Mechenbier [1] and W. Schulz [2]

1) Lehrstuhl für Energieanlagentechnik
   Ruhr- Universität Bochum
2) Vereinigte Elektrizitätswerke Westfalen AG
   Dortmund

1. INTRODUCTION

The emissions of nitrogen oxides are much higher with the combustion of fossil fuels containing organic bound nitrogen compounds than with clean fuels like natural gas and light distillate oil. During combustion the fuel-bound nitrogen can be converted partly or totally into nitrogen oxides, these causing the higher emissions.

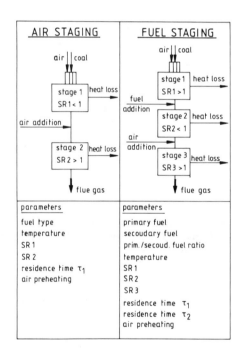

Legislation limiting the specific emission rates of nitrogen oxides has led to intensive efforts to achieve these goals technically by controlled combustion only. Staged combustion has been found the most effective measure to reduce NOx emissions with combustion of fuel nitrogen containing coals and heavy fuel oils /1/. The influence of different combustion parameters on the reduction of NOx emissions from pulverized coal flames by air staging was investigated previously to our own experiments /2,3/ by other authors /4 to 7/. Based on these results and the gained experience the possibilities of fuel staging are at present systematically investigated. This paper deals with available results on air and on fuel staging with coal combustion.

Fig. 1 is showing both staging principles as they may be applied with low NOx burners, assuming that all stages normally are losing a certain portion of energy by radiation. However, it also might be possible that, e.g. in the reducing stage, with air-staged combustion

Figure 1. Principles of staged combustion of pulverized coal.

heat is added from the flame circumference. There are several parameters which are of importance for the final flue gas NOx concentration.

2. FORMATION OF NO FROM FUEL NITROGEN

Coal is containing from 0.5 to 2 % fuel nitrogen which consists mainly from
aromatic compounds like pyridine, pyrole or amines /8/. Fuel nitrogen is
oxidized to NO after having been reduced previously to hydrogen cyanine
(HCN) or ammonia (NH_3) respectively. This statement is supported by many
results from investigations with fuels which were doped with nitrogen additives like pyridine, NH_3 etc. and with fuel nitrogen containing fuels. Fenimore /9,10/ has found that prompt NO in rich hydrocarbon flames is formed
via the same nitrogen compounds.

With coal combustion one has to differentiate among NO from gaseous fuel
nitrogen and NO from char. A part of the NO formed can be reduced to nitrogen at the carbon surface and form molecular nitrogen N_2 after recombination. Fuel nitrogen reactions are much faster than thermal NO formation.

2.1. NO from gaseous fuel nitrogen

NO formation from gaseous fuel nitrogen occurs along a complex reaction
path which up to now can not be described quantitatively. Fuel nitrogen is
converted to fairly simple compounds like amines and cyanides. With growing
temperature during pyrolysis of nitrogen compounds the proportion of HCN is
increasing and becoming the dominating one with very high temperatures. HCN
is reacting further and forming N atoms containing radicals which can react
rapidly with oxygen containing components /11/. However, this finding is not
supported by the results of the authors.

The cyanides are converted to either molecular nitrogen or NO by forming
amines as intermediates /12/. Similar reaction paths seem to exist for
ammonia. If coal particles during combustion are not heated too fast they
devolatilize prior to ignition and lose a great part of the bound nitrogen.
These products are converted during the combustion of the volatiles. The
devolatilization itself is dependent on the temperature and has an effect
upon the formation of fuel NO.

Bound nitrogen is devolatilizing mostly (90 %) along with the tar and high
boiling components at temperatures above 900 K, i.e. in the second range of
coal devolatilization. With low pyrolysis temperatures the nitrogen is remaining mostly in the char, at high temperatures from 70 to 90 % of the
nitrogen can evolve. This is depending on the thermal stability of the
nitrogen compounds. Because of their fairly late escape from the coal particles they are usually not burning in the premixed phase of substoichiometric combustion but later in the diffusion-controlled combustion stages
/13/.

2.2. NO from fuel nitrogen in the char

The conversion of fuel nitrogen in the char is less well understood than
that of the volatiles /11/. It is known, however, that the rate of char
nitrogen conversion to NO is smaller than that of the volatile nitrogen
compounds. The amount of volatiles evolved from the coal particles is depending more on the final temperature than on the heating rate. But if the
latter is extremely high then the devolatilization step can be skipped and
the combustion of volatiles and of char occurs simultaneously /7/.

This heterogeneous conversion is dominated by the chemical reaction of the
nitrogen containing components in the interior and at the outer surface of

the particles. Song et al. /11/ found out that the oxidation of nitrogen and carbon is not selective which is corroborated by the results of
this investigation. During char combustion the devolatilization of nitrogen
is continuing and diminishing the N/C ratio. The formation of NO from char
is very little depending on the combustion process except on the overall
excess air level. Therefore it can hardly be influenced by proper control
of the combustion process.

3. REDUCTION OF NO BY STAGED COMBUSTION

Depending on conditions with pulverized coal combustion, NO formed initially can be reduced later during carbon burnout by means of homogeneous and
heterogeneous reactions, the latter involving unburnt char. Also heterogeneous catalytical surface reactions at ash and other solid particles are
known to be effective. The fairly low NO formation can apparently be explained by indirect NO reduction at available char particles. The heterogeneous reduction of NO at the char surface can be described by the following gross reactions

$$C + 2\ NO \longrightarrow CO_2 + N_2 \tag{1}$$

$$C + NO \longrightarrow CO + 1/2\ N_2 \tag{2}$$

However, this reaction mechanism is of minor importance for the reduction
of gas phase NO with staged combustion.

3.1 Nitrogen oxide reduction with air staging

With air-staged combustion of nitrogen containing fuels homogeneous as well
as heterogeneous catalytic reactions are of importance. The presence of CO
causes a fairly rapid diminution of NO which can be explained by catalytically boosted heterogeneous reactions /14/. These can occur directly among
CO and NO, catalyzed by the ash, or indirectly via chemisorbed oxygen which
converts CO to CO_2. At the free active surface NO is reduced by chemisorp-
tion of an oxygen atom, according to the following reactions

$$CO + NO \longrightarrow CO_2 + 1/2\ N_2 \tag{3}$$

$$CO + C(O) \longrightarrow CO_2 + Cf \tag{4}$$

$$NO + Cf \longrightarrow C(O) + 1/2\ N_2 \tag{5}$$

The decreasing predominance of heterogeneous catalytical reactions with
higher temperatures can be explained by a weaker adsorption of oxygen

$$C(O) \longrightarrow CO \tag{6}$$

The heterogeneous catalytical NO reduction is dependent on the ash content
and on the ash composition, but no information on the decisive components
is available.

Under reducing conditions the fuel nitrogen can be present in the components NO, HCN, NH_3, and is partly contained in the char /11,15,16/. Nitrogen monoxide as well as HCN and NH_3 can be reduced to molecular nitrogen
/8/. The reduction of NO is occuring with sufficiently low stoichiometric
ratios, preferably homogeneously, via the ammonia compound NH_2

$$NH_2 + NO \longrightarrow N_2 + H_2O \tag{7}$$

which is equilibrated with ammonia NH_3

$$NH_3 + OH \longrightarrow NH_2 + H_2O \tag{8}$$

Rising temperatures cause higher NH_2 concentrations and speed up the NO
reduction.

3.2 Nitrogen oxide reduction with fuel staging

A new type of combustion, incorporating fuel staging, was derived from flue gas treatment with reducing fuel components /7/. In fuel-rich flames NO is diminished by fuel radicals /17,18/. As Myerson /18/ has shown, reactions involving CH and CH_2 are responsible for the reduction of NO

$$CH + NO \longrightarrow HCO + N \tag{9}$$

The nitrogen atoms formed can be consumed in fast consecutive reactions

$$N + NO \longrightarrow N_2 + O \tag{10}$$

With the experimental investigations HCN, plotted versus the stoichiometric ratio, showed a maximum value which can be attributed to the following reaction

$$CH + NO \longrightarrow HCN + O \tag{11}$$

Under minimum NO formation conditions only little HCN is formed which in the presence of atomic oxygen is rapidly oxidized to NO. According to Shaub /10/ not the fuel radicals CH and CH_2 are decisive, but the NO reduction is occurring indirectly via higher concentrations of atomic nitrogen following reaction (10). These results also confirm the importance of the reactions (9) and (10) involving the CH radical. After this investigation the reduction of NO by NH_2 and by NH is fairly weak.

Fuel staging, also called "reburning" or "in-furnace NOx reduction", is involving a splitting of the fuel into two or more fuel streams which are feeding two combustion stages being followed by a burnout zone. The second combustion stage can be considered the NO reduction stage. It is of great importance to operate this second stage substoichiometric /19/. The NOx reduction is influenced by stoichiometry, the reduction zone temperature, and the residence time in the first stage (between both fuel additions). It was found that a stoichiometric ratio of 0.9 and a temperature of above 1200 °C lead to minimum NO concentrations. If the reducing fuels are containing bound nitrogen, combustion at too low a stoichiometric value will lead to low NO concentrations but also to increased HCN and NH_3 concentrations which in the final burnout zone, according to Chen et al. /19/, can again cause higher NO concentrations. The reduction of NO can be caused by homogeneous as well as heterogeneous reaction mechanisms. The composition of the fuel and that of the products of conversion, besides thermal effects, have to be considered during combustion. The technical realization of fuel staging requires at least three stages, with the first stage being operated under excess air or under substoichiometric conditions.

The "MACT" process /20/ can be considered as a combination of air and of fuel staging.

4. EXPERIMENTAL INVESTIGATIONS

4.1. Experimental coal combustor

The experimental investigations were performed at a quasi one-dimensional pulverized coal flame. The experimental furnace consists of a vertical ceramic combustion tube of 150 mm internal diamater which is split into two sections. Electrical resistance heating by "Crusilite" heating rods control the temperature of each section in three control groups each. The Combustion tube is provided with opposite holes along the furnace, each couple 100 mm apart, for the introduction of thermocouples and gas composition probes. The thermocouples can be moved across the tube and serve to measure gas and wall temperatures. The primary fuel-air mixture is introduced at the head of the furnace, the secondary fuel and/or air streams between the upper and the lower furnace sections.

The maximum coal throughput of the test furnace is about 2 kg/h. With a total length of 2.3 m the resulting residence time of the combustion

308

gases can be varied between 1 and 3 seconds, comparable to industrial boiler combustion chambers.
With the test series the combustion chamber wall temperatures and the air preheating temperatures were preset. The stoichiometric ratios were changed by adjusting the air and coal dust flow rates. Because of the coal combustion in the primary zone there is a rapid increase of gas temperature near the burner, exceeding usually first the wall temperature. Since no heating takes place in this part of the combustion tube the surplus heat is irradiated to the walls and dissipated towards the colder periphery. Thus gas and wall temperatures are approaching quickly and are about constant in the reducing zone.

4.2 Analytical techniques

The flue gas composition was measured largely by continously operating physical measuring devices (CO_2, CO, O_2, SO_2, and H_2). In earlier investigations water-cooled probes were used: water-quenching with substoichiometric combustion allowed to determine the concentrations of NH_3 and HCN by means of calibrated ion-sensitive electrodes. Now only air-cooled probes are applied. It is planned to replace this time consuming step by the application of infrared analyzers for both components.
The proportion of unburnt hydrocarbons in the flue gases is determined by a flame ionisation detector (FID). All components with high water-solubility and condensable hydrocarbons were measured after passing electrically heated probe lines; the other components were determined after having passed a drying stage. The probe openings along the combustion tube sections serve to obtain axial profiles of temperature and composition.

	Medium Volatile Bituminous		Ballast Coal		High Volatile Bituminous		Anthracite		Lignite	
	Coal	Char	Coal	Char	Coal	Char	Coal	Char	Coal	Char
Proximate Analysis										
Moisture [%]	2,75		0,99		2,13		0,85		21,8	
Ash [% d]	13,6	14,3	23,7	25,2	9,02	13,4	9,70	10,60	6,04	10,3
Volatile Matter [% daf]	34,9		9,6		36,8		9,40		56,2	
Char %	67,9		91,8		65,1		90,70		36,9	
Ultimate Analysis										
C	71,40	80,0	69,2	72,2	73,5	81,3	82,4	84,8	63,3	81,6
H	4,40	0,81	1,91	0,25	4,28	0,89	3,41	0,86	3,04	1,22
N [% d]	1,84	1,26	0,80	0,77	1,49	0,96	1,72	1,22	0,90	0,78
O	8,72	1,80	3,30	0,68	10,77	2,50	1,93	1,37	23,10	5,98
S	0,83	0,64	0,72	0,75	0,73	0,52	0,76	0,73	0,20	0,20
Heating Value [kJ/kg d]	29697	28691	26252	25453	29857	28753	31724	30273	25384	29574

Table I: Analyses of the five coals used in the combustion experiments.

5. RESULTS ON AIR STAGING

The results obtained with 5 different coals (see Table 1) are similar to those obtained by Wendt et al. /21/ who operated an autothermal combustion chamber. At the end of the first, substoichiometric zone the nitrogen oxide concentration has dropped, for example, with a subbituminous imported coal at SR1=0.7 from 950 vppm to 120 vppm, while there is an increase of NH_3 and HCN to 210 and 50 vppm respectively, the mean gas temperature being 1200°C. In the following burnout stage with a stoichiometric ratio of SR2=1.35 the nitrogen oxide concentration rose again to 250 vppm. CO vanished nearly completely.

5.1. Residence time

The residence time in the primary zone exerts a strong influence as was demonstrated earlier /4/. Fig. 2 shows for an ash rich "ballast" coal, with a volatility like anthracite, that beyond a residence time of 0.75 seconds there is no further reduction of final NO in the second stage. This is caused by the char nitrogen fraction which is partly converted to NO in the final burnout zone. With West German brown coal (called lignite in this context) there was no such limitation demonstrating the importance of char reactivity for the carry-over of unburnt char from the reduction stage to the burnout stage.

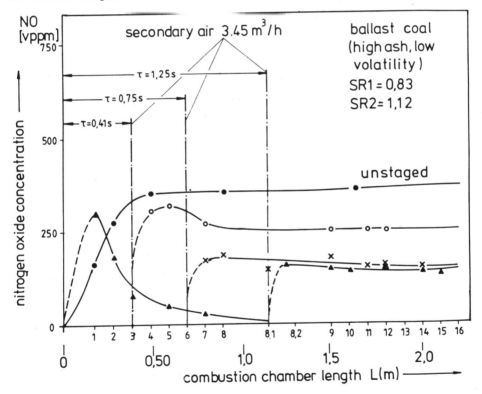

Figure 2. Influence of residence time on the NOx emission from the one-dimensional combustion chamber for an ash rich,low volatile "ballast" coal.

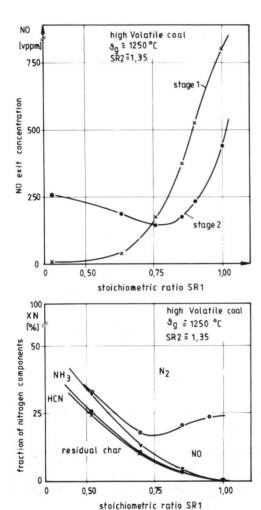

5.2. Primary zone stoichiometric ratio

The effect of the primary zone stoichiometric ratio SR1 on the NO concentration at the end of stage two can be seen from Fig.3 for a high volatile bituminous (imported) coal showing the plotted primary stage NO concentration without secondary air addition and the NO concentration at the end of the second stage. The experiments show also that for substoichiometric combustion the nitrogen content of the char is increasing rapidly, at the same time there is a marked increase mainly of NH_3 concentration, while the HCN concentration is lower for most combustion temperatures. The nitrogen content of the residual char is not selective, meaning that carbon and nitrogen contents are rising with falling primary air in the same manner.

5.3. Temperature

The experiment 1 set-up was operated under nearly isothermal conditions which allowed to assess the influence of temperature in a better way than with conventional autothermal combustion. It was found that the reduction of NO formed at the beginning of the reducing zone was faster with higher temperatures, but paid by higher initial NO concentrations. For low values of SR1 this could be over-compensated by the reduction effect. There is a definite influence of coal quality, as can be seen from Fig. 4, showing the effect of air staging at different wall temperatures (1000 to 1300 °C) for the imported high volatile bituminous coal and the lignite coal. The important finding is that lignite

Figure 3. Influence of primary zone stoichiometric ratio SR1 on NO emission for a high volatile , bituminous (imported) coal.

because of its high char reactivity, is vanishing faster by combustion and gasification reactions than older coals, thus transporting less char nitrogen into the burnout zone in which therefore less fuel NO can be formed.

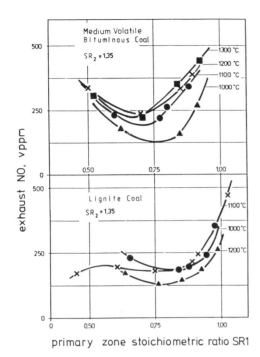

primary zone stoichiometric ratio SR1

Figure 4. Influence of temperature on the overall NOx emission for two coals with air staging.

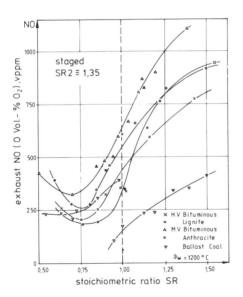

stoichiometric ratio SR

5.4. Coal quality
Also coal quality has a definite effect on NO formation which is demonstrated by Fig.5 /13,16/. It was found with premixed combustion, that the lowest NO concentrations were achieved with lignite, the highest with h.v. bituminous coal. This also has to be compared with the total fuel nitrogen content of the different coals.

6. RESULTS ON FUEL STAGING
6.1. Methane as reducing agent
With pulverized coal combustion the same coal can be used as reducing agent, but also fuels like other coals, coke, oil or fuel gases can be used. In these experiments first methane was used as secondary fuel. It can be expected that CH radicals are of major importance for the reduction efficiency. One of the goals of the experiments was to find out the coal-methane feeding ratios needed to comply with the German legal emission limits for different overall stoichiometric ratios (defined as the fractions of air/fuel ratio divided by its stoichiometric value).
Fig. 6 shows the distributions of concentrations of the main components and of the mean gas temperature along the furnace. This graph also demonstrates that the temperatures along the test furnace can be kept nearly constant, i.e. isothermal conditions can be maintained. The NOx concentration rises in the

Figure 5. Dependence of NOx emission with air staging on coal quality.

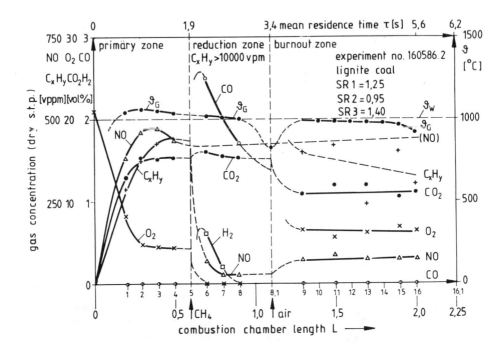

Figure 6. Axial distribution of stable species concentrations in the one-dimensional combustion chamber with Methane as reducing fuel for lignite combustion.

first stage with the stoichiometric ratio SR1=1.25 to a maximum value of 470 vppm and drops to 435 vppm at approaching near-stoichiometric conditions. Methane was introduced downstream via probe opening no. 5 through a multibore tube (at length 0.6 m, residence time t = 1.9 seconds). Thus in the second zone reducing conditions, corresponding to SR2=0.95, were created, causing NO to fall below 25 vppm. Here also the maximum carbon monoxide (2,5%) and hydrogen (0,6%) concentrations were measured.

At probe opening no. 8,1 (1.1 m, t=3,4s) 127 °C tertiary air was added leading to a total stoichiometric ratio of SR3=1.4 in the final burnout zone. In this zone the NOx concentration reached the final value of 75 vppm. It should, however, be pointed out that the hydrocarbon concentration measured at the end of the combustion chamber was about 300 vppm which can be attributed to the fairly low gas temperature (1000 °C) and possibly non-perfect mixing.

The following graph (Fig.7) demonstrates the influence of the secondary stage stoichiometric ratio on the NOx emission of the final burnout stage (SR3=1.4). Even for SR2=1.1 there is a reduction of NOx by two thirds as compared to the non-staged case. The shaded area indicates somehow unstable conditions due to incomplete mixing of the methane.

In Fig. 8 the NOx concentration is plotted versus the stoichiometric ratio of the second (reduction) stage showing that lower overall NOx values

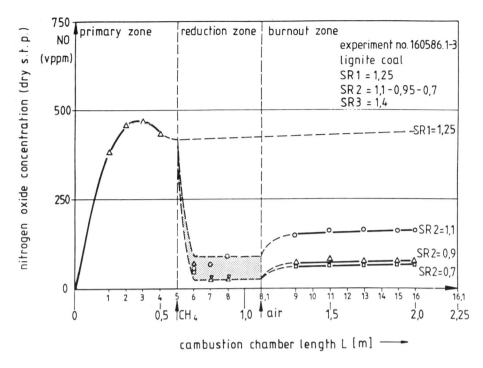

Figure 7. Influence of the secondary stage stoichiometric ratio on NOx emission of the final burnout stage (SR3=1.4) burning lignite with Methane as reducing fuel.

are obtained for higher primary zone stoichiometric ratios. Also plotted in this diagram is the ratio of methane and methane plus coal enthalpies pointing out that, e.g. for a primary zone stoichiometric ratio of SR1=1.1, the methane contribution to the total energy only has to be about 10% in order to keep the NOx concentration below the momentary German legal limit of 200 mg NOx/m³ (6% O_2) for large coal-fired power station boilers. It is expected that higher reduction zone temperatures, whose influence will be investigated in the near future, may lead to even more advantageous results.

6.2. Pulverized coal as a reducing agent

In other experiments fuel staging was investigated with pulverized coal as a reducing agent. Fig.9 contains the results of fuel staging using lignite in the first two zones. It can be seen that for a wall temperature of 1200 °C, SR1=1.31, SR2=0.88, and SR3=1.20 a final NOx concentration of under 200 mg/m³ can be achieved. The NOx reduction in the second stage occurs, as discussed earlier, by fuel radicals and later by ammonia compounds. For lower SR2 values the reduction can be attributed to reactions with NH_2, being equilibrated with NH_3. If SR2 approaches one, CO is the only reducing component available.

The concentration distributions at the start of the second stage show that there is even a diminution of NOx if neither NH_3 nor CO are present. It is supposed to be caused by fuel radicals from the combustion of the volales, emanating in the beginning of the second zone after fuel injection.

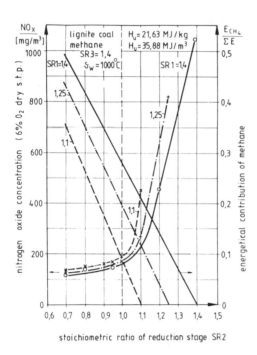

stoichiometric ratio of reduction stage SR2

Figure 8. NOx concentration and specific energy consumption with lignite combustion, using Methane as reducing fuel.

Figure 9. Influence of the secondary stage stoichiometric ratio on NOx emission of the final burnout stage (SR3=1.4) burning lignite with lignite as reducing fuel.

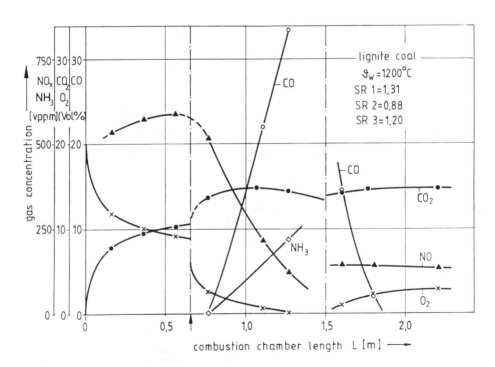

combustion chamber length L [m] ⟶

No detailed mechanisms are known for a quantitative description of these processes involving high molecular weight hydrocarbons. The further course of reduction follows similar lines as with methane as a reducing medium. Immediately after addition of the coal dust there is no oxygen shortage until the volatile matter and part of the char is burnt. Progress of combustion leads to the formation of the reducing components NH_3 and CO. The decrease of nitrogen oxides in the fuel-rich zone shows a development similar to the case of air staging. Using fuel-staging with bound nitrogen containing coal may lead again to the same problems as with air staging. The optimum stoichiometric ratio of the reducing stage is determined by the primary zone excess air and by the kind of fuel used. The primary zone stoichiometric ratio is responsible for the amount of nitrogen oxides, gaseous nitrogen components and residual char nitrogen. The absolute value of the NO concentration is less important in this context. If the stoichiometric ratio of the primary zone is too low then also a low NO concentration will be obtained, besides HCN and NH_3 as additional nitrogen components. In the following reducing step gaseous nitrogen components, except NO, are only converted very slowly. The reducing fuel will increase the concentration of the nitrogen components which in the final burnout zone will be converted again to nitric oxides. On the other hand a high amount of excess air in the primary zone is requiring a large consumption of secondary fuel in the following zone in order to achieve reducing conditions. The bound nitrogen can also contribute to the formation of nitrogen oxides.

Operating the first stage under near-stoichiometric conditions appears to be the most favorable case to use little secondary fuel in the reducing stage. Bound nitrogen compounds should be converted to nitrogen oxides or molecular nitrogen before entering the second stage. If one burns coals with low reactivity a certain amount of residual char containing some bound nitrogen will unavoidably leave the first stage. Fig. 10 shows the influence of four different primary zone stoichiometric ratios on the NO emission rate of the final burnout zone of h.v. bitum, coal as a function of the secondary zone stoichiometric ratio Rising primary zone stoichiometric ratios cause higher NOx emissions. Optimum reducing zone stoichiometric ratios are about 0.85 for h.v. bituminous coal and 0.90 for lignite coal. Also Chen et al. /19/ arrived at the same value with their investigation of fuel staging As with air staging, also with fuel staging there seems to exist a shift of the optimum stoichimetric

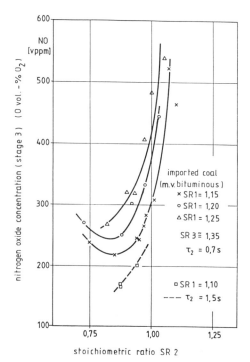

Figure 10. NOx emission with different primary and secondary zone stoichiometric ratios burning high volat. bituminous coal.

316

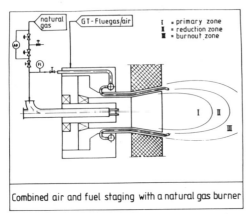

Combined air and fuel staging with a natural gas burner

Figure 11.Realisation of air and of fuel staging with a boiler natural gas burner.

ratio towards higher values with increasing coal reactivity. In addition to this there is an influence of temperature and of residence time in the reducing zone. With fuel staging the resulting NOx emissions are lower, since also bound nitrogen of the residual char of the primary zone can be converted to N_2. At the moment there are systematic investigations under way to study the mechanisms of fuel staging in greater detail.

6.3. Technical application of fuel staging

The findings of fuel staging can generally also be applied to natural gas combustion systems. In the following the remodeling of an existing combined cycle natural gas power plant is shortly discussed. The combustion products of the gas turbine, containing about 15 to 17% oxygen, are used as combustion air for the natural gas boiler.

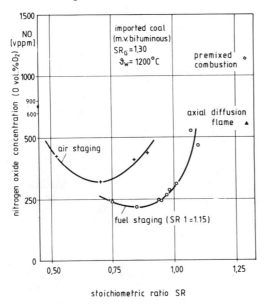

stoichiometric ratio SR

Figure 12.Comparison of air and of fuel staging withethe combustion of a high volatile bituminous coal,showing also results of one-staged premixed and diffusion flame combustion.

As Fig. 11 is demonstrating it has been possible to apply the principle of fuel staging with the type of natural gas burner used by introducing special nozzles to inject additional natural gas approximately at the boundaries of the primary flame, thus creating a reducing zone in which nitrogen oxides from the primary zone are reduced. Another change was made by inserting an air deflector between the primary and secondary circumferential air flows. This leads to some kind of air staging in the primary zone,resulting in lower mean flame temperatures. Because of different flow directions of secondary combustion air and additional fuel (20% of the total amount seems to be preferable) a sufficient amount of air is available to provide the required complete burnout. Since only one burner among others at an existing boiler could be tested only qualitative results have been obtained so far. All other burners are retrofitted in the same way giving hope to obtain satisfactory results from all the burners in the near future.

7. CONCLUSION

Air and fuel staging are both interesting and powerful means to reduce NOx emission from coal-fired power plants and other applications. While air-staging is more established than fuel-staging it obviously is less effective to achieve very low NOx levels, mainly because of the carry-over of char nitrogen form the first to the burnout stage. Fuel-staging with excess air combustion in the first stage can lead to better burnout in the first stage and therefore has a greater overall potential to reduce NOx emission. But also combinations of both methods can be applied.
A comparison of the effect of air and of fuel staging is shown in Fig. 12 for a medium volatile bituminous (imported) coal. The nitrogen concentration, plotted versus the stoichiometric ratio of the reducing zones (primary with air staging, secondary with fuel staging) is based on 0% oxygen. The mean wall temperature was maintained at 1200 °C with these experiments. This figure also contains two single results from premixed and diffusion flame combustion without staging, being obviously a lot higher than the lowest NOx emission rates with staging. In all cases the stoichiometric ratio of the burnout zone was kept at 1.3. A near-stoichiometric operation of the primary zone seems to be favorable to obtain the lowest NOx emissions, because NO emanating from the char burnout can be reduced to molecular nitrogen. Quite opposite to this behavior with air staging at too low primary zone stoichiometric ratios bound nitrogen from residual char is carried over into the final burnout zone and converted to nitrogen oxide there. However, it must be mentioned that char burnout can become a problem with boiler combustion chambers if one approaches stoichiometric composition in the primary zone too closely.

The financial support of this investigation by the Stiftung Volkswagenwerk is gratefully acknowledged by the authors.

REFERENCES

1. Kremer, H., Schulz,W.: Minderung der NOx-Emissionen durch verbrennungstechnische Maßnahmen. VDI-Bericht No. 495,pp.133/142, 1984.
2, Kremer,H., Schulz, W.: Reduzierung der NOx-Emissionen durch Stufenverbrennung. VDI-Bericht No. 574,pp. 413/437, 1985.
3. Schulz, W.: Bildung von Stickstoffoxiden in Kohlenstaubfeuerungen und deren Unterdrückung. VGB-Kongreß "Kraftwerke 1985", Essen.
4. Wendt, J.O.L.: Fundamental Coal Combustion Mechanisms and Pollutant Formation in Furnaces. Progr. Energy Comb. Sci.6, pp. 201/222, 1980.
5. Pohl, J.H., Chen, S.L., Heap, M.P. and Pershing,D.W.: Correlations of NOx Emissions with Basic Physical and Chemical Characteristics of Coal. Proceedings of the 1982 Joint Symposium on Stationary Combustion NOx Control, Palo Alto 1982.
6. Glass, J.W.: Fuel Nitrogen Conversion During Fuel Rich Combustion of Pulverized Coal and Char. Doctorate thesis, University of Arizona, 1981.
7. Kelly, J.T.,Pam, R.L., Suttmann, S.T.: Fuel Staging for Pulverized Coal Furnace NOx control. Proceedings of the 1982 Joint Symposium on Stationary Combustion NOx control.Amer.Flame Res. Comm.Sympos., Newport Beach, California, October 26-28, 1982.

318

8. de Soete, G.: Physikalisch-chemische Mechanismen bei der Stickstoffoxid-
   bildung in industriellen Flammen. gas wärme intern.30,pp.15/23,1981.
9. Fenimore,C.P.: Formation of Nitric Oxide in Premixed Hydrocarbon Flames.
   13th Symposium (Int.) on Combustion, The Combustion Institute, Pittsburgh, pp. 373/380, 1972.
10. Fenimore, C.P.: Reactions of Fuel Nitrogen in Rich Flame Gases. Com-
    bustion and Flame 26,pp.249/256,1976.
11. Song,Y.H., Pohl,J.H., Beér,J.M., Sarofim,A.F.: Nitric Oxide Formation
    During Pulverized Coal Combustion. Comb. Sci. and Technol.28,1982,pp.
    177/183.
12. Okazaki,K.,Shishido, H.,Nishikawa,I. and Ohtake,K.: Separation of the
    Basic Factors Affecting NO Formation in Pulverized Coal Combustion.
    Twentieth Symposium (Int.) on Combustion. The Combustion Institute,
    Pittsburgh, 1984.
13. Schulz,W., Kremer,H.: Bildung von Stickstoffoxiden bei der Verbrennung.
    Brennstoff Wärme Kraft, 37,pp.29/35, 1985.
14. Chan, L.K., Sarofim, A.F., Beér, J.M.: Kinetics of the NO-Carbon
    Reaction at Fluidized Bed Combustor Conditions. Combustion and
    Flame 52,pp. 37/45, 1983.
15. Chen, S.L., Heap,M.P., Pershing,D.W.: Bench-Scale NO Emissions Testing
    of World Coals: Influence of Particle Size and Temperature. Proceedings
    of the 1982 Joint Symposium on Stationary Combustion NOx Control. Amer.
    Flame Res. Comm. Sympos., Newport Beach, Cal.,October 26/28, 1982.
16. Schulz,W.: Experimentelle Untersuchung der Bildung von Stickstoffoxiden
    bei der Kohlenstaubverbrennung. Doctorate thesis Ruhr-Universität
    Bochum 1985.
17. Myerson, A.L.: The Reaction of Nitric Oxide in simulated Combustion
    Effluents by Hydrocarbon-Oxygen Mixtures. 15th Symp.,Int. on Combustion
    The Combustion Institute, Pittsburgh,pp. 1085/1092,1974.
18. Shaub,W.M., Bauer,S.H.: The Reduction of Nitric Oxide During the Com-
    bustion of Hydrocarbons: Methodology for a Rational Mechanism. Com-
    bustion and Flame 32,pp. 35/55, 1978.
19. Chen,S.L., Clark,W.C., Heap,M.P., Pershing,D.W., Seeker, W.R.: NOx
    Reduction by Reburning with Gas and Coal-Bench Scale Studies. Procee-
    dings of the 1982 Joint Sympos. on Stationary Combustion NOx Control.
    Electric Power Research Institute, Palo Alto, July 1983.
20. Takahashi, Y., Sakai,M., Kunimoto, T., Ohme,S., Haneda, H., Kawamura,
    T., Kaneko, S.: Development of the "MACT"In-Furnace NOx Removal Pro-
    cess for Steam Generators. Proceedings of the 1982 Joint Sympos. on
    Stationary Combustion NOx Control.Electric Power Research Institute,
    Palo Alto, July 1983.
21. Wendt, J.O.L., Pershing, D.W., Lee, J.W., Glass, J.W.: Pulverized Coal
    Combustion: NOx-Formation Mechanisms under Fuel Rich and Staged Com-
    bustion Conditions. 17th Symposium (Int.) on Combustion. The Combustion
    Institute, Pittsburgh,pp. 77/87, 1978.

OPTICAL DIAGNOSTICS

OPTICAL DIAGNOSTICS FOR IN-SITU MEASUREMENTS
IN PULVERIZED COAL COMBUSTION ENVIRONMENTS

DONALD R. HARDESTY and DAVID K. OTTESEN

SANDIA NATIONAL LABORATORIES
LIVERMORE, CA 94550

1. INTRODUCTION

Recent progress in understanding pulverized coal combustion is attributable, in part, to the use of new optical diagnostics for in-situ measurements of gas temperature, concentrations of major and trace gas species, particle size distribution, and gas and particle velocity fields. There is great potential for more widespread application of these techniques in the future, particularly in laboratory bench-scale studies. In addition, advanced methods which are now undergoing laboratory development and validation hold promise for in-situ measurement of the composition of both entrained particulates and deposits on material surfaces exposed to coal combustion environments. This paper emphasizes the features of optical diagnostics that have the greatest potential for near-term application to detect the properties of gaseous or particulate species in small-scale pulverized coal (PC) combustion experiments. These experiments contain many of the hostile features of practical combustors (e.g. high temperatures, high particulate loadings, flow turbulence and windows) that limit optical path lengths and signal-to-noise. Extensive references to the earlier literature are included in recent reviews [1,2].

2. DIAGNOSTICS REQUIREMENTS

Perhaps the key ingredient in defining the applicability of any optical technique for PC studies is the degree of temporal and spatial resolution required. All other requirements are derived from the specific parameters to be measured and depend upon the diagnostic system itself. To order this discussion we define two classes of PC combustion flows as shown schematically in Figure 1. Class I includes all wall-bounded, laminar and turbulent duct flows, while Class II refers to mixing, ducted, laminar or turbulent shear flows. Unconfined, premixed or diffusion flames can be considered special cases of Class I or Class II respectively. With the notable exception of suspended or electrodynamically-levitated particle experiments, these represent the two limiting cases of the myriad of possible laboratory- scale PC combustion studies. Clearly some methods are relevant to some applications, but have little advantage over more conventional methods (eg. intrusive probes for thermometry, and gas or solids extraction) in others. Where the measurement needs suggest that an optical technique be applied, consideration of the key features of the two classes of flows in Figure 1 helps to define diagnostics requirements.

2.1. Characteristics of Class I Flows

The important features of Class I flows include: a) well mixed gases; b) temporal steadiness; c) negligible pressure gradients, and; d) plug flow, where reaction zones extend along the space coordinate or are quite small or zero (e.g. in the post flame zone). Mean radial gradients in

322

temperature, velocity, or particulate loadings are either negligible
(distribution curve A) or at most are on the order of the transverse duct
dimension (distribution curve B) due to, for example, a nearby bend in the
duct. In some cases, smaller characteristic dimensions and steeper
gradients arise due the presence of surfaces inserted in the flow.

The range of gas phase parameters for Class I flows is: temperature,
1100 to 2000 K; major species concentrations (in mole percent) from 0.03
(H_2O) to 0.8 (N_2); minor species concentrations from a few to a 100 ppm
for SO_3 and less than 2000 ppm for NO and SO_2. Concentrations of CO due to
incomplete combustion may range from a few hundred ppm to several percent.
In addition, at combustion temperatures volatilization of some mineral
species such as vanadium, sodium and potassium occurs. For typical amounts
of these species in the unreacted coal, the highest concentrations to be
expected are on the order of 25 ppm for sodium and potassium and 1 ppm for
vanadium. Concentrations of these mineral species in the tens to hundreds
of ppb will occur after cooling and condensation.

As noted, an important feature of any coal-derived flow is the high
loading density of entrained particulates including coal, char, free
mineral matter, soot and fly ash. Mass loading densities are typically
weighted toward the large particle sizes. Of special importance for
optical measurements is the wide distribution of both size and particle
number densities. Figure 2 summarizes several recent observations of the
number-size distribution of particulates in a variety of typical steady
combustion environments. Particle velocity is a function of the mass and
size of the particle and may range from a few to a hundred meters/second.
In general particles of different sizes move at different velocities due

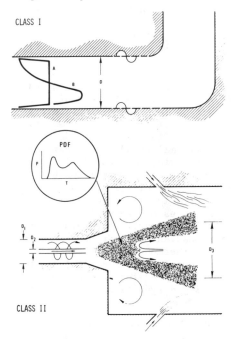

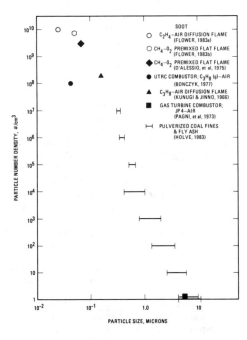

FIGURE 1. Two principal flow regimes for PC
combustion. Class I: Wall-bounded duct flows
Class II: Mixing, ducted shear flows

Figure 2. Typical particle number density and
particle size for soot, PC fines and fly ash
in combustion flows. [citations in Ref. 1]

to slip with the gas stream.

2.2. Diagnostics Considerations for Class I Flows

For these steady flows requirements on temporal resolution are minimal; time-averaged measurements are adequate. An exception is the case where both the mean value and turbulent fluctuations of velocity are required, eg. for heat transfer estimates. Similarly, requirements on spatial resolution are generally not severe; e.g. none for flat (A) profiles or about one-tenth of the duct diameter for a skewed (B) profile. In view of these relaxed requirements on resolution, intrusive isokinetic probes (where sampling velocities match the mean gas flow velocity) may be valid diagnostic candidates. The alternative choice of an optical method would have to be based on other considerations such as the inherent errors and flow disturbances introduced by intrusive probes. For Class I flows both line-of-sight as well as spatially-resolved optical methods are generally applicable. Either may yield the desired information for a flat radial profile (A). Note, however, that most of the advanced optical methods require tight focusing of laser beams either in principle or to overcome natural signal to noise limitations--a spatially resolved measurement is obtained whether it is needed or not.

2.3. Characteristics of Class II Flows

The important features of Class II flows include: a) locally unmixed gases and particulates with potentially steep gradients in all properties; b) locally unsteady flow with substantial turbulent fluctuations; c) finite mean pressure gradients and three dimensional flow both in the mean and in the turbulent fluctuations; d) particle motion may have significant inertial components, and; e) the range of mean values for the concentrations of gaseous and particulate species is from zero to the levels indicated for Class I flows. Local zones of high temperature and high rates of chemical reaction can generate high levels of intermediate species such as the free radicals OH and C_2.

2.4. Diagnostics Considerations for Class II Flows

The unsteady, three-dimensional, "steep gradient" character of such flows generally precludes the application of line-of-sight optical diagnostics. The exceptions to this include visualization techniques which may be of use in "whole flame" observation and line-of-sight indicators of gross flame behavior. The existence of regions of steep gradients and high turbulent intensities imposes severe constraints on diagnostics. High spatial resolution is required to measure local properties (on the order of 1 mm^3 for small geometries and perhaps 1 cm^3 for large flows). As indicated by the schematic of a typical local probability density function (PDF) of, e.g., gas temperature in the reacting shear layer, Class II flows have regions where fluctuations about the local mean value are bimodal; the conventional definition of the mean value may be very misleading. In such regions, high temporal resolution is required in addition to high spatial resolution. Adequate time resolution for Class II flows is likely to be on the order of 50 microseconds. In other words, single pulse measurements with pulse widths less than 50 microseconds are dictated; the PDF must be constructed from an ensemble of such single pulse data. In some cases, such as for a single particle counting diagnostic system, this occurs naturally--the number of counts of particles in each size bin is stored. In other cases, such as in any of the Raman scattering techniques, the diagnostic must be specially configured to acquire single pulse information. In still others, such as in laser Doppler velocimetry an ambiguous jumble of information may be generated due to the difficulty of distinguishing true turbulent fluctuations from variations due to the size dependence of particle velocity. Locally high concentrations of reaction intermediates (such as OH and C_2) can be exploited in some diagnostics to sense local temperature or zones of incipient soot formation. On the other hand, the presence of C_2 and other reaction-generated hydrocarbon species can have strong adverse "noise-inducing" consequences on laser-based diagnostics.

In Class II flows, probe measurements of local mean values of gas temperature, species concentrations and particulate loadings are highly questionable from virtually every technical viewpoint. Without exception, it would be advantageous if a suitable, nonintrusive optical diagnostic having the requisite spatial and temporal resolution, were available for measuring each important local property in Class II flows. As discussed in the following sections, in principle, this is the case. In fact, however, there have been few validated measurements using advanced optical methods in such flows where coal or coal slurry fuels have been used.

3. GAS PHASE TEMPERATURE MEASUREMENT

In this and in the next section promising, principally laser-based, optical methods for measuring the temperature and concentrations of major and minor gas species in Class I or II PC combustion are summarized.

3.1. Raman Scattering Techniques

Temperature measurement in flames using Raman spectroscopy has been an active area of research for more than a decade. For most applications of Raman scattering in combustion, temperatures are calculated from vibrational or rotational population distributions in N_2, which is in high concentration in nearly all regions of air-fed flames. In addition, it has been demonstrated that the rotational and vibrational modes of N_2 are close to equilibrium with the translational mode for the great majority of combustion situations. Raman scattering has the important advantage that the temperature can be calculated from relative intensities of scattering from different transitions.

3.1.1 Spontaneous Raman Scattering

Spontaneous Raman scattering (SRS) is a proven temperature measurement technique, and has been widely applied in clean flames. Incident photons (typically from a fixed frequency laser) are scattered at frequencies which are shifted by the energy difference between two vibration-rotation levels. Temperature measurements are typically performed by ratioing the intensities of the Stokes (lower frequency) and anti-Stokes (higher frequency) signals or by spectrally resolving the Raman signal (either Stokes or anti-Stokes) and fitting theoretical curves to the data. Application of spontaneous Raman scattering to luminous flames has been complicated by the weakness of the Raman signal relative to background emission. The presence of particles in combustion gases introduces severe problems for any SRS method. Incandescence from hot particles can swamp the weak SRS signal. In some cases, background-subtraction techniques, combined with high-power pulsed lasers and gated detection, may circumvent this continuous background. However, excessively high energy laser pulses heat particles above the gas temperature [3], causing additional radiation and fluorescence modulated at the laser frequency, and this cannot be eliminated by gated-detection and background-subtraction techniques. It appears that only one rather unique approach [4] is suited for making time-averaged SRS measurements in luminous particle-laden flows. The system uses a cavity-dumped argon-ion laser of intermediate peak power (50-100 W) and short pulse length (20 ns) combined with a high repetition rate (10^6/s) to obtain moderate average power (0.5-1.0 W). The synchronous detection sampling gate is 25 ns. This system improves average signal-to-

background ratio by a factor of 10 relative to a similar system using a
5-W continuous (cw) laser. Flower [5] validated the feasibility of
time-averaged temperature measurements in highly luminous flows by this
method. Application of the technique is restricted to environments which
are steady over the time period required for the measurement (1 to 10
minutes using an optical multichannel detection system).

3.1.2 Stimulated Raman Scattering Techniques

Stimulated Raman scattering techniques have important advantages over
spontaneous Raman techniques for probing luminous, particle-laden or
turbulent flows. In the most widely developed method, coherent anti-Stokes
Raman scattering (CARS), the Raman scattered signal is emitted coherently,
i.e., it has laser-like directionality. Thus much greater discrimination
against background luminosity is possible. Further, signal levels are
usually sufficiently high that single laser pulse temperature measurements
are possible for both Class I and Class II flows. CARS requires two
lasers--a pump laser and a Stokes laser. A CARS experiment is depicted
schematically in Figure 3. When the difference frequency between the pump
laser (ω_p) and the Stokes laser (ω_s) corresponds to a Raman resonance (ω_s)
of the probed species, a signal at the anti-Stokes frequency $2(\omega_p - \omega_s)$
is generated. The Stokes laser must be either tunable or broadband,
and is typically a dye laser. The response of the molecule to the
pump and Stokes beam can be described in terms of the third-order
resonant susceptibility. The susceptibility can be divided into a
resonant part which has a strong frequency dependence and a nearly
frequency-independent nonresonant component. The resonant and nonresonant
signals interfere and the disappearance of the resonant signal into the
nonresonant background determines the sensitivity limit for CARS. As
described below, there are essentially two approaches, scanned CARS, a
time-consuming method, and broadband CARS, a rapid and more promising
approach for PC combustion studies. Much current research in broadband
CARS is devoted to improving the accuracy and precision of single pulse
temperature measurements. The precision of the technique is presently
limited by frequency noise in the broadband dye laser spectrum.

Accurate modeling of CARS spectra is necessary to extract quantitative
temperature and concentration information. Such modeling requires
molecular linewidth and transition frequency information. The molecular
susceptibility must be convolved with the laser linewidth in order to
compare theoretical and experimental spectra. Temperature is deduced by
least-squares fitting of computer-generated spectra to obtain the best
match to experimental spectra. Such
least-squares fitting is a time-
consuming procedure, and various
fast algorithms have been developed
for single pulse broadband CARS
temperature measurements.

3.1.3 Application of SRS and CARS to Pulverized Coal Combustion

Although background flame
emission is rarely problematical in
CARS measurements, the interaction
of the high intensity laser beams
used for CARS with particles in the
flame causes difficulties. As discussed above, the laser radiation
can heat particles to very high
temperatures causing laser-modulated

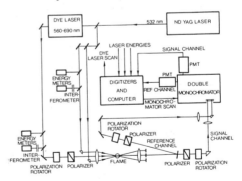

Figure 3. Typical experimental schematic for
scanned CARS diagnostics [7]: 2 channel, in
situ-referenced with background subtraction

particulate incandescence. Other, more subtle effects occur in particle-laden flows, particularly in the case of soot-laden environments. When soot particles are rapidly heated by laser absorption, C_2 concentrations increase considerably in the gas phase surrounding the particles. Interferences can arise in the CARS measurements due to laser-induced C_2 fluorescence. In flows seeded with coal slag, Beiting [6] observed a coherent background signal in the region of the nitrogen spectrum which interferes with CARS thermometry. The source of the background has not yet been fully explained.

Recently, Lucht [7] completed the most systematic investigation to date of the applicability of CARS for gas phase temperature and species concentration measurements in a PC combustion environment. Using a well-controlled, coal-seeded laminar flow reactor, CARS spectra of oxygen and nitrogen were successfully acquired and temperatures were determined from theoretical fits to nitrogen spectra. Significant spectral interferences were observed in the oxygen spectra due to laser-induced particle breakdown at high laser powers. Lucht used the system shown in Figure 3. The pump and Stokes beams were generated by a Molectron Nd:YAG laser and a Quanta-Ray dye laser. The 532 nm, frequency-doubled, 10 Hz output of the Nd:Yag laser serves both as a CARS pump beam and as a pump for the dye laser (bandwidths are 0.1 and 0.2 cm^{-1} respectively). The pump beam is split into two equal intensity beams; all three beams are focused to a common probe volume (4 x 0.2 x 0.2 mm) by a 238 mm focal length lens. In this work the narrowband dye laser frequency was scanned across the Raman resonances of oxygen or nitrogen to generate the CARS signal, which is focused onto the entrance slit of a 1.5 m monochromator that is scanned synchronously with the dye laser. A photomultiplier (PM) tube detects the CARS signal, which is digitized and stored on a PDP 11/24 minicomputer. This time consuming approach (30 minutes to acquire 600 frequency data points and 30 laser shots averaged at each point) has now been eliminated by using a broadband dye laser to obtain the entire CARS spectra in a single shot with an optical multichannel analyzer (OMA) for detection rather than a PM tube; data acquisition is reduced to a few seconds. The measurements were performed in a densely coal-seeded laminar flow reactor (100 x 25 mm) enclosed with quartz windows. The laser beams traversed the 100 mm length, thus maximizing any effects, such as absorption, which arise from propagation through the medium.

Typical results for oxygen and nitrogen spectra are shown in Figure 4. With high pulse energies (26 mJ total pump and 5 mJ Stokes) significant interference (random lines) were observed in the spectrum (Fig 4b) compared to a coal-free spectrum (Fig 4a). These spectral interferences are most likely due to laser-induced breakdown at the surfaces of coal particles at random positions along the laser beam paths. Two approaches, use of a longer focal length focussing lens or use of lower pulse energies, can circumvent this problem. In a large system the sacrifice of spatial resolution with a longer lens may be acceptable. In this work lower pulse energies (8 mJ pump and 1 mJ Stokes) greatly reduced the breakdown, but at the expense of diminishing the CARS signal (Fig. 4c). Lucht successfully implemented a simple conditional sampling system to detect off-axis emission during breakdown and to reject those shots on which breakdown occurs. Time-gating the detection electronics, synchronously with the lasers, discriminates against noise due to particle luminosity. Since CARS signals from nitrogen are 100 times stronger than from oxygen, interferences are much reduced at the same laser energies. Fig. 4d shows a nitrogen CARS spectrum obtained at high pulse energies with a very good theoretical fit superimposed. In any application single-shot broadband CARS will eliminate additional systematic uncertainties in such slow scans due, for example, to unsteadiness in the combustion conditions and variation in window transparencies.

In summary, the applicability of Raman scattering techniques, both SRS and CARS, to PC combustion environments is determined by characteristics of the flow such as: (1) the particle size distribution and number density; (2) the intensity, intermittency and scale of flow turbulence; (3) the size and geometry of the flow, and; (4) the temperature, pressure and composition of the flow. Heavy loadings of luminous particles will be most detrimental to SRS because of the weakness and isotropic nature of the signal; CARS will be most affected by turbulence and combustor size because of beam overlap requirements, and of course by beam attenuation at larger path lengths in particle-laden flows.

There is little doubt that, for a variety of Class II flows in laboratory combustors, CARS has potential for achieving local measurements of the probability density function of temperature from which valid mean and fluctuations in gas temperature can be inferred. The limits of applicability of CARS in such flows have not been thoroughly investigated. Beam attenuation due to heavy particle loadings and beam steering due to turbulence are expected to be the major limitations on CARS applicability in Class II flows. Beam steering (especially for long optical path lengths) will cause the three laser beams to diverge from the required common focal volume. If conditional sampling techniques can be devised to distinguish large beam steering effects from the real temperature fluctuations due to turbulence, the main effect of beam steering will be to decrease the data rate, a relatively minor consequence. Only for very

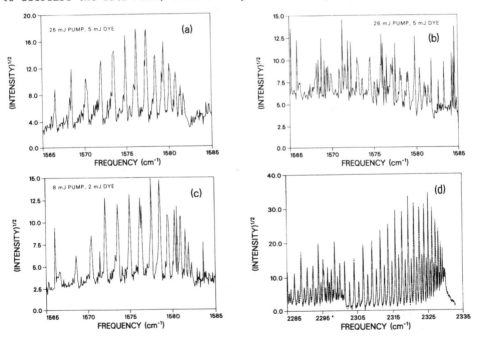

Figure 4. Typical CARS spectra obtained in laminar flow reactor seeded with 50 to 70 micron-sized, bituminous coal particles. a) O_2, clean flow. b) O_2, high laser energy. c) O_2, low laser energy. d) N_2, high energy; solid line is data, dashed is theoretical fit for 1790 K. [7]

clean Class II flows can the same be said for SRS. To date a systematic demonstration and validation of CARS for local temperature measurements in Class II coal-fired flows has not been done; this is, in fact, one principal objective of current research at the Sandia Livermore Combustion Research Facility.

For Class I PC combustion flows the prospect for successful application of CARS is good. For such flows time averaging is acceptable because of the low level of temperature fluctuations. Application of SRS is severely limited by the background luminosity level, but time-gating and signal averaging may help.

3.2 Fluorescence and Absorption Techniques

Laser-induced fluorescence (LIF) and absorption methods (using laser diodes or other infrared sources) have been applied extensively to determine gas phase temperature in relatively clean and in soot-laden combustion environments. Only recently, Fourier transform infrared spectroscopy (FTIR) has emerged as a potentially useful technique for application to bench-scale coal combustion studies (e.g., see Refs. 8-10). Laser fluorescence and absorption methods require many of the same ingredients as the CARS apparatus shown in Figure 3. The laser-pumped tunable dye laser (for LIF) or FTIR source is required, with detection at 90 degrees to the incident beam for fluorescence or along the beam for absorption. In these measurements, the variable frequency light source, is tuned to a real electronic transition of a molecule or atom. The species absorbs laser radiation and is excited to a higher-lying electronic state. Excited state species can decay back to the ground state by spontaneous emission (fluorescence), a process in which a photon is emitted isotropically. The fluorescence or absorption signal is proportional to the excited state population of the molecule or atom. Unfortunately, the excited state population which results from laser excitation depends not only on the rates of laser absorption and spontaneous emission, but also on collisional transfer rates which must be measured or estimated.

Numerous studies in clean flames have been performed in recent years using LIF (including various special derivatives, such as two-line fluorescence) from the hydroxyl (OH) radical to measure temperatures in flames. Laser-excited OH is attractive as a temperature diagnostic because of signal strength considerations; i.e., OH is typically present in high concentrations in flames, and several vibrational bands of the first excited electronic transition lie at the frequency-doubled wavelengths of high efficiency rhodamine laser dyes. In addition, the frequency and radiative transition rates of OH rotational transitions are unusually well-characterized. Fluorescence from vaporized atomic species (assumed to be in equilibrium with the combustion gases) can also be used to infer gas temperature. Five methods have been outlined [12] for measuring flame temperature using atomic fluorescence. Because numerous possible seed atoms (e.g. lithium, sodium, potassium) have electronic transitions in the visible region of the spectrum, where ring dye lasers operate very efficiently, such techniques offer the possibility of temperature measurements at kHz data rates.

In a similar way, temperatures of molecular gas phase species in flames have been determined from absorption spectra. Infrared sources are generally used for absorption measurements because all molecules except homonuclear diatomics have strong absorption bands in this spectral regime. As with LIF, temperatures are determined from the assumed equilibrium Boltzmann population distribution among the vibration-rotation energy levels.

3.2.1 Application of LIF and FTIR to Pulverized Coal Combustion

In general, application of LIF or absorption techniques for gas phase temperature measurement in pulverized coal combustion environments will be feasible in situations where SRS is only marginally applicable. The strength of the LIF signal will depend on the type of species probed. Because transient species such as OH will be used for the LIF temperature measurement in unseeded flows, LIF will be useful as a temperature diagnostic only in relatively high temperature regions. In such regions, where transient species concentrations can be very high (typically 1000 ppm for OH), the strength of the LIF signal is likely to be orders of magnitude higher than the Raman signal. Thus, in situations where temperature and particle loadings are high and SRS cannot be used, LIF temperature measurements may be possible. LIF is an attractive alternative to CARS in some instances due to the less complicated apparatus and data reduction procedures. Single-pulse LIF temperature measurements will require either a fast-scanning dye laser or laser excitation with two or more distinct frequencies for so-called two-line methods. No obvious technical obstacles exist for single-pulse, two-line temperature measurement methods, although no such measurements have been reported in flames to date. Seeding the flow with species such as atomic sodium, will extend the temperature range of applicability of LIF at the cost of added experimental complexity to ensure uniform seeding, and possibly some uncertainty about the effect of the seed species on local reaction processes. In many coal-derived flows sufficient atomic absorbers may already be present. An important consideration for application of LIF in PC combustion studies is the effect of fluorescence trapping, or re-absorption of fluorescence emission as it traverses the medium between the probe volume and collecting lens. In axisymmetric Class I flows, the symmetry (or lack of symmetry) of the fluorescence temperature profile should be an excellent indication of such effects. In Class II flows it may be much more difficult to detect the influence of fluorescence trapping, and serious temperature errors could result.

The feasibility of FTIR absorption thermometry in PC combustion experiments is largely due to the fact that modern infrared sources are much brighter than the infrared emission of the hot gases in the flame. This is especially true for applications in bench-scale experiments. However, even in situations where the infrared emission from the flame is comparable to that of the infrared source, modulation of the infrared source provides discrimination against the infrared emission of the flame so long as the combined infrared intensities do not saturate the infrared detector. Further comments on this approach are included below.

4. GAS PHASE SPECIES CONCENTRATION MEASUREMENT

In this section we comment briefly on the prospects for application of both Raman scattering techniques and laser-induced fluorescence or absorption methods for detection of gas phase species concentrations in PC combustion environments.

4.1. Raman Scattering Techniques

Spontaneous Raman scattering (SRS) is a well-developed technique for measuring concentrations of major species in clean flows, including combustion and turbulent mixing flows. However, SRS has not been applied to measurements in environments laden with coal particles, soot or gas phase hydrocarbons due to interferences from particulate luminosity and broadband fluorescence from the gaseous organic species. As described above, CARS has been applied for gas thermometry in a wide variety of combustion systems, but CARS has not been as widely applied for concentration measurements. Concentrations can be calculated from CARS signals either from the signal intensity or by ratioing the resonant and nonresonant CARS signals. The disadvantage of calculating concentrations from signal intensity is that the CARS signal strength is very sensitive to factors such as beam overlap and the temporal structure of the laser pulse which can vary substantially over the course of an experiment or even from one laser pulse to the next. In principle, more precise and accurate concentration measurements can be obtained by ratioing the resonant CARS signal from the species of interest with the frequency-independent nonresonant background. Such measurements can be performed by using polarization analysis to separate the CARS signals into resonant and nonresonant signals, separately recording the intensities of each signal, and then recording the ratio spectrum. Variations in CARS signal strengths from pulse to pulse are reflected in both the resonant and nonresonant signals, and are approximately cancelled by ratioing the signal. Calculation of species concentrations from the ratios is straightforward provided that the nonresonant susceptibility is accurately known. Alternatively, broadband CARS spectra can be directly analyzed to obtain species concentrations when the resonant and nonresonant signal magnitudes are approximately equal.

There have been very few attempts at obtaining species concentration measurements using CARS in heavily particle-laden combustion and gasification environments. A few preliminary demonstrations have been done for Class I flows [7, 13, 14]. We are not aware of any, even preliminary, validations of CARS for species concentration measurements in coal-fired, Class II flows. While the few results for the Class I flows have been encouraging (ie. a CARS signal was detected and a concentration estimated) from the standpoint of diagnostics proof and validation, much remains to be done even for well-defined high temperature environments. In view of the aforementioned uncertainties in interpreting CARS spectra (including such phenomena as pressure narrowing) considerable fundamental diagnostics research on the method is required.

4.2. Fluorescence and Absorption Techniques
4.2.1 Laser-induced Fluorescence

In recent years considerable progress has been made in defining the limits of accuracy of various laser induced fluorescence (LIF) techniques for detecting species concentrations. Species of interest in PC combustion flows which can be measured using LIF can be divided into three basic categories: (1) Radical species such as O, OH, and NH which are important in reaction kinetics and pollutant formation; (2) Pollutant species such as NO, CO and SO_2, and; (3) Metal atoms such as Na, K and V and compounds such as NaS which are important in corrosion, fouling and slagging.

The use of LIF techniques as a diagnostic for species concentrations in PC combustion has not been actively pursued. One reason for the lack of activity is that the detection of free-radical molecular species, which has been the primary object of fluorescence investigations in combustion media to date, has not been of particular interest to coal combustion scientists. However, recent fluorescence investigations of turbulent diffusion flames have shown that the radical pool is far from equilibrium in certain portions of the flame. Consequently, accurate predictions of processes such as NO formation and CO burnout may require accurate knowledge of radical concentrations. In addition, there is indication that OH may be responsible both for oxidation of coal volatiles as well as for direct attack on the residual char. Therefore, we expect to see more applications in PC combustion but, as noted above, this will be difficult when particle loadings are high.

Quantitative 'fluorescence diagnostics of pollutant species such as NO and CO, while certainly feasible, have not yet been demonstrated even in laboratory flames. Qualitative fluorescence methods may be of use in locating high zones of soot and hydrocarbon formation. Most hydrocarbons, including soot precursors have strong absorption bands in the visible and near ultraviolet, and produce broadband fluorescent emission in the visible portion of the spectrum. Fluorescence techniques for the measurement of Na have been actively pursued for over a decade; in fact, Na has served as a model atom for the development and demonstration of numerous fluorescence techniques. LIF techniques for the measurement of other metal atoms are less well-developed, but it is anticipated that techniques developed for Na will also be applicable to numerous other metal atoms. LIF diagnostics of species such as NaS are handicapped by the lack of a well-developed spectroscopic data base.

4.2.2 FTIR Absorption Techniques

The commercialization and wide spread use of Fourier Transform infrared (FTIR) spectrometers has enabled their use for quantitative gas analysis. This has been aided by increasingly accurate values for infrared line strengths being reported in the literature. Furthermore, virtually all gases (except homonuclear diatomics) have at least one strong infrared absorption band in the mid-infrared region ($400 - 4000$ cm^{-1}) that is easily accessible to most FTIR instruments. With many different species present in the sample, the mid-infrared region may become congested with many absorption lines. However, in most cases the features due to different species can be sorted out if the spectra are taken at sufficiently high resolution (0.5 cm^{-1} or better) since each species has a unique infrared absorption spectrum. Also, the task of sorting out the spectrum is much easier if only low molecular weight species (those having fewer than 5 atoms of atomic weight greater than hydrogen) are present in the sample. Clearly, lower gas concentrations can be measured if higher signal-to-noise (S/N) spectra are obtained. With FTIR, a number of factors influence the maximum attainable S/N. For example, averaging many spectra together improves the S/N by a factor equal to the square root of the number of spectra averaged. The price one pays for improved S/N is the increased time required to acquire the data.

In general FTIR spectroscopy is an effective technique for detection of many different species simultaneously over a wide spectral bandwidth when conditions permit time-averaged measurements along a line-of-sight (e.g. in laminar or turbulent, plug-flow, flames and reactors). Determination of gas phase species concentrations using absorption spectroscopy is based on the application of the Beer-Lambert Law of absorption which relates the intensity of the light transmitted through the sample to the incident light intensity, the transition line strength, the transition line shape function, the concentration of the absorbing species and the path length of the light through the sample.

Recently two investigations, [8-10] have explored the use of FTIR for gas phase species concentration measurements in PC combustion studies. Solomon's work is reviewed elsewhere in this Workshop [11]. Ottesen's recent work at Sandia [8] has been of the nature of technique demonstration and validation. In this work he has extended FTIR measurements into the high resolution domain in order to assess the technique's sensitivity for the determination of the concentration and rotational temperature of various molecules in a PC combustion environment. Infrared absorption measurements were made with uniformly sized coal particles entrained in a laminar flow reactor.

332

Past attempts to make high resolution absorption measurements have been plagued by the infrared emission of hot coal particles passing through the detector field-of-view. Solomon [10] has attempted to make use of these emissions by inverting the order of the combustion reactor-interferometer apparatus, and by extracting the average emittance of the hot particulate ensemble for a given measurement time. These particulate emission signals, however, act as an additional noise source for absorption measurements, and are mitigated only somewhat by laborious time-averaging. This has prevented the measurement of spectra at a resolution greater than 0.5 cm^{-1}. However, this particle generated "noise" occurs at Fourier frequencies lower than the pertinent bandwidth containing infrared spectral information. By electronically filtering the detector output, Ottesen successfully reduced this particulate noise, and obtained the first reported measurements of pulverized coal combustion products at a spectral resolution of 0.08 cm^{-1}.

Although these measurements are more time consuming than those at lower resolution, they allow a more accurate determination of both the molecular rotational temperature and concentration. This is due principally to the lessened interference of nearby absorption lines from other species (mainly water vapor molecules) and higher excited state transitions. These measurements also showed that the measured line widths of CO and CO_2 transitions are considerably greater than the reported values of 0.04 cm^{-1} at temperatures around 1300 K. Measured widths at half maximum are 0.105 cm^{-1}, and it is postulated that the additional width is caused by collisional broadening with the large quantity of water present (15 mol %). A portion of the CO P-branch rotation-vibration band produced during the combustion of a pulverized western Kentucky bituminous coal is shown in Figure 5. The CO rotational temperature determined from the relative intensities of the absorption lines was 1218 +/- 28 K and was in good agreement with a temperature of 1247 +/- 22 K derived from the CO_2 ν_3 R-branch rotation-vibration during the same experimental measurement. Reduction of the noise caused by particulate emission increased the sensitivity for many small infrared-active molecules (CO, CH4, NO, HCN, H_2S) to about 100 ppm for a 10 cm path length at combustion temperatures in the presence in coal particles. Current efforts at detecting minor products of combustion in situ have been hampered by the serious overlap of water absorption lines with N- and S-containing species of interest.

5. PARTICLE SIZE, NUMBER DENSITY AND TEMPERATURE MEASUREMENT

Among the spectrum of diagnostic requirements for examining PC combustion perhaps the most serious is the need for in-situ techniques for real-time detection of the loading and size distribution of entrained particulate matter, including solid coal, char, fly ash, and soot material and liquid droplets or slurry mixtures. In fundamental bench-scale studies of the reactivity of condensed phase fuels, there is need to obtain simultaneous information on the size, temperature and residence time of reacting particles larger

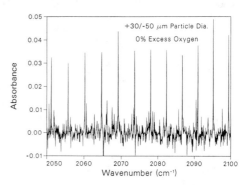

Figure 5. CO P-branch spectrum obtained by FTIR spectroscopy in a coal-seeded laminar flow reactor; rotational T = 1218 K. [8]

than 10 microns. In situ optical sizing techniques offer great promise for
meeting many of these measurement needs. Detailed chemical analysis will
probably continue to require extractive sampling, but a novel and
promising, in situ optical method called laser spark spectroscopy is
discussed in the final section.

Optical methods can be characterized as imaging or light scattering. In
the case of imaging methods individual particles are resolved while in the
latter they are not. As indicated in Figure 6, both methods can be further
subdivided into ensemble or single particle techniques. In all cases the
analysis of particle size relies on the assumption of independent light
scattering which should be valid down to particle separations on the order
of four times the particle diameter and at number densities up to $10^{10}/d^3$
per cm^3 where d is the particle size in microns. Because of the need for
spatial resolution, real time analysis, and mass resolution (mass is
concentrated in the upper end of the size distribution) the emphasis here
is on the features of two of the most successful single particle counting
(SPC) methods; one is a light scattering method, the other an imaging
technique.

5.1. Single Particle Counter (Scattering) Instruments
5.1.1 User Requirements

Hardesty [1] has summarized typical user requirements for application
of SPC instruments to PC combustion environments. For many applications
both number and mass loading densities are of interest. The latter can be
inferred from SPC measurements of the former if the material density of
the particulates is known. Note that the sizing range required for most

studies probably has a lower bound of 0.1 μm, although soot formation and
ash condensation effects would require monitoring down to 0.01 μm.
However, in all likelihood the practical limit of single particle counting
in high number density flows is about 0.1 μm. Minimum sensitivity to
particle refractive index ($m = n_1 + in_2$) and particle shape variations is
desirable in coal-fired flows where particle properties range from
irregularly fractured carbon particles ($n_2 \sim .5$) to fused silica fly ash
($n_2 \sim .001$). In comparison, liquid fuel droplets are in general
transparent ($n_2 \sim 0$) and spherical. If particle properties and shapes are
known and invariant, these restrictions on instrument design become less
stringent.

5.1.2 Design Constraints

While many discussions refer to particle measurements as "particle
sizing," the correct reference is to "particle size distribution"
measurements. Measurements of particle size alone, without accurate
counting of the number in each size class, is insufficient for
characterizing mean diameters (number, area, mass) or their integrated
values. Many discussions have emphasized accurate size characterization

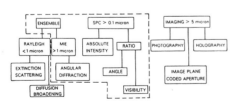

Figure 6. Ensemble and single particle
optical particle measurement techniques

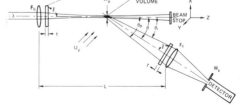

Figure 7. Schematic diagram of a typical SPC
device employing near-forward Mie scattering

with only limited discussion of proper number counting. In practice, accuracy of frequency distribution measurements and derived averages are equally dependent on sizing and number counting accuracy.

The fundamental relationships which constrain the design of SPC diagnostics based on Mie scattering are well known. Of particular concern here are the theoretical results which lead to the variety of in situ SPC diagnostics. First, a monotonic relationship of the scattered light amplitude on the particle size is required, with little dependence on the particle index of refraction or shape. Second, confining all measurements to the forward diffraction lobe for particles larger than about 0.1 µm is the most advantageous. Consistent with these considerations, Figure 7 shows a schematic diagram of a typical SPC instrument with symbols indicating the primary instrument design variables. A laser of wavelength λ and beam f-number f_b is focused through a window of thickness t and angle γ to a beam waist W_o. Scattered light from particles of velocity U_y is collected by a lens of open aperture $(\Theta_i - \Theta_o)$ or f-number F_c at an angle Θ and focused onto a detector slit assembly of width W_s. The available working space between the two lenses is denoted by L. The sample volume from which scattered light is detected is determined by the intersection of the beam focus and the image of the slit. The single particle signatures at the detector are processed by a minicomputer to provide amplitude–frequency distributions.

Given this general configuration, there are two basic methods of relating particle size to a scattering signal. One approach [15] is based on absolute scattering and requires accurate knowledge of the distribution of laser intensity in the sample volume. The second approach [16] uses the ratio of two independent absolute scattering signals from the same particle. Although ratio methods characterize particle size independently of the illumination intensity, the absolute magnitude of each ratio signal still depends on the local illumination intensity and the absolute response function. In Figure 8 a theoretical comparison is shown of the typical absolute scattering response functions for the two classes of SPC methods. In the top half of the figure, the dashed curves refer to the angle ratio method with different scattering angle pairs, and the solid curve characterizes a typical visibility response curve with fringe spacing. In general all such response curves are multivalued, which places undesirable limits on the dynamic range (in particle size) or requires multiple angle ratios to determine the correct particle size. The lower half of Figure 8 shows typical absolute scattering response curves for both absolute and ratio scattering methods.

In our original analyses of both types of scattering methods we considered in detail the basic relationships among the elements of the typical SPC optical system, the particles being measured, and various instrument or hardware aspects. These include: a) the dynamic range; b) the sample volume size; c) the particle number density; d) resolution and accuracy; e) laser intensity in the focal volume; f) windows; g) beam steering, and; h) electronics. The key points are as follows:

a) dynamic range – Enlarging the collection solid angle reduces the dynamic range. The required dynamic range in the signal for a factor of 10 in particle size approaches 10^4.

b) sample volume – All SPC methods require knowledge of the absolute scattering response function and the distribution of laser intensity in the sample volume. Particles of the same size which pass through different regions of the focal volume give rise to different scattering signatures.

c) particle number density – The maximum particle number density in the flow that can be accommodated without having two or more particles in the sample volume simultaneously is given by (N < 4P/V) where P is the probability of having two or more simultaneous scattering events in the sample volume (V). The requirement for a small volume dictates a small beam waist Wo, which produces a nonuniform laser intensity distribution in the sample volume.

d) resolution and accuracy - A 95 % confidence level in the particle count rate requires more than 1000 counts. All SPC instruments require time to acquire sufficient counts in each size bin to meet the accuracy requirement; the time required is a function of the particle loading density and will range from a few seconds to minutes for very lightly loaded flows. Thus although data is being sensed in a "single pulse" mode for each particle, the measurement of a size distribution is inherently time-averaged.

e) illumination intensity and depth of field - There are simple interrelationships among the laser wavelength, the focal length of the focussing lens, the beam waist size, the depth of field, the size of the largest sensible particle and the maximum number density of the smallest particles that can be sensed [1]. In practice, these factors dictate that at least two laser beams are required to measure particles from the submicron to 100 micron range; one tightly focussed to measure the small, higher number density particles and the second with a larger focus to measure the larger particles.

f) windows - Most PC combustion experiments require windows for optical access. Displacement of the focal volume along the incident laser direction is proportional to the thickness and index of refraction of the window material; lateral displacement will occur if the beam is not perpendicular to the window. The incident beam must be monitored to account for beam attenuation due to window fouling.

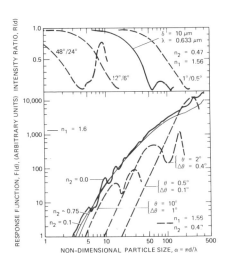

Figure 8. Comparison of response functions [2] for ratio and absolute SPC scattering methods

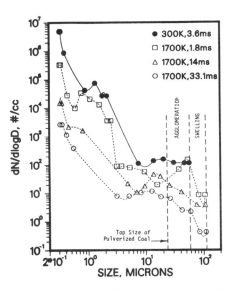

Figure 9. Evolution of a bituminous coal/water slurry particle size distribution as a function of residence time in a laminar flow reactor obtained using the SPC/ID system. [18]

g) beam steering - Beam steering, primarily due to gradients in the index of refraction of the combustion gases which are transverse to the laser and scattered beams, can be appreciable and unsteady but can be compensated for by enlarging the entrance slit of the detector.

i) electronics - Because of the large dynamic range required of the detection and amplification circuit logarithmic circuits or staged amplifiers are required with at least a 10 mHz frequency response.

5.1.3 The Single Particle Counting Intensity Deconvolution Method

The only single particle counting optical method that has been used reliably for measuring particles over the size range of 0.1 to 100 microns with number densities ranging from 10^2 at the large sizes to 10^7 at the small sizes is a dual-beam absolute intensity deconvolution (SPC/ID) method [17]. A key feature of this method is the use of dual beams to achieve the required small sample volume size to accommodate the increased particle loadings at the small end of the particle size distribution. The system uses near forward scatter as in Figure 7 with $\Theta_i < 6°$ to minimize refractive index and particle shape sensitivity, with the choice of inner angle light collection, Θ_i, dependent on the size range of interest. The response functions for $\Theta_i = 0.6°$ are the solid curves in Figure 8.

Due to the variation of the laser beam intensities in the sample volume the scattered signal is particle trajectory dependent. A numerical inversion scheme and calibration procedure allows unfolding of the distribution of signal amplitudes and yields an indicated size distribution which eliminates the dependence on trajectory. The scattered light from particles passing through each of the two sample volumes is collected by a single lens and divided by a beam splitter and focused onto the 'large' and 'small' slits respectively.

Extensive calibration and validation experiments have been conducted by Holve [15]. Figure 9 illustrates recent results obtained for the evolution of particle size distribution in a laminar flow reactor study of the combustion of pulverized coal and coal/water slurries [18].

5.2 Imaging Methods for Single Particle Counting

All imaging methods in principle permit some degree of particle size distribution measurement capability. With reference to Figure 6, there are three basic types of systems: photographic, holographic, and image-plane coded aperture methods. Unlike scattering techniques, all are applicable only to particles larger than the diffraction limit of in situ optical systems--typically 5 to 10 microns. The first two have been extensively reviewed and will not be discussed here. The third method is relatively new. Among the three candidates it is uniquely amenable to straight-forward, real-time, spatially resolved measurements under certain coal combustion conditions. Under the right conditions however (see the discussion below), it appears to be an extremely powerful method, e.g. for basic studies of the reactivity of pulverized coals, with clear advantages over scattering methods.

5.2.1 Image-Plane-Coded-Aperture Single Particle Methods

Several image plane coded aperture systems for application to measurements of both luminous and nonluminous particles during PC combustion have recently been developed at Sandia [19]. These systems have been evaluated in detail for application to flows where the particle loading density of particles larger than about 10 microns does not exceed $10^4/cm^3$. In contrast to the light scattering methods, the image plane methods permit direct and simultaneous measurement of the size and velocity of particles. In addition, if the particles are sufficiently hot, their temperature can also be measured. In contrast to light scattering methods, the image plane technique permits more direct discrimination against background "noise" due to the presence of smaller (unsized) particles in the focal volume. In other words, if it is sufficient for a particular application to measure information only on particles larger than say 10 microns and if these occur at number densities no greater than about $10^4/cm^3$, then the usual higher number densities of smaller particles may not prove limiting since reflected or emitted, not diffracted, light is employed.

5.2.2 The Multiple Slit Method

In this method [19], the signal that contains the particle size information is generated by imaging the particle onto a physical mask containing a grid of three slits and detecting the transmitted light. Particles may be illuminated or, in the case of hot or reacting particles, direct emission from the particles may be used to obtain size and velocity information. In general, deconvolution is required to extract information from any image plane coded aperture system. It is possible, however, to design the aperture so that the particle size can be retrieved using only two values in the output waveform. Such an aperture is shown in Figure 10. It is composed of two slits for encoding the particle size information. A third slit is used in the implementation of a laser trigger to eliminate edge effect errors and to further distinguish and minimize the size of the focal volume. The large aperture is made bigger than the largest particle to be measured; the small aperture smaller than the smallest. As the image of the particle scans across the mask the ratio of the signals from detectors behind the apertures is proportional to the diameter of the particle. In addition to size information, the velocity of the particle can be easily obtained by measuring the particle transit time between the two apertures. Since size and velocity of a particle can be measured simultaneously, velocity-size correlations are readily derived. The method is reasonably insensitive to particle shape. For irregular particles the dimension measured is approximately equal to the particle length in the direction normal to the slits. While the system is insensitive to the particle index of refraction, the particle must be a diffuse surface in order to be sized correctly. This condition is particularly important when coherent illumination is used. In general, white light illumination yields better signals for nonideal particles. It may be possible to size specular particles by back lighting and operating the system in a schlieren mode.

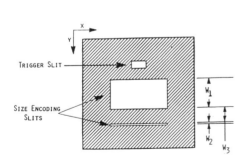

Figure 10. Aperture mask used in the image plane coded aperture (SPC) method. [19]

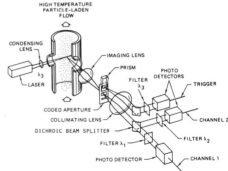

Figure 11. Integrated single particle imaging system for simultaneous particle sizing, temperature and velocity measurement. [19]

5.2.3 Particle Sizing Pyrometry Based on the Multiple Slit Method

While all of the image plane techniques suffer cannot measure particles smaller than 10 microns, they do yield an important bonus capability in situations where the particles are sufficiently hot such that their incandescence can be used to obtain the size information. Under these conditions, by virtue of its simplicity and ease of implementation in bench scale PC combustion environments, the multiple-slit technique is particularly amenable to integration with a multi-wavelength optical pyrometer system. The integrated system, shown in Figure 11, has been developed for the simultaneous detection of particle size, particle velocity, and particle temperature [19]. In addition, by measuring the absolute magnitude of emitted radiation, once the particle size and temperature have been determined its emissivity may be estimated. Analysis of the particle size is done by ratioing the single particle signatures from the large and small slits. The particle temperature follows by ratioing the signals from the large slit at two discrete wavelengths (assuming that the particle emits as a gray body). By the use of reflecting optics and extension of the detection system into the near infrared, the lower limits in terms of particle size and temperature may be significantly extended. Particle velocity is determined by timing the signals from the two slits.

The ability of the method to simultaneously detect the size, temperature, and velocity of reacting particles is proving invaluable in recent studies of PC combustion. The system is now being routinely applied by Mitchell [20] to measure these properties for a variety of coals, chars and other reactive solids (and thereby to directly determine global particle reactivities) in a laboratory flow reactor.

5.3 Ensemble Methods for Analysis of Particulates

Our discussion has emphasized single particle counting methods because: (1) they are generally applicable to Class I and Class II flows; (2) they yield a direct measure of the particle size and number density, and; (3) they provide a measure of the flux of particles through the focal volume (by virtue of the fact that the transit time of each particle is sensed). Ensemble methods are inherently line-of-sight measurements, with one exception--the method of diffusion broadening spectroscopy for local measurement of the total volume of submicron particles in the focal volume of the incident laser beam. As shown in Figure 6, ensemble methods fall into two classes--those that work in the Rayleigh limit of small particles (generally less than a few tenths of microns in practice) and those that work in the Mie regime, generally larger than 2 microns in practice.

5.3.1 Ensemble Methods for Large Particles

For measuring the mean size of particles larger than 1 micron, two commercial instruments are available, the Malvern instrument [21] and Leeds and Northrup instrument [22]. Both devices operate on the principle of detecting the Mie scattering from an ensemble of particles or droplets in a relatively large sample volume defined by the width of the duct and the cross section of an incident expanded laser beam. The use of these instruments and their inherent limitations with regard to determining particle size distributions have been discussed elsewhere [1]. In general the ensemble methods inherently yield less information than single particle methods. In addition to the lack of spatial resolution in the ensemble methods, they provide no measure of velocity or of particle flux. However, if the phenomenon being observed is inherently unsteady an ensemble device must be used since all SPC instruments average over relatively long times.

5.3.2 Ensemble Methods for Measuring Particles < 0.1 Micron in Size

High number densities of fine mineral matter particles, less than 0.1 micron in size, are common in most PC combustion environments (Fig. 2). These arise both from the liberation of solid mineral matter during combustion and as a result of condensation of vaporized inorganic species in the combustion and post flame zones. In addition high number densities of fine soot particles can be expected in regions where temperatures are high and fuel rich conditions predominate. Of the various possible methods for obtaining information on particles less than 0.1 micron in size, only one--laser Doppler broadening spectroscopy (also called diffusion broadening spectroscopy, DBS) permits the kind of spatial resolution required in Class II flows. The DBS technique is unfortunately applicable only to low velocity laminar flows where particle diffusion velocities due to Brownian motion dominate particle transport through the laser focal volume. The method is extremely useful in bench scale measurements in laminar flames and flow reactors [23]. It would appear, however, that there are no near-term candidate methods for obtaining spatially resolved sub-0.1 micron particle size measurements in highly structured Class II flows.

For Class I flows where a flat (A) profile of local particle loading density is ensured, the conventional line-of-sight absorption method may be combined with a scattering technique to infer the total particle volume loading. The method relies on the classical assumptions and results of Rayleigh and Mie scattering and light extinction. In the Rayleigh limit the method applies for particles which are small compared to the wavelength of light (for visible light, strictly for particles smaller than about 0.05 microns). In this regime, it can be shown that the ratio of the intensities of light scattered at 90 degrees (to an incident beam direction) to transmitted light is a strong function of the volume fraction of particles. For larger particles, in the size range of 0.05 to 0.1 microns, Mie scattering can be used to advantage.

6. VELOCITY MEASUREMENTS IN PARTICLE-LADEN FLOWS

As described above in all single particle counting diagnostic methods the velocity of particles passing through the sample volume is measured. The measurement is essentially one of transit-timing. More precisely only the magnitude of the component of the particle velocity projected into the plane perpendicular to the axis of the collection optics is sensed. A true velocity is measured only for well-ordered flows where the axis of the collection optics can be arranged to be perpendicular to the flow direction. The inversion of the image plane aperture method has also been used to directly measure transit times; two laser beams are focussed in close proximity within the combustion zone and the time for a particle to transit the intervening distance is measured.

Clearly transit timing methods are, in principle, applicable to Class I flows. For more complex flows, and in most cases where turbulent intensity information is sought, the method of choice for direct velocity measurement is laser Doppler velocimetry (LDV). The method is not new; fundamental descriptions of the technique are numerous. In most cases, however, the emphasis in previous work is on the measurement of gas velocity and turbulent fluctuations associated with the gases. Small particles of uniform size and spherocity in the 1 micron size range are typically used to minimize slip between the gas and the particles. Here we comment briefly on some of the special considerations which pertain to velocity measurements using LDV in coal-derived flows. In this discussion, basic familiarity with the conventional LDV method is assumed.

6.1 Application of Laser Doppler Velocimetry to Coal-Derived Flows

Despite the ready availability of commercial LDV systems, interpretation of LDV data from particle-laden, coal-derived flows is anything but a simple matter. Care is required. The two features of such flows which have the greatest impact on straightforward interpretation of LDV signatures are: (1) the need to operate with long focal length optics in large coal combustion geometries, and; (2) the presence of high number densities of irregular particles over (typically) a very wide size range. For most bench-scale experiments the latter effect will dominate.

6.1.1 The Effect of Long Focal Length Optics

Large combustion zone dimensions produce long optical path lengths, which require long focal length optics. As the focal length of the incident lens(es) increases, the length of the sample or probe volume increases in the direction of the input laser beams. Consequently, off-axis (as opposed to backscatter) collection is desirable. For example, for a 50 mm beam spacing, with a 120 mm (4.7 inches) focal length lens the probe volume length is 0.35 mm; with a 600 mm (2 feet) focal length lens the probe volume enlarges to 8 mm. This loss of spatial resolution must be considered. While this degree of resolution may be acceptable, the probability of multiple particles in the probe volume poses problems. Another immediate effect is the decrease in efficiency of backscatter due to the decreased solid angle of collection.

Multiple particles of the same size (hence presumably at the same velocity) have the effect of degrading the signal to noise. A higher pedestal in the LDV signature occurs due to the increased scattered light intensity, but less modulation occurs since the scattered light from different particles will be somewhat out of phase. If multiple particles of different velocities (sizes) are present simultaneously the velocity differences will appear as random modulation of the Doppler signal.

The presence of particles in the line of sight between the detector and the probe volume can introduce problems with both signal level and noise. The problems are accentuated with longer focal length systems. Light which is scattered from a large particle, which crosses one of the two laser beams or which passes through the probe volume, can undergo secondary scattering off particles in the (increased) line of sight. Finally, the possibility of beam refraction or steering is increased with the longer lever arms involved. Uncrossing of the incident beams causes signal dropout, and relative motion between the two beams introduces a bias error in both the mean velocity and the turbulence intensity.

6.1.2 The Effect of Different Particle Sizes

There are several additional adverse consequences of increasing particle number densities and of an increasing range of particle sizes in the flow. First, the dynamic range of the photodetector is limited. When the detection threshold is set to follow the weak signal from a small particle, the detector may be saturated by the signal from a large particle. A solution may be to use two detectors, set at different levels to monitor velocities of small and large particles independently. Secondly, the quality of the Doppler signal in terms of the modulation depth will most likely vary greatly. There is, in principle, an optimum ratio of particle size to fringe spacing; the result is that particles at the small and large ends of the size spectrum may be difficult to detect above the increased noise level. Finally, the single largest cause of low signal to noise in an LDV system is flare, usually from surfaces and windows. Large particles which cross the incident laser beams create flare directly or due to scattering off surfaces or other particles.

In summary, the necessity for careful analysis cannot be over-emphasized. Laser Doppler velocimeter systems, whether packaged or home-grown will invariably produce Doppler signals--whether they mean a great deal, especially from a PC combustion flow, is another matter.

7. PARTICULATE COMPOSITION MEASUREMENT BY LASER SPARK SPECTROSCOPY

One of the most challenging areas for coal combustion diagnostics is that of determining the composition of entrained coal, char, and mineral matter particles in the the combustion zone. At present the only means for accomplishing this is to extract a sample by insertion of a generally bulky, water-cooled quench probe and to subject the sample to a battery of off-line analytical techniques. In a new diagnostic research effort, Ottesen [24] is extending earlier work at Sandia Laboratories [25] on a technique called laser spark spectroscopy (also referred to as laser-induced breakdown spectroscopy [26]) to obtain in-situ measurements of the elemental composition of particles in PC combustion environments. Although the technique is about six years old, Ottesen's results are sufficiently encouraging to warrant an extended comment in this review.

A schematic diagram of the experimental apparatus [24] is shown in Figure 12. The measurement sequence for single particles uses two continuous (cw) lasers to provide particle size measurements and to trigger the laser breakdown process. A He-Ne laser beam is focused to a 50 micrometer waist size, and the 90 degree Mie scattering by individual particles passing through this focal volume is used as a trigger pulse. A co-linear, Argon-ion laser beam with a waist size of 250 micrometers is also scattered by particles passing through the focal volume, and the intensity of the near-forward scattered light is used as a measure of mean particle size. These measurements are calibrated using precision pinholes and uniform liquid droplets; the particle sizing method is discussed elsewhere [27]. Incorporating particle size information into the present emission measurements will help to remove the scatter in data caused by differences in particle size, will be useful in observing systematic trends in mineral matter composition during combustion as a function of

particle size, and may also permit an absolute elemental mass measurement for each particle.

Immediately following the measurement of particle size, the He-Ne laser trigger pulse is used to initiate the laser spark sequence. A Q-switched

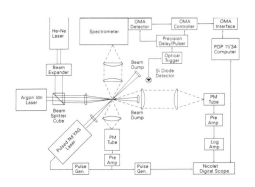

Figure 12. Schematic diagram of Sandia laser spark spectroscopy diagnostic system. [24]

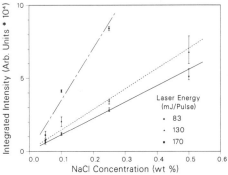

Figure 13. Dependence of Na emission on solution concentration and laser energy after laser breakdown of single droplet. [24]

Nd:YAG laser with maximum pulse energies of 170 mJ and pulse widths of 7 nsec at the frequency doubled wavelength of 5320 A is used to vaporize part or all of the coal particle. This rapid deposition of energy breaks down the molecular structures in the coal particle, and ionizes the resulting atomic species forming a high temperature plasma [26]. The intense emission lines from the plasma are viewed with a 0.5 m Spex monochromator equipped with an optical multichannel anayzer capable of time resolution to 0.1 microsecond following the laser pulse. Ottesen's results confirm the expectation that the emission spectrum is extremely complex immediately following the formation of the plasma [25]. This is due to the presence of highly ionized species and results in a poorly characterized spectrum. By delaying the starting time of the observation, we find that a well defined line spectrum dominated by neutral and singly ionized atoms occurs approximately 2 microseconds after the pulse. The results discussed here were obtained during a 2 to 4 microsecond window following this initial delay.

A central concern during the development of any new diagnostic technique is the issue of calibration. To date a piezo-electric droplet generator has been used to produce uniform diameter liquid droplets which contain a known concentration of material. For example, breakdown spectra of uniform droplets generated from dilute aqueous solutions of NaCl.

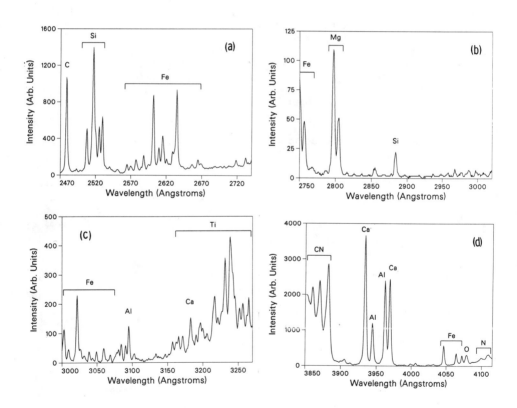

Figure 14. Representative emission spectra following laser breakdown on four single coal particles. [24]

Excellent straight line plots of the sodium 5890 A emission line versus
salt concentration were obtained. The reproducibility of the observed
emission intensities is on the order of +/- 10-20% for salt solutions of
0.05 to 0.5 % by weight, and droplet sizes on the order of 70 micrometers;
this is illustrated in Figure 13. The dependence of emission intensity on
laser energy is not as straightforward. Although the Na emission
monotonically increases with laser energy, no clear functional dependence
is observed. It is speculated that this is mainly due to a complex
increase of plasma temperature with increasing laser energy, and the
incomplete vaporization of the 70 micrometer droplets, even at the maximum
laser energy [26]. This has not been verified.

In addition to such calibration studies, preliminary investigations of
LASS applied to raw coal particles have been completed. Samples of a high
volatile bituminous Kentucky No. 11 coal with an ASTM ash content of 10%
were used in the range of +30/-50 micrometers. A spectrochemical analysis
of a bulk sample yielded the following composition ranges for the observed
mineral species: > 1000 ppm (Al, Fe, Ca, Si), 100 - 1000 ppm (K, Mg, Na,
Ti), < 100 ppm (Mn, Cr). To date, emission lines using LASS have been
recorded for all of these species except Mn and Cr. Representative spectra
for four single particles are shown in Figure 14. Additional species
identified in the spectra include C, N, O, and CN. CN (cyanogen) is
observed as a recombination product of carbon atoms from the organic
matrix with nitrogen (the entrainment gas). Nitrogen and oxygen emissions
are also observed, although quantitative analysis is not possible for
these species since they originate primarily from the entrainment gas and
the ambient laboratory atmosphere.

Detection of Na is best done using the well known emission lines near
5890 A. Figure 15a shows four successive single particle emission spectra.
These spectra are representative of a much larger data set, and were taken
sequentially in approximately 8 seconds total real time. The small
variation in line intensity in Figure 15a is typical of the 80 particles

which were measured. If these differences in observed intensity
are due primarily to variations in particle size, then one inference from
this preliminary data might be that the sodium content is rather evenly
distributed in these unreacted coal particles. A similar study for
potassium content evidenced much larger differences in emission line
intensities (Figure 15b).

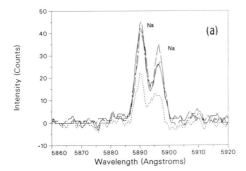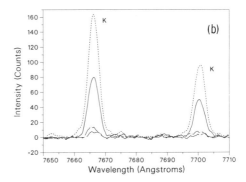

Figure 15. Spark emission spectra obtained after laser breakdown
of four single coal particles. a) Sodium. b) Potassium. [24]

344

Also of great interest is the correlation of different emission line intensities in a given spectroscopic region for a set of single particle spectra. This has been done for two iron lines (2601 A and 2634 A) as shown in Figure 14a, and the results are plotted in Figure 16a for 80 successive particles. To a first approximation this should yield a straight line with a slope equal to the ratio of the line strengths for these two iron transitions. This is clearly the case for these preliminary data, and some of the observed scatter can be accounted for by variations in plasma temperature (caused by differences in particle size and shape, the absorption of laser energy, and self-absorption of emission lines). Since these transitions do not originate from the same energy levels, changes in plasma temperature will affect their observed intensity ratio. The intensity of the stronger iron transition (2601 A) is correlated with the intensity of a carbon transition at 2475 A (as shown in Figure 14a) for the same set of 80 particles. These results are shown in Figure 16b. The large scatter in the data points illustrates the great variability in Fe distribution in the raw coal particles, possibly due to various iron-containing mineral inclusions. A much more consistent set of points, however, is observed for those particles with Fe transitions of less than 200 counts intensity. These points lie near the baseline of Figure 16b and may be due to iron bound in the organic matrix of the coal, which would be much more homogeneously distributed than the mineral inclusions.

Although these early studies have shown several very interesting and exciting qualitative results, much work remains to be done in developing LASS as a diagnostic technique, and in quantifying the limits of its accuracy under a variety of conditions. Current efforts at Sandia Laboratories are being directed to the quantification of these emission spectra in the areas of particle size and plasma temperature measurement and emission intensity calibration.

Finally, note that laser spark spectroscopy may also be used to sample the gas phase in a combustion environment including very small particles which may not produce a measurable Mie scattering trigger signal. While some adjustment in laser energy may be required to compensate for the reduced breakdown threshold of the gas or fine aerosol [26], in principle, one needs merely to conditionally sample the spark emission spectra and to distinguish among those obtained with and without a sensible Mie signal in

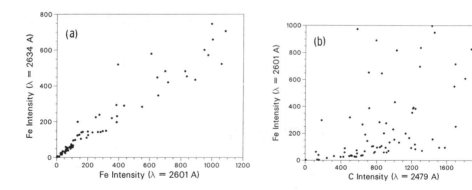

Figure 16. Correlation of spark emission intensities on a particle-by-particle basis in a series of 80 single particles. a) for two Fe emission lines. b) for a single Fe and a single C emission line. [24]

the particle detection channel. This complementary data is potentially
quite valuable in determining the composition of soot, mineral matter
fumes, and condensing particulates in combustion environments.

REFERENCES

1. Hardesty, DR(ed): Assess. of opt. diag. for in situ meas. in high
    temp. coal comb. and conv. flows, Sandia Tech. Rpt. SAND84-8724, 1984.
2. Holve, DJ, et al: Design criteria and recent dev. of opt. SPC,s for
    fossil fuel systems, Optical Engineering, 20, 4, 1981.
3. Flower WL and Miller JA: An analy.of particle-temp modul. induced by
    pulsed-laser sources, Sandia Tech. Rpt. SAND79-8607, Feb. 1979.
4. Mulac, AJ, et al: Appl. Opt. 17, 2695, 1983.
5. Flower, WL: Raman scattering gas temperature measurements in partic.
    laden flows, Sandia Technical Report, SAND81-8608, July 1981.
6. Beiting, EJ: Coher. interfer. in multiplex CARS meas.: nonres. suscep.
    enhance. due to laser breakdown, App. Opt. 24, 18, Sept.1985.
7. Lucht, RP: CARS meas. of temp. and species concen. in coal comb.
    flows, Proc. 1985 Intl. Conf. Coal Science, Sydney, Oct. 1985.
8. Ottesen, DK: In situ studies of PC combustion by FT-IR spect., Proc.
    1985 Intl. Conf. on Coal Science, Sydney, October 1985.
9. Solomon, PR et al: Proceedings of the 19th International Symposium
    on Combustion, The Combustion Inst., p. 1139, 1982.
10. Solomon, PR, et al: Amer. Chem. Soc. Preprints 31, 141, 1986.
11. Solomon, PR: FT-IR Spectroscopy Diagnostic Techniques, this Workshop.
12. Bradshaw, JD, et al: Five laser-excited fluoresc. methods for meas.
    spatial flame temp.: Theor. basis, Appl. Opt., 19, 1980.
13. Taylor, DJ: CARS concentration and temperature measurements in
    coal gasifiers, Los Alamos Nat. Lab. Rpt. LA-UR-83-1840, 1983.
14. Ferrario, A, et al: Real-time CARS spectr. in a semi-industrial
    furnace, pres. at Conf. on Lasers and Electro-Optics, Baltimore 1983.
15. Holve DJ and Self, SA: Optical particle sizing for in situ
    measurements, parts I and II, J. App. Op 18, 10, May 1979.
16. Hirleman, EO: Optical technique for particle characterization in
    combustion; multiple ratio particle counter, PhD Thesis Purdue, 1977.
17. Holve, DJ: An SPC sys. for meas. fine partic. at high numb. dens. in
    resear. and industr. appl., Sandia Tech. Rpt. SAND83-8246, Oct. 1983.
18. Holve, DJ: Compar. comb. studies of coal/water slurries and PC, 7th
    Intl. Symp. Coal Slur. Fuels Prep. and Util., New Orleans, May 1985.
19. Tichenor, DA, et al: Simultan. in situ meas. of partic. size, temp.
    and vel. in comb. envir., Proc. 20th Int. Symp. on Comb., 1984.
20. Mitchell, RE: Experimentally determined overall burning rates of
    coal chars, subm. to Comb. Sc. and Tech., January 1986.
21. Swithenbank, J: A laser diag. tech. for meas. droplet and particle
    size distribution, Proc. AIAA 14th Aerosp. Sc. Mtg., 1976.
22. Wertheimer, AL and Wilcox, WL: Light scattering measurements of
    particle distributions, Applied Optics, 15, p. 1616, 1976.
23. Flower, WL: Measurements of the diffusion coefficient for soot
    particles in flames, Phys. Rev. Let., 51, p. 2287, 1983.
24. Ottesen,DK: private communication of unpublished results.
25. Schmieder, RW: Techniques and applications of laser spark
    spectroscopy, Sandia Tech. Rpt. SAND83-8618, March 1983.
26. Radziemski, LJ et al: Time-resolved laser-induced breakdown
    spectrometry of aerosols, Anal. Chem. 55, p. 1246, 1983.
27. Wang, JCF and Hencken, KR: In situ particle size meas. using a two-
    color laser scat. tech., Sandia Tech. Rpt. SAND85-8869, Dec. 1985.

DISCUSSION

Z. Habib

     Concerning the mineral matter emission lines in the visible range and wich can affect the two color temperature, I agree with Dr Hertzberg that the best way to avoid such problem is to develop multiple-wavelengths pyrometers such as four, six and more. At University of Rouen, we have developped a pyrometer wich allows an instantaneous record of the total visible spectrum i.e. 0.5-1 μm and which allows a very accurate temperature measurements. We have also detected the presence of mineral matter lines like potassium, sodium, lithium wich are easy to avoid, but also some unexpected lines like rubidium at 0.78 and 0.7947 μm which in spite of their weakness introduce an error up to 100 degree in the two color temperature measurement. In our case, the two wavelengths wich give accurate results are 0.65 and 0.9 μm. But one must be careful because the nature of the mineral matter varies from a coal to another, and reliability of the measure is improved with the number of wavelengths.

IN-SITU FT-IR EMISSION/TRANSMISSION DIAGNOSTICS IN COMBUSTION

Peter R. Solomon, Robert M. Carangelo, Philip E. Best,
James R. Markham, David G. Hamblen, and Po Liang Chien

Advanced Fuel Research, Inc., 87 Church Street, East Hartford, CT 06108

INTRODUCTION

Many chemical and energy conversion processes involve multi-phase feed or product streams. Streams can consist of mixtures of solids, liquids and gases and in some cases, the important species may be in transition between phases (e.g. liquid fuel combustion which involves liquid fuel droplets, vaporized fuel and soot). Process monitoring requires the measurement of parameters for the separate phases. There are requirements for monitoring feedstocks and samples of the process stream as well as for in-situ monitoring.

Fourier Transform Infrared (FT-IR) absorption spectroscopy has been used previously as an in-situ diagnostic for both gas species concentration and gas temperature determinations (1-9). Recently, a new method was developed for on-line, in-situ monitoring of particle laden streams to determine chemical composition, concentration, particle size, and temperature of the individual components. The technique uses a FT-IR spectrometer to perform both emission and transmission (E/T) spectroscopy along a line of sight through the reacting medium. The theoretical foundation for the method was considered by Best et al. (10). Applications of E/T spectroscopy were recently reported for: measuring particle temperatures in non-combusting media (10-13); measuring particle emittance (10,14); measuring temperature and concentration of gases and soot in both combusting and non-combusting media (15); and quantitative infrared analysis of gas suspended solids (16).

This paper considers the application of FT-IR E/T spectroscopy to study particles in flames. Examples are presented for a hexane spray flame and coal water fuel (CWF) flames.

The work described in this paper was carried out to demonstrate the feasibility of applying the FT-IR E/T diagnostics to the measurement of composition, particle size, and temperature in moderate size flames. It was not meant to provide a complete analysis of the flames.

EXPERIMENTAL

FT-IR Spectrometer

The apparatus employed in the experiments consists of a FT-IR spectrometer coupled to a reactor, such that the FT-IR beam passes through the sample stream as shown in Fig. 1. Emission measurements are made with the movable mirror in place. Transmission measurements are made with the movable mirror removed. The Fourier transform technique, in contrast to wavelength dispersive methods, processes all wavelengths of a spectrum simultaneously. For this reason it can be used to measure spectral properties of particulate flows, which are notoriously difficult

348

to maintain at a constant level. The emission and transmission can be
measured for the same sample volume. The technique is extremely rapid; a
low noise emission or transmission spectrum at low resolution (4 cm^{-1}) can
be recorded in under a second. Also, radiation passing through the
interferometer is amplitude modulated, and only such radiation is detected.
Because of its unmodulated nature, the particulate emission passing
directly to the detector does not interfere with the measurements of
scattering or transmission.

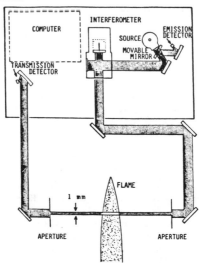

Figure 1. FT-IR Configuration for Measuring Flame Properties.

In an ideal emission-transmission experiment both measurements are
made on the identical sample. In our case the emission and transmission
measurements are made sequentially in time along the same optical path, for
a sample flowing through the cell in a nominal steady condition.

Spray Combustion Facility

A spray facility for testing the combustion characteristics of liquid
fuels has been constructed as shown in Fig. 2. Both hydrocarbon liquid and
coal water sprays have been produced and burned in this system. The spray-
down geometry was chosen so that the high density residues do not re-enter
the flame under the influence of gravity. A down-spray flame into **still**
air is not suitable for study, of course, because such a flame has a
turning point to become an upward-going flame, due to buoyancy forces. We
avoid this transition by embedding the nozzle-flame system in a relatively
high velocity, downward flow of air. The furnace is in an "air-tight"
enclosure, 6' x 16'x 8 1/2' high, which has the floor and rear walls lined
with steel sheet. The ceiling in the vicinity of the flame is lined with
non-flammable Homosote. The high velocity downward flow of air about the
flame results from the location of the opening into the enclosure being
directly above the suction vent.

The FT-IR spectrometer is also located within the furnace room (Fig.
2). This spectrometer is on a movable hoist table, so that its 0.4 cm
diameter beam can be positioned to probe any line-of-sight across the
flame. The IR probe beam is taken from the interferometer to the flame
region by mirrors which are mounted on the end of beams. The spectrometer

used in this work is a Nicolet model 20SX, which employed 4 cm^{-1} resolution. Most spectra were measured with a total of 128 scans, which take 1.5 minutes to record.

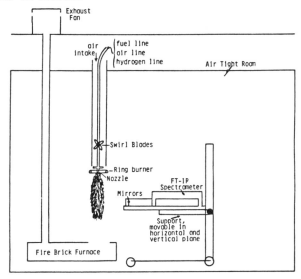

Figure 2.   Schematic of Down Fired Spray Flame Facility.

The facility employed a spray-producing nozzle, which was of the air-assist type, Parker Hannifin model number EDL 6850661ml. Liquid fuel passes through an orifice, which is in a removable plug so that the orifice size can be varied. High pressure air enters through an annular tube surrounding the orifice. This nozzle was modified for the CWF work to narrow the spray cone. All air and fuel supplies, as well as the FT-IR spectrometer controls, reside outside of the furnace room.

The hexane flow through the nozzle was induced by constant pressure over the liquid reservoir. A Robbins-Myers Ramoy #61011 slurry pump was used to pump the CWF at rates between 2 and 5 cm^3 s^{-1}.

ANALYSIS

For multi-phase reacting systems, measurements are made of the transmittance and the radiance, and from these a quantity called the normalized radiance is calculated. The analysis, which follows Siegel and Howell (17), has been presented previously (10). The relevant equations for a homogeneous medium are presented below.

Transmittance, τ_ν

The transmittance, τ_ν, is defined in the usual manner,

$$\tau_\nu = I_\nu / I_{0\nu},$$ (1)

where $I_{0\nu}$ is the intensity transmitted in the absence of sample, while I_ν is that transmitted with the sample stream in place. The geometry for the transmittance measurement is illustrated in Fig. 3. With a particle in the focal volume, energy is taken out of the incident beam by absorption and scattering. The figure illustrates scattering by refraction of energy at the particles surfaces. Scattering will also be caused by reflection and diffraction. The terms are defined in Table I.

TABLE I

NOMENCLATURE

Particle Optical and Other Properties

ε_ν	emittance
F^a_ν	absorption efficiency ($= \varepsilon_\nu$), defined as the absorption cross-section divided by the geometric cross-section.
F^t_ν	total efficiency (extinction), for scattering out of the angular acceptance aperture of our instrument plus absorption.
F^s_ν	scattering efficiency, for scattering out of the angular acceptance aperture of our instrument
$F^{s'}_\nu$	scattering efficiency, for scattering wall radiation into the acceptance aperture of our instrument
N	particle number density
A	geometrical cross-sectional area
D	particle diameter
D_s	Sauter-mean-diameter
α^g_ν	absorption coefficient for gases
α^s_ν	absorption coefficient for soot
f_v	soot volume fraction
b, q	parameters of Rosin-Rammler distribution

Material Optical Constants

m_ν	complex index of refraction ($= n - ik$)
n_ν	real part of the index of refraction
k_ν	imaginary part of the index of refraction

Optical Measurements and Derived Quantities

I_ν , $I_{o\nu}$	detected intensity
τ_ν	transmittance ($= I_\nu / I_{o\nu}$)
R_ν	sample radiance
R^n_ν	Normalized radiance ($= R_\nu/(1 - \tau_\nu)$)
$R^b_\nu(T_w)$	Black-body radiance
W	instrumental response function
S	observed emission spectra from sample (corrected for background)
ν	wavenumber
L	The optical pathlength through the sample stream
T	Temperature

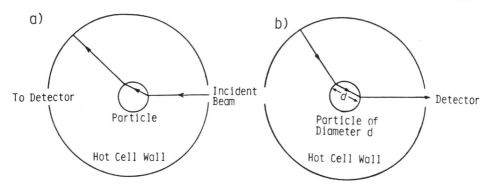

Figure 3.   Measurement Geometry.   a) Geometry of Transmission Measurement
showing the Incident Beam Refracted by the Particle Out of the Beam; b)
Geometry of Radiance Measurement showing Wall Radiance Refracted through a
Particle to the Detector.

For a medium containing gases and soot with absorption coefficients α_ν^g and α_ν^s and particles of spherical total additional area A at a density of N particles cm^{-3},

$$\tau_\nu \;=\; \exp\,(-(\alpha\,{}_\nu^{\,s} + \alpha\,{}_\nu^{\,g} + NAF_\nu^t)\,L) \tag{2}$$

where F_ν^t is the ratio of the total cross-section (extinction) to geometric
cross-section, and L is the path length. τ_ν is sometimes plotted as a
percent.

Radiance, R_ν
To measure the sample radiance R_ν, the radiative power emitted and
scattered by the sample with background subtracted, S_ν, is measured, and
converted to the sample radiance,

$$R_\nu = S_\nu / W_\nu \tag{3}$$

where W_ν is the instrument response function measured using a black-body
cavity.   The radiance measurement detects both radiation emitted, as well
as radiation scattered or refracted by the particles as illustrated in Fig.
3.
R_ν is given by (10,17),

$$R_\nu =$$

$$\frac{\left[\alpha\,{}_\nu^{\,s}R_\nu^b(T_s)+ \alpha\,{}_\nu^{\,g}R_\nu^b(T_g)+NA\,\varepsilon_\nu\,R_\nu^b(T_p)+NAF_\nu^{s'}R_\nu^b(T_w)\right]\left[1-\exp(-(\alpha\,{}_\nu^{\,s}+\alpha\,{}_\nu^{\,g}+NAF_\nu^t)L)\right]}{\alpha\,{}_\nu^{\,s} + \alpha\,{}_\nu^{\,g} + NAF_\nu^t} \tag{4}$$

where $R_\nu^b(T_g)$, $R_\nu^b(T_s)$ and $R_\nu^b(T_p)$ are the black-body emission spectra at the
temperatures of the gas, soot and particle, respectively. ε_ν is the
particle's spectral emittance and $F_\nu^{s'}$ is the particle's cross section for
scattering radiation into the spectrometer.   In deriving this expression we
have included the scattering terms $F_\nu^{s'}$ and F_ν^t for particles but assume no
scattering for soot particles in the IR region.

Normalized Radiance, R_ν^n

The normalized radiance, R_ν^n, is defined as,

$$R_\nu^n = R_\nu / (1 - \tau_\nu).$$ (5)

From Eqs. 2 and 4, the normalized radiance is

$$R_\nu^n = \frac{\alpha_\nu^s R_\nu^b(T_s) + \alpha_\nu^g R_\nu^b(T_g) + NA\epsilon_\nu R_\nu^b(T_p) + NAF_\nu^{s'} R_\nu^b(T_w)}{\alpha_\nu^s + \alpha_\nu^g + NAF_\nu^r}$$ (6)

Examples

Several examples of the method and spectra for components in flames are considered below. We first consider the case of soot. Figure 4 shows $100(1 - \tau_\nu)$, R_ν and R_ν^n for soot produced by the pyrolysis of butane. The transmittance spectrum shows a sloping continuum, typical of soot particles, with the absorption bands for gases, mainly CO_2, acetylene, methane, ethylene, and PAH's, superimposed on the continuum. From Eq. 2, in the absence of particles (N = 0), $\tau_\nu = \exp(-(\alpha_\nu^s + \alpha_\nu^g)L)$, where α_ν^s is proportional to the volume fraction of soot particles.

The radiance spectrum presented in Fig. 4b shows non-gray-body continuum radiation from soot and from the gas bands.

The soot temperature is determined from the measured spectrum, R_ν^n, Fig. 4c. In regions where α_ν^g is zero, $R_\nu^n = R_\nu^b(T_s)$ (from Eq. 6) a result identical to that used by Vervisch and Coppalle (18). A theoretical black-body curve for $T_s = 1280$ K yields the best match to the experimental curve in both spectral shape and amplitude in the regions away from the gas bands. The uncertainty in determining R_ν^n is typically within $\pm 5\%$. At 1280 K this results in ± 30K uncertainty in temperature. Note that R_ν^n matches the black-body curve even in regions of the gas bands, so the gases and soot are at the same temperature.

A second example illustrates the spectra obtained from other particles important in combustion. Figures 5a and 5b illustrate results for fine (less than 20 micron diameter) mineral particles. The spectra show the characteristic absorption bands of the minerals (peaks in the 500-1500 wavenumber region). The spectra also illustrates the reduced attenuation at large wavelengths (i.e. low wavenumbers) characteristic of fine particles. The silica particles (Fig. 5a) which have a very tight size distribution show an increase in attenuation at 3500 wavenumbers (~ 3 microns). This is a resonance effect where the wavelength of the radiation matches the diameter of the particle. In this case, because the silica is very porous, its effective diameter is smaller than the observed diameter of 11 microns. The resonance is not so apparent for the ash (Fig. 5b) because it has a very broad size distribution. Figure 5c shows the results for a sample of coal with large size particles (i.e., ~ 40 micron diameter). The characteristics of such large particles are that they scatter or absorb essentially all of the light incident on them. Any radiation hitting the particle which is not absorbed will be scattered. Consequently, the attenuation is proportional to the area of the particle and no absorption effects can be seen. There is an increase in attenuation at long wavelengths due to diffraction effects.

Figure 6 illustrates the use of theoretical reference spectra to interpret the observed spectra. The reference spectra were generated using Mie theory. The code was taken from Bohren and Huffman (19). Mie theory is a general solution of Maxwell's equations for an isotropic, homogeneous sphere (diameter d) with a complex index of refraction m = n - ik in a medium of different index of refraction. The complex index of refraction

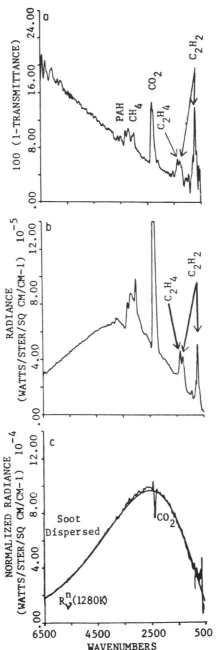

Figure 4. Butane + CO_2 Pyrolyzed in Nitrogen in the Entrained Flow Reactor with 66 cm Reaction Distance. a) $100(1-\tau_\nu)$, b) Radiance, and c) Normalized Radiance. A Suction Pyrometer Temperature of 1295 K was Measured at the Optical Focus.

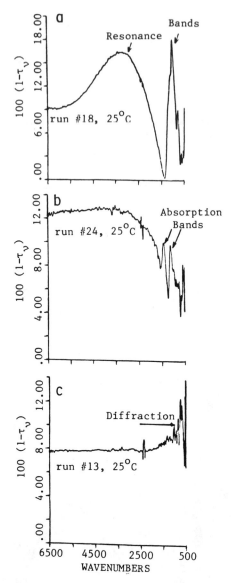

Figure 5.   Infrared Extinction $100(1-\tau_\nu)$ for Particles Exiting the Heated
Tube Reactor at an Asymptotic Tube Temperature of 25°C. a) Silica 1.4 g/min
feed rate),   b) Fly Ash (3.4 g/min feed rate), and   c) Ohio #6 Coal,
(2.0 g/min feed rate).

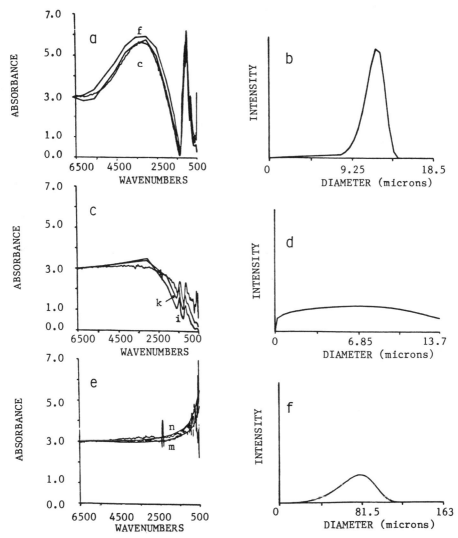

Figure 6. Infrared Extinction Efficiencies for Particles. The Spectra are
Arbitrarily Scaled to give the Same Amplitude at 6500 Wavenumbers. a)
Comparison of Measured and Computed Extinction Efficiency for Silica.
Theory Curve c is for D = 5 microns, Solid Particles; f is for D = 12
microns with a Pore Volume of 65%. b) Particle Size Distribution for
Simulation f. c) Comparison of Measured and Computed Extinction Efficiency
for Ash. Theory Curve i is for D = 3 microns, Solid Particles; k is for D
= 7 microns with a Pore Volume of 57%. d) Particle Size Distribution for
Simulation k. e) Comparison of Measured and Computed Extinction Efficiency
for Coal Particles of Two Size Cuts (200 x 325 and 325 x 400). f) Particle
Size Distribution for simulation m.

contains all of the electromagnetic properties of non-magnectic materials. The complex index of refraction has been determined using a transmission method for the samples of silica, ash and the Ohio #6 coal (for which spectra are shown in Fig. 5) as described in Refs. 12 and 14. The absorbance used in Fig. 6 is defined as $(-\log_{10} \tau)$.

When Mie theory is used to calculate the extinction properties of monodisperse spheres, it is found that surface resonances and other interference effects dominate the spectra. Such features are not observed in our data, and the scattering spectra should be compared with those calculated for scattering by a distribution of particle sizes, as done by Bohren and Huffman (19). That is the approach we have used.

The particle size distribution we have chosen is a two parameter (b and q) distribution, where the number of particles, P, having diameters between D and D + dD is given by

$$P = \frac{6b}{\pi} q \ D^{q-4} \ \exp(-bD^q) \ dD \tag{7}$$

which is the Rosin-Rammler function (20).

Calculated extinction spectra for several characteristic size distributions are shown in Figs. 6a, c, and e. The size distribution parameters (b, q, the diameter of maximum intensity, D (peak), and width ΔD at half the intensity) are presented in Table II. The spectra are normalized to give the same amplitude at 6500 wavenumbers and the value of F^t_{6500} is given in Table II. The observed spectra normalized to the same amplitude at 6500 wavenumbers can then be compared to the reference spectra.

TABLE II

PARTICLE DISTRIBUTION PARAMETERS

	b	q	ΔD	D(peak)	Porosity	F^t_{6500}
Silica						
c	9.36×10^{-11}	14	1.3	5	0	1.96
f	4.45×10^{-16}	14	2.4	12	.58	1.53
Ash						
i	8.23×10^{-4}	5	3.1	3	0	2.64
k	1.35×10^{-5}	4.2	13.5	7	.57	1.95
Coal						
m	2.04×10^{-14}	7	40	80	--	1.12
n	1.53×10^{-13}	7	30	60	--	1.10

Figure 6a shows the results for silica. The reference spectra are for a tight size distribution as measured for this sample. The best fit between the spectrum and the reference spectrum is for an average diameter of 5 microns (spectrum c). This is substantially smaller than the observed diameter of 11 microns. The discrepancy is caused by the material's porosity($\sim$ 65% void fraction). This changes the effective index of refraction. Calculations were made using an index of refraction based on an average between the silica and air (the voids). This average employs the Garnett relation for inhomogeneous materials given in reference (19). The resulting spectrum f calculated for 12 micron diameter spheres

(58% voids) is now in excellent agreement with the observed spectrum.

Figure 6c presents the results for ash. Here a wide distribution was used in agreement with the SEM results. The same problem of void fraction exists here as the powder has a density of 0.75 g/cm^3 which suggests a void fraction of 57%. While the best fit for solid material is for a diameter of 3 microns, the fit for 57% void fraction is 7 microns, in good agreement with the observed distribution.

Results for coal are presented in Fig. 6e. For particles larger than 20 micron diameter, the spectra are reasonably smooth with a slight upward slope going toward low wavenumbers and with a sharper rise approaching 500 wavenumbers. The variation with particle size is not drastic, so the sensitivity of the fitting procedure is much lower than for the small particles. Nevertheless, the observed spectrum for the two size fractions of coal (200 x 325 and 325 x 400) do show the expected difference, and can be used to infer relative size. Figure 6e compares the measured spectra for the two size fractions of Ohio #6. These are expected to have an average difference of 18 microns in diameter. The reference spectra which best fit are for 60 and 80 micron diameters. These do show the right trends but were for a larger average radius than expected.

The determination of temperature or emittance of particles from the normalized radiance is discussed in Refs. 10-14. The interpretation of gas spectra is discussed in Ref. 15.

RESULTS

Hexane

Non-Combusting - The initial work was performed on sprays without combustion to demonstrate the measurement of droplet size distributions. This is done by comparing the transmittance with theoretically determined reference spectra as described above. The shape of the transmittance provides information on the particle size and composition, while the amplitude is proportional to concentration.

The transmittance of the spray for two pressures of atomizing air (34 and 60 psi, respectively), was measured at a number of on-axis position, and at cross-section positions at a depth of 12" below the nozzle.

We start by discussing the on-axis result measured at a pressure of 60 psi, plotted as transmittance in Fig. 7a. Line absorption features due to the hexane can be seen, superimposed on a smooth background (with noise) which increases from 6500 cm^{-1} to 1500 cm^{-1}; and subsequently decreases toward 500 cm^{-1}. By comparing this spectra with calculated spectra it can be shown that the line absorption features in Fig. 7a are due to hexane vapor, rather than liquid. The Mie calculations for each particle size distribution were made with the aid of a program supplied by Bohren and Huffman (19). The best fit theoretical spectrum is overlayed on the measured spectrum in Fig. 7a. In Fig. 7b we show the droplet size distribution which gave rise to the best fit. This distribution has a maximum at D_{peak} = 16 microns, while D_s, the Sauter-mean-diameter, is equal to 21.5 $\pm$ 1.5 microns. If we narrow or broaden the distribution we see poorer fits; while the same width distribution moved to smaller diameter or greater diameter also give a visibly poorer fit. For this size range the extinction measurement is a remarkably sensitive gauge of particle size distribution: i.e. slight differences in the particle size distribution show noticeable differences in the FT-IR spectra. We note that, to our knowledge, this is the first time this measurement method has been applied to the problem of spray droplet size determination.

At lower pressure, where droplets are larger, there is a little more

358

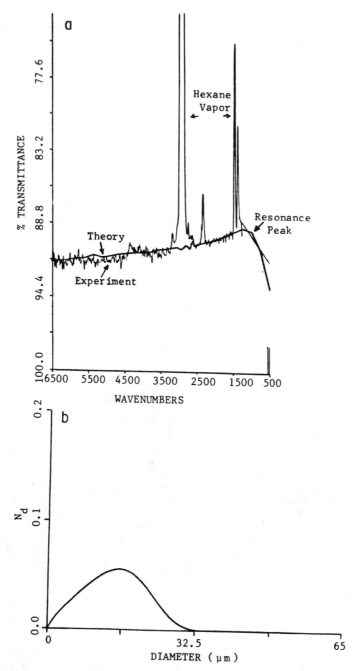

Figure 7. Comparison of Theoretical and Experimental Transmittance
Spectra. a) Spectra, b) Size Distribution used for Calculation.

uncertainty in the measured value of D_s. For this larger sized distribution, the resonance peak in the spectrum, seen in Fig. 7a for example, has moved to smaller wavenumbers, out of the range of our present instrument. The comparison for this case is shown in Fig. 8. The Sauter-mean-diameter, D_s, is estimated to be 39 ± 3 μm. The variation of droplet diameter with atomizing air pressure is as expected for this nozzle.

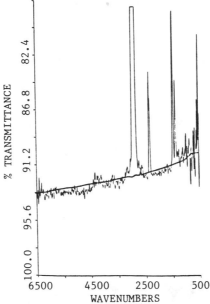

Figure 8. Comparison of Theoretical and Experimental Transmittance Spectra for Larger Droplets ($>$30 μm diameter).

The production of smaller droplets with increasing atomization pressure is visually shown in Figs. 9a and 9c, which show back-lighted sprays at two different atomizing air pressures. The higher pressure clearly produces the smaller droplets.

Combusting - The remaining work for hexane was on spray flames. Both emission and transmission measurements were obtained to determine concentrations of hexane pyrolysis and combustion products and their temperatures. The hexane spray flame was ignited by a hydrogen pilot flame, which was then turned off: the flame is of hexane burning in air. Photographs of the flames (Figs. 9b and 9d) shows them to be about 70 cm in length, with real differences visible due to change in the pressure of atomizing air. The FT-IR E/T diagnostics were applied to these hexane flames.

For an atomizing air pressure of 34 psi, measurements of the combustion species temperatures and concentrations were obtained for 12 axial positions, as well as 18 off-axis positions at depths of 20 and 30 cm. Examples of the species which can be detected by FT-IR spectra are shown in transmittance and normalized radiance spectra presented in Figs. 10 and 11. They include:

- Unburned hexane in both liquid and vapor phases.
- Pyrolyzed hexane as methane, polycyclic aromatic hydrocarbons (PAH), and soot.
- Combustion products, H_2O, CO_2, and CO.

360

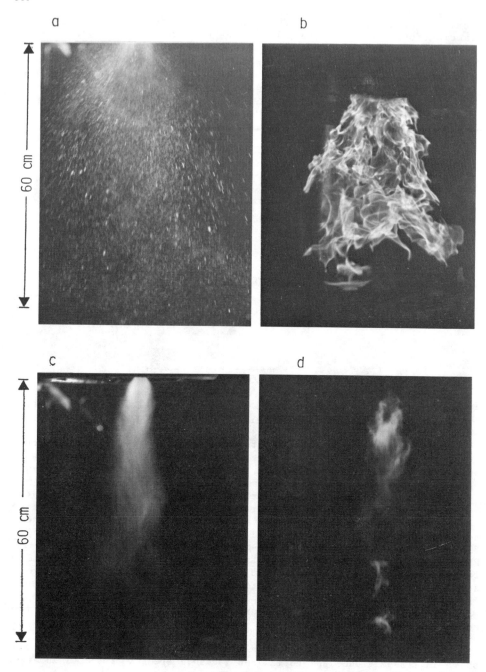

Figure 9. Hexane Spray Pattern and Flame for 34 PSI Atomizing Air Pressure (a and b) and 60 PSI Atomizing Air Pressure (c and d). a and c) Spray – No Flame, b and d) Flame at 1/1000 sec. Shutter Speed.

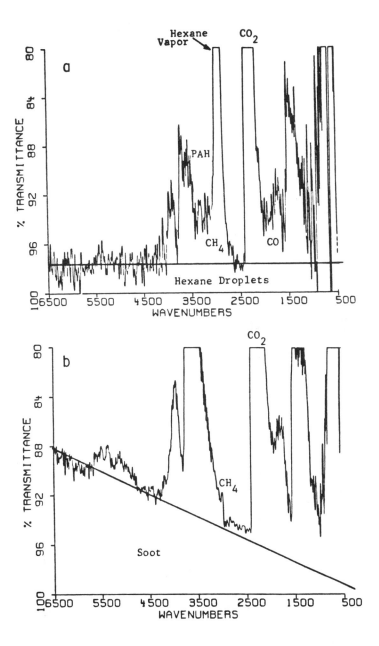

Figure 10. % Transmittance, τ_ν, for Hexane Flame (34 PSI Atomizing Air Pressure, 0.1/min Fuel Flow). a) 3.5 cm Below Nozzle, b) 30 cm Below Nozzle.

362

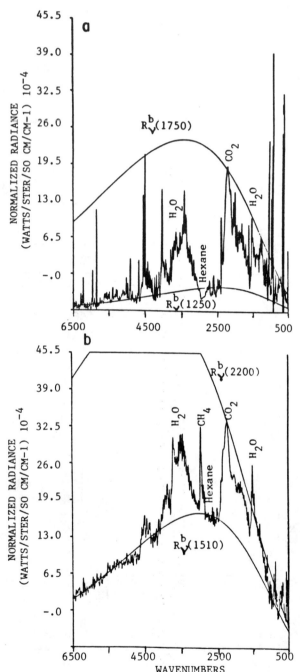

Figure 11. Normalized Radiance, R_ν^n, for Hexane Flame (34 PSl Atomizing Air Pressure, 0.1/min Fuel Flow). a) 3.5 cm Below Nozzle, b) 30 cm Below Nozzle.

It is interesting to note that in this spray flame where the combustion is controlled by mixing, the various components (fuel, soot, CO_2, H_2O, etc.) can be at different temperatures and these can be seen when compared to theoretical black-body curves in the normalized spectra, e.g. the low fuel temperature compared to the CO_2 in Fig. 11a, and the high methane and CO_2 temperatures compared to the lower soot temperature in Fig. 11b. The range of temperatures measured for soot and CO_2 are in agreement with those determined for the diffusion flame discussed in Ref. 15.

Figures 10 and 11 show transmission and normalized emission spectra taken through the axis at depths of 3.5 and 30 cm respectively. The flat, non-zero base line of the transmission spectrum at the depth of 3.5 cm, Fig. 10a, indicates that hexane droplets exist in the flame here. Intermediate spectra between 3.5 and 30 cm show that the flat non-zero base line disappears at 7 cm indicating that there are no liquid droplets left in this flame past 7 cm depth. Gas lines can be readily identified on top of the continuum.

The continuum spectrum, in Fig. 10b, is made up of a small flat part due to particle blockage, and a sloping part that intercepts the flat part at 0 cm^{-1}, due to soot. Here we ignore the spectral slope of the droplet extinction spectra due to diffraction effects. The soot concentration is measured by the amplitude of the triangular "wedge" (Fig. 10). In Fig. 11, the temperature for soot and gas components are obtained by comparison with the black-body reference spectra.

The concentrations of hexane, methane, soot and CO_2 that have been derived from transmittance spectra are summarized in Fig. 12a together with temperature information obtained from normalized emission spectra. A flame photograph for these conditions is presented in Fig. 12b. The sequence of hexane combustion along the flame axis is very clearly shown in these data. Close to the nozzle, the combustion process is just beginning (low temperature), and hexane can exist in both the liquid and vapor phase. As the combustion continues, liquid hexane becomes completely vaporized between the depths of 3.5 and 7 cm, while methane, a pyrolysis product of hexane, is initially detected in this region by the appearance of its characteristic emission line at 3002 cm^{-1}, as a shoulder on the hexane peak. The maximum concentration of methane is found to be at the depth of 15 cm. At this point the hexane concentration has decreased to a rather low value. However, polycyclic aromatic hydrocarbon (PAH), a known soot precursor, identified by the line at 3080 cm^{-1}, was identified in the spectra at this height. These observations imply that in our flame the pyrolysis of hexane is an important step. The PAH emission intensity is seen to increase until a depth of 34 cm is reached. The soot concentration also peaks in this region. Beyond this point, all of the fuel concentrations, including hexane, methane, PAH and soot, decrease with increasing depth. This observation implies that oxidation reactions predominate in the lower half of the flame. The CO_2 concentrations increase reaching a maximum at 50 cm at which point all the products have been consumed. The fuel, soot, and CO_2 temperatures are also shown in Fig. 12a. These temperatures increase as the fuel is consumed.

Coal Water Fuels (CWF)

Measurements in Non-Combusting CWF Sprays - When measuring spray properties without a flame, conditions were maintained that would later be used in the flame work. These conditions include hydrogen in the ring burner (27 psi) and natural gas in the atomizing air flow. The natural gas and air were fed through one-way valves. Only the pressure of the atomizing air is reported below.

364

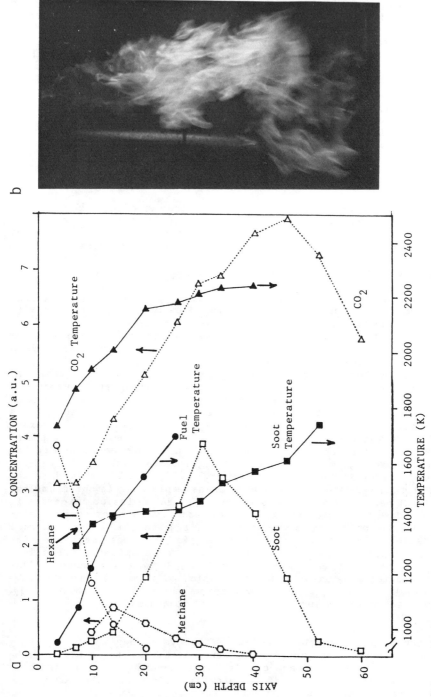

Figure 12. Characteristics of the Hexane Flame. a) Concentrations and Temperatures of Species along the Flame Axis as a Function of the Depth Below the Nozzle (34 PSI Atomizing Air Pressure, 0.1 ℓ/min Fuel Flow). b) Photograph of the Flame.

In a manner similar to that described for hexane, FT-IR transmission measurements were made through the coal slurry spray for a variety of conditions. Comparisons between these spectra and those calculated for coal particles using Mie theory are reported. The results suggested that Mie calculations for water droplets, or coal and water mixed, would have been more appropriate for the comparison.

Droplet size distributions were obtained from the FT-IR transmittance measurements. An example illustrating the precision of the method in this case is shown in Fig. 13 which is a transmission measurement taken through the CWF spray on-axis at a depth of 30 cm and air pressure of 34 psi. The data show better agreement with the calculated curve for 64 microns, than for 84 microns (Sauter-mean-diameters). The 64 $\pm$ 10 microns that we would quote for this measurement agrees with 57 $\pm$ 17 microns determined from a micrograph of droplets collected on filter paper.

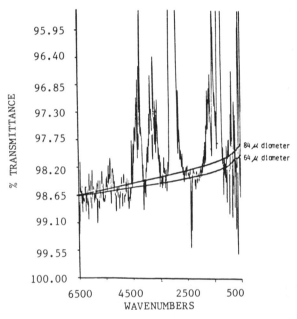

Figure 13. Comparison of Experimental Transmittance with Mie Theory Predictions for Two Drop Size Distributions.

Droplet size distributions were also determined for the modified nozzle. These measurements are on-axis measurements. The transmission spectra were recorded at different atomizing air pressures, on axis, at a depth of 12". Atomizing air pressures of from 30 to 90 psi with a constant CWF flow rate were used to obtain the data of Fig. 14, which compare the transmittance spectra on a common scale. As can be seen, both the shape and amplitude change with pressure. The amplitude of the extinction increases with pressure. Smaller droplet sizes and narrower spray cones, expected with high pressure, will both tend to increase $(1-\tau_\nu)$, while higher velocity would decrease $(1-\tau_\nu)$. Apparently the former effects dominate.

Sauter-mean-diameters were obtained by best visual fits to theoretical spectra, as before. The 30 and 50 psi data were fit with Sauter-mean-

366

diameters of $D_s = 100$ and $D_s = 52$ microns, respectively. The trends are similar to those for hexane.

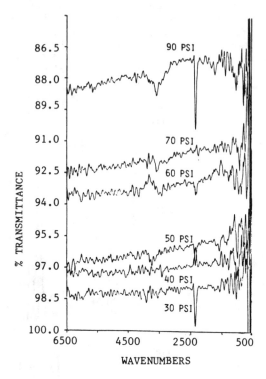

Figure 14.   Effect of Atomizing Air Pressure on Transmittance.

Measurements of Combustion Performance – To characterize the CWF sprays, FT-IR diagnostics were applied to monitor combustion. Measurements were made along the axis as a function of depth below the nozzle.   From these measurements, the following quantities can be determined:
- Gas concentrations and temperatures for CO_2, H_2O, CH_4.
- Concentrations of particles and soot.
- Particle temperatures.
- Percent of particles ignited.
- Flame radiation intensity from individual components (particles, soot, gases).

To illustrate the measurements, Fig. 15 shows the transmittance, radiance, and normalized radiance spectra obtained from the combustion of a CWF at a depth of 25 cm.

The concentration of CO_2 can be estimated by measuring the peak height of the absorbance at 2297 wavenumbers. The CO_2 temperature is obtained from the normalized radiance spectrum (Fig. 15c) with the correction for soot and coal particle emissions (as discussed in Ref. 15).

The concentrations of particles and soot are determined from Fig. 15a. The base line extinction represents the optical blockage of soot and CWF droplets. Since the droplets have a relatively large average size, their blockage in the (transmittance) spectrum is taken to be independent of wavenumber. The concentration of coal droplets is proportional to the

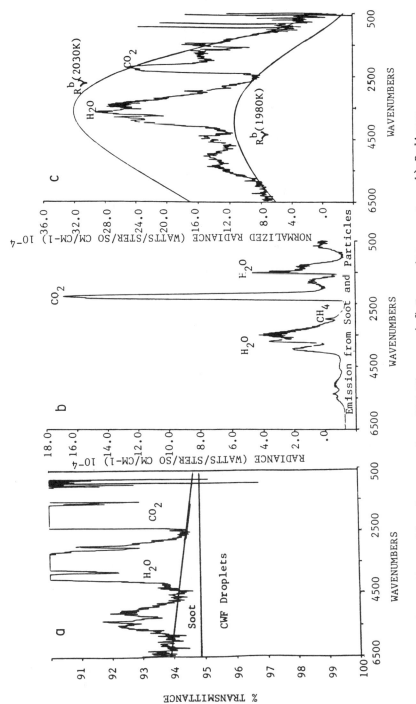

Figure 15. Spectra for CWF Flame. a) % Transmittance, τ_ν, b) Radiance, R_ν, and Normalized Radiance R_ν^n.

amplitude of the horizontal line which intersects the sloping baseline at zero wavenumbers, as shown in Fig. 15a. The amplitude of the sloping base line above this horizontal line, at 6500 cm^{-1} wavenumbers, represents the concentration of soot.

Particle temperatures were determined from Fig. 15c by comparison with a theoretical black body distribution, $R_\nu^b(T)$. This curve is fit in the region where there is no interference from gas lines. While the **shape** of the experimental curve matches R_ν^b for a temperature of 1980 K, the **amplitude** of the experimental curve is too low. The explanation for this observation is that a significant fraction of coal particles were not ignited during the combustion process. The measured particle concentrations from the transmittance (Fig. 15a) contain both hot and cold coal particles. If the blockage of unignited coal particles in the transmittance spectrum is taken out, the normalized particle emission spectrum should fit the theoretical black body curve in both shape and amplitude. By a trial and error method we found the fraction of coal particles that was burning; that fraction for which the shape and amplitude of the normalized emission spectrum matched R_ν^b. The emissivity of the ignited char particles is assumed to be 0.7 for this determination.

The use of normalized radiance for particle and soot temperature determinations has considerable advantages over the use of simple radiance. In particular, when simple radiance spectra of non-grey bodies (e.g. soot) are interpreted as grey body spectra, significant errors can occur in temperature determination. The soot and hot coal particle temperature, which can be obtained from the fitting of black body curves to the **normalized radiance spectrum** (Fig. 15c), will, however, be correct if the soot and char particles are at the same temperature. If soot and char are at different temperatures, a more complicated fitting procedure is required.

Figures 16 to 18 summarize the results of measurements of combustion of the CWF spray formed by the modified nozzle. For these measurements the atomizing gas consisted of 22.5 psi of natural gas and 17.5 psi of air, with 27 psi of hydrogen through the ring burner. Figure 16a shows a photograph of the flame.

Figure 16b contains the information concerning the fraction of ignited coal particles, plus the energy radiated by particles and soot during the combustion. The increase in the fraction of ignited coal particles in the first 10 inches of the flame implies that the atomized coal slurry is increasingly ignited by the gas flame with increase in residence time. In this flame the maximum fraction of ignited particles at any one time reached only 50%. The subsequent rapid decrease of the fraction of ignited droplets can be attributed to burnout of the fine particles. The hot coal particle concentration is fairly constant in the last stage of combustion. The measured radiation compares qualitatively with the appearance of the flame in the photograph which is on the same scale.

In Fig. 17, the on-axis temperatures of CO_2, soot, and hot coal particles, (determined by axial scans) are shown as a function of depth. Consider first the CO_2. As the combustion progresses from the nozzle to the 5 inch depth, the contact of unignited CWF droplets with CO_2 apparently causes the CO_2 temperature to dip lower than that at the depth of 1 inch. At depths from 5 to 10 inches, the increase in the CO_2 temperature suggests that most of the small slurry particle droplets have already dried out, and volatiles from these are already burning. Beyond the depth of 15 inches, a decrease of CO_2 temperature is observed again. This decrease of CO_2 temperature indicates that the combustion of both the natural gas and the coal volatiles is essentially complete at this depth.

369

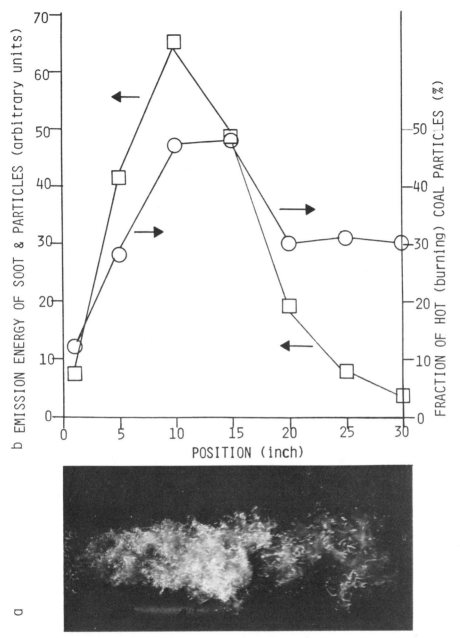

Figure 16. Characteristics of Lingan CWF Flame as a Function of Axis
Position with Modified Nozzle, the Atomizing Gas was Composed of 22.5 PSI
of Natural Gas and 17.5 PSI of Air, with 27 PSI of Hydrogen in the Pilot
Flame. a) Photograph of Flame, and b) Radiance of Soot and Ignited
Particles, and Fractions of Ignited Particles.

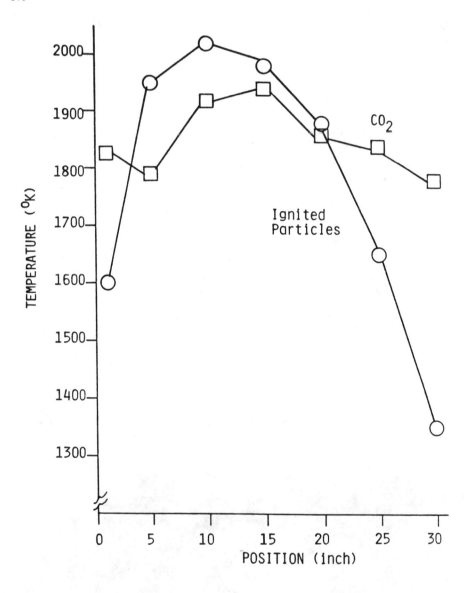

Figure 17. Characteristics of Lingan CWF Flame as a Function of Axis Position with Modified Nozzle, the Atomizing Gas was Composed of 22.5 PSI of Natural Gas and 17.5 PSI of Air, with 27 PSI of Hydrogen in the Pilot Flame. Concentration of CO_2, Soot, Total Particles, and Ignited Particles.

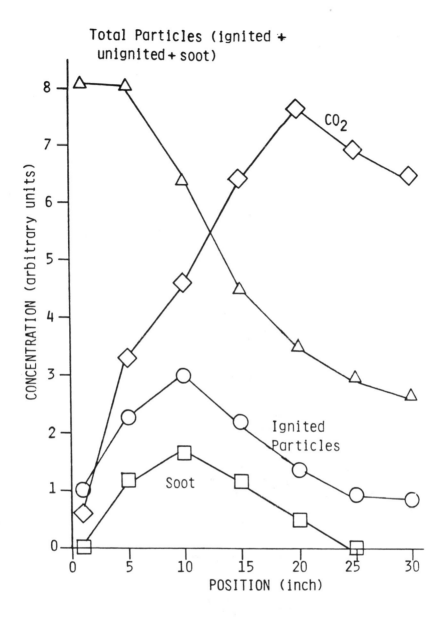

Figure 18. Characteristics of Lingan CWF Flame as a Function of Axis Position with Modified Nozzle, the Atomizing Gas was Composed of 22.5 PSI of Natural Gas and 17.5 PSI of Air, with and 27 PSI of Hydrogen in the Pilot Flame. Temperature of CO_2, Soot, Total Particles, and Ignited Particles.

The soot and hot burning coal particle temperatures are also shown in Fig. 17. The temperatures are found to increase from 1600 K at a depth of 1 inch to 2020 K at 10 inches, gradually decreasing after that. The initial increase of the temperature is due to the heating up by the gas flame (particularly the H_2 pilot). Once ignited the hot coal particles will generate energy themselves, and because they do not contact the unignited droplets, their temperature can be greater than that of CO_2, which does contact colder (unignited) droplets. Once these particles leave the region of burning gas, their temperatures gradually decrease from 2070 to 1350°K.

The concentrations of CO_2, the total particles (ignited, unignited and soot), the ignited coal particles, and the soot are shown in Fig. 18. The CO_2 concentration monotonically increases for the first 70% of the flame, then decreases slowly. The first stage of rapid increase of CO_2 concentration is due to generation by simultaneously combusting natural gas and coal. The decrease of CO_2 concentration in the last stage of the flame can be attributed to a generation rate which is lower than the out-diffusion rate.

The above observations can be well correlated with the soot production of the flame. Because of the continuous ignition of the atomized coal slurry, soot generated from the pyrolysis of the volatile matter from coal is observed to increase to a maximum value at a depth of 10 inches, Fig. 18. From 10 to 20 inches the depletion of coal volatile matter results in a decreasing soot concentration. At this stage, the appearance and removal of soot from the flame is important in controlling the flame radiation properties.

CONCLUSIONS

An FT-IR emission and transmission (E/T) spectroscopic technique has been employed for in-situ diagnostics in moderate size laboratory spray combustion flames of hexane and coal water fuels. The E/T spectra are shown to yield extensive information on particle or droplet size, composition and temperature, on soot density and temperature and on gas composition, concentration and temperature. The results demonstrate the feasibility of employing the FT-IR E/T technique in practical flames. Additional work is needed to refine the technique, develop tomographic reconstruction techniques to obtain point data, and to validate the measurements in flames by comparison to measurements using other techniques.

ACKNOWLEDGEMENTS

Funding for aspects of this work were provided under contracts #DE-AC21-81FE05122 from the Morgantown Energy Technology Center of the Department of Energy and #DE-AC01-85ER80320 from the Department of Energy under the Small Business Innovative Research (SBIR) program. The authors wish to thank Dr. Richard Goodman of Carbogel, Inc. for supplying the CWF samples and for advice during the research.

REFERENCES

1. Liebman, S.A., Ahlstrom, D.H., and Griffiths, P.R., Applied Spectroscopy, **30**, #3, 355, (1976).
2. Erickson, M.D., Frazier, S.E., and Sparacino, C.M., Fuel, **60**, 263, (1981).

3. Solomon, P.R., Hamblen, D.G., Carangelo, R.M., and Krause, J.L., Coal Thermal Decomposition in an Entrained Flow Reactor; Experiment and Theory, 19th Symposium (International) on Combustion, 1139, The Combustion Institute, Pittsburgh, PA, (1982).

4. Solomon, P.R., Hamblen, D.G., and Carangelo, R.M., Applications of Fourier Transform Infrared Spectroscopy in Fuel Science, ACS Symposium Series, 205, 4, 77, (1982).

5. Ottensen, D.K. and Stephenson, D.A., Combustion and Flame, **46**, 95 (1982).

6. Ottensen, D.K. and Thorne, L.R., 1983 International Conference on Coal Science, Pittsburgh, PA, pg. 621, (Aug. 15-19, 1983).

7. Solomon, P.R., "Pyrolysis", in Chemistry of Coal Conversion, R.H. Schlosberg, Editor, Plenum Publishing, New York, NY, (1985) in press.

8. Solomon, P.R. and Hamblen, D.G., Measurements and Theory of Coal Pyrolysis, A Topical Report for U.S. Department of Energy, Office of Fossil Energy, Morgantown Energy Technology Center, Contract No. DE-AC21-81FE05122, (1984).

9. Solomon, P.R., Hamblen, D.G., and Carangelo, R.M., "Analytical Pyrolysis", K.J. Voorhees, (Editor), Butterworths London, Chapter 5, pg. 121, (1984).

10. Best, P.E., Carangelo, R.M., Markham, J.R., and Solomon, P.R., "Extension of Emission - Transmission Technique to Particulate Samples using FT-IR", accepted for publication in Combustion and Flame (1986).

11. Solomon, P.R., Serio, M.A., Carangelo, R.M., and Markham, J.R, Fuel, **65**, 182, (1986).

12. Solomon, P.R., Carangelo, R.M., Best, P.E., Markham, J.R., and Hamblen, D.G., ACS Div. of Fuel Chem. Preprints, **31**, #1, 141, (1986).

13. Best, P.E., Carangelo, R.M., and Solomon, P.R., ACS Div. of Fuel Chem. Preprints, **29**, #6, 249, (1984).

14. Solomon, P.R., Carangelo, R.M., Best, P.E., Markham, J.R., and Hamblen, D.G., "The Spectral Emittance of Pulverized Coal and Char", accepted for the 21st Symposium (Int) on Combustion Meeting (1986).

15. Solomon, P.R., Best, P.E., Carangelo, R.M., Markham, J.R., Chien, P.-L., Santoro, R.J., and Semerjian, H.G., "FT-IR Emission/Transmission Spectroscopy for In-Situ Combustion Diagnostics", accepted for the 21st Symposium (Int) on Combustion Meeting, (1986).

16. Solomon, P.R., Carangelo, R.M., Hamblen, D.G., and Best, P.E., "Infrared Analysis of Particulates by FT-IR Emission/Transmission Spectroscopy", accepted by Applied Spectroscopy for publication (1986).

17. Siegel, R. and Howell, J.R., "Thermal Radiation Heat Transfer", McGraw Hill, NY, (1972) (Section 20-5).

18. Vervisch, P. and Coppalle, A., Combustion and Flame, **52**, 127, (1983).

19. Bohren, C.F. and Huffman, D.R., "Absorption and Scattering of Light by Small Particles," John Wiley and Sons, NY (1983).

20. Chigier, N., Energy Combustion and Environment, McGraw Hill, NY, 261, (1981).

DISCUSSION

A. Sarofim

Can you comment on the apparent discrepancy between your emissivities for coals and those inferred from optical constants of coals as measured for example by Blokh ?

P.R. Solomon

This subject has been considered by Brewster and Kunimoto (1) and by Solomon et al. (2). Brewster and Kunimoto believe that the surfaces of coal can not be prepared with sufficient perfection to insure specular reflection required by the method of Blokh. They used a transmission method to determine the imaginary part of the index of refraction, k and got values much lower than those of Blokh and others. We have used a similar transmission technique and get similar low values of k. If we calculate the emissivity using our values of k using Mie theory, we get agreement with our measured values of emissivity using the Ft-IR E/T method. We don't get agreement using Blokh's value of k.

(1) M.Q. Brewster and T. Kunitomo Tran. ASME 106, 678 (1984).
(2) Solomon, P.R., Carangelo, R.M., Best, P.E., Markham, J.R. and Hamblen D.G. "The spectral emittance of pulverized coal and char", 21st Symposium (Int) on Combustion, 1986.

A. Williams

Please will you comment on the influence of self-absorption effects by cold flame components (egCO$_2$) at the interface of the hut flame with its cold surroundings gases on your measurements of the concentration of the flame species.

P.R. Solomon et al.

Regions of the spectrum where cold species have absorption should be avoided. One can make measurements on CO$_2$ by using regions of the spectrum which are absorbing and emitting only for hot CO$_2$. Radiation at such frequencies are not affected by cold CO$_2$. The same idea applies to H$_2$O.

F. Kapteijn

Along this conference various values for the emissivity of coal or char particles have been mentioned. The presented radiance curve vs wavenumber for a Kentucky 9 coal cannot be fitted with a black body radiance curve.

What values of emissivity must be used in temperature and radiactive heating calculations ?

Or alternatively : What is the true story about emissivity ?

R. Solomon

The results are reported in a paper "The spectral Emittance of pulverized coal and char" for the 21st Symposium
(Int) on Combustion. Bituminous coals and lower rank coals of
pulverized coal size are non-gray body. They have high values
of emittance only in regions where there are strong absorption
bands. The emittance increases with particle size and with
increasing rank. Anthracites appear to be gray body with ε ≃ 9.
With pyrolysis, coals become more gray body with reaching a
maximum of 0.9. At high temperatures as the chars become more
"graphite-like" is reduced to approximitely 0.7.

E.M. Suuberg

You have raised the question of temperature measurements
in systems for studying coal pyrolysis kinetics a number of
times and it may be worthwhile to more quantitatively discuss
the point. Most heated grid experiments tend to be performed
with < 100μ particles at heating rates of 1000 K/S in the
presence of an inert gas. A number of unpublished calculations
suggest that this technique may be off, in actual particle
temperature, by 10 to a few 10'S of degrees. I agree completely that the calculations should be published to once and for
all establish how good these measurements arc. My question is
for comparison with these presumably soon to be published
analyses.

A - How accurate is the FTIR based technique ?
B - What is the limiting factor in accuracy ?
C - The FTIR is not a bulk temperature technique but
provides a "SURFACE" temperature. From what depth in the particle does radiation originate ?

D - At what particle size does one have diffuculty associated with approach of particle size to the wavelength of
measured radiation ?

P.R. Solomon

A - We estimate that the FT-IR temperature measurement
is good to within ± 50°C in general and ± 30°C for gray bodies.

B - The limiting factors are : 1) The accuracy of the
black body standard. 2) The uniformity of the ensemble of
particles being measured. 3) The accuracy with which the em-
missivity is known. 4) Instrumental alignment so that the
transmittance and radiance are measured for the same sample
volume. 5) The constancy of conditions between the transmittance and radiance measurement. 6) Possible saturation of the
detector. 7) Avoidance of radiation from the sample in the
transmittance measurement.

C - The depth is dependent on what coal and what frequency. For lignite at wave numbers larger than 1700 cm^{-1} the

penetration depth may be as large as 200 microns. At 1400 cm^{-1} the penetration depth is on the order of 40 microns. For char, the penetration depth can be much less.

D - Not a problems. Such effects can be calculated using Mie theory.

THE MIE THEORY AND THE NATURE OF ATTENUATION OF THERMAL RADIATION IN
PULVERIZED COAL FLAMES

Z.G. HABIB and P. VERVISCH

Laboratoire de Thermodynamique - U.A. CNRS 230 - Faculté des Sciences
et Techniques de Rouen - B.P. 67 - 76130 Mont-Saint-Aignan (France)

ABSTRACT

The scattering effect of thermal radiation (0.5-10μm) on absorption
measurements is theoretically investigated by using the Lorenz-Mie theory.
Non-dimensional monochromatic correction coefficients χ_λ^a are determined by
calculating the scattered power collected by the detector when assuming a
single scattering. χ_λ^a, included between 1 (absorption situation) and 1-ω
(extinction situation) where ω is the single particle albedo, are to be
used in order to estimate the absorption contribution to an experimental
optical depth.

Theoretical values of χ_λ^a are presented for a wide range of particle
size parameters(0 1000) and for a useful range of optical apertures
(0-20 deg). The interest of these coefficients is that no information
is needed on particle concentration provided that the assumption of a
single scattering holds.

Particle temperature measurements by the emission absorption method at
4μm are presented and show a clear underestimation due to scattering. Temperature values after scattering correction using χ_λ^a and ranging between
1500 and 2200 K are more realistic compared to the two color temperature
measured in the visible range. The use of 1-ω, a common pratice for
narrow solid angles of measurement, leads to unrealistic results in some
situations.

Measurements were performed in a coal-Air-Argon flames generated in a
cylindrical furnace at different oxygen mole fractions (0.16-0.33) and for
different size graded samples of a high V.M. bituminous coal : 20-40,
40-63, 63-80 and 80-100μm.

1. INTRODUCTION

Reliability of spectral absorption measurements in pulverized coal
flames depends upon the accuracy of the assumptions made concerning the
scattering process occuring when large particules are present in the bulk gas.

The very wide range of particle dimensions(0-200μm) encountered in coal
flames and the large spectral range (0.4-10μm) generally investigated by
infrared experimentalists do not enable us to adopt a general solution.
The same particle of a given diameter will scatter differently at different wavelengths, since scattering depends on particle size parameter. Hence, a monochromatic treatment of scattering has to be considered.

In general, the objective of transmission measurements is to determine
the monochromatic emissivity ε_λ of a slab of depth L through its optical
depth $k_\lambda L$ where k_λ is the monochromatic absorption coefficient.

In presence of radiation scatterers, the attenuation of a beam of radiation passing through the slab will be partly due to scattering. Then
considerable care must be taken to avoid misinterpretation of the transmittance data(1,2). Soot radiative characteristics have been widely studied

(3,4,5,6). Scattering by soot particles whose maximum dimensions are less than 0.1μm can be ignored since it is insignificant compared to absorption.

For $\alpha \ll 1$ scattering can be neglected, but it rapidly becomes comparable to absorption for $\alpha > 1$. Particles for which $\alpha \approx 1$ (cenospheres, char, fly ash) scatter in a nearly isotropic manner, and for large particles ($\alpha \gg 1$), scattering is so predominentely forward (7,8,9) that the bulk of the scattered radiation is contained in a narrow angle around the incident direction. Qualitative information on scattering effect during a transmittance experiment can be obtained by comparing the aperture of the optical system to the angle of the forward lobe of the scattering pattern (7,10,11). This approach does not give the order of magnitude of the scattered intensity included in the solid angle of measurement but can be sufficient in the case of large apertures (>15 deg) where more than 90% of the scattered light might be collected by the detector. For very narrow apertures (<1 deg) the attenuation is generally identified to extinction so that the absorption contribution is estimated by using theoretical values of the single particle albedo ω.

For the intermediate range of apertures, precise information is needed. The calculations presented in the next section were carried out for that purpose.

2. THEORETICAL DEVELOPMENT

The interaction between a particle and radiation impinging on it depends on its complex refractive index, $n-iK$ and size parameter α. For a spherical particle of radius a, $\alpha = 2\pi a/\lambda$. This interaction occurs in two different ways, i.e. scattering and absorption. For an incident intensity I_0 the total scattered power is given by

$$P_{sca} = I_0 C_{sca}$$

C_{sca} is the scattering cross section and is related to the geometrical cross section πa^2 by means of the scattering efficiency Q_{sca}

$$C_{sca} = \pi a^2 Q_{sca}$$

In the same way, C_{abs} and Q_{abs} are defined for absorption. The sum of scattering and absorption is referred to by extinction,

$$C_{ext} = C_{sca} + C_{abs}$$
$$Q_{ext} = Q_{sca} + Q_{abs}$$

Scattering is not necessarily isotropic and the total scattered power is written, at a distance r

$$P_{sca} = \int_{4\pi} I_{sca}(\theta,\Phi) \, r^2 d\Omega \qquad (1)$$

$I_{sca}(\theta,\Phi)$ is the scattered intensity distribution, θ and Φ are the common angles of a spherical coordinate system intrinsic to the incident beam.

P_{sca} of course is independent of r since $d\Omega = dS/r^2$. In consequence, the scattered power received by a surface Σ at a distance r from the particle is

$$P = \int\int_\Sigma I_{sca}(\theta,\Phi) \, d\Sigma \qquad (2)$$

Let us consider a detector of radius R centered on the incident beam axis at a distance D from an isolated particle. This determines a total angle of collection given by $2 \, \mathrm{Arc}\, \tan(R/D)$, and the collected scattered power in the forward direction is given by eq. 2 in which Σ is the detector area.

The extinction relation can be rewritten as follows

$$P_{ext} = I_0 C_{ext} - \int\int_{\Sigma_d} I_{sca}(\theta,\Phi) \, d\Sigma_d \qquad (3)$$

or

$$C_{ext} = C_{abs} + C_{sca}$$

then $P_{ext} = I_o[C_{abs} + C_{sca}(1- \dfrac{\sum\iint_d I_{sca}(\theta,\Phi)d\Sigma_d}{I_o\ C_{sca}})]$

or $P_{ext} = I_o[C_{abs} + C_{sca}(1-R_\lambda)]$ (4)

where $R_\lambda = (\dfrac{1}{I_o\ C_{sca}\ \Sigma_d}) \iint I_{sca}(\theta,\Phi)d\Sigma_d$ (5)

R_λ is the normalized collected scattered power, subscript d denotes the detector. R_λ must vary between 0 and 1 standing respectively to extinction and absorption. It is dependent on the size parameter and for a fixed particle diameter, R_λ will be dependent on wavelength.

For a monodispersed particle suspension and in absence of multiple scattering, the extinction coefficient is the sum over individual particles of the extinction cross sections(9)

$k_{ext} = N\ C_{ext}$

where N is the particle concentration. For large particles k_{ext} is usually assumed to be independent of wavelength. In terms of C_{ext} and hence of R_λ, the apparent extinction will depend on wavelength by means of R_λ

$k'_{ext,\lambda} = N[C_{abs} + C_{sca}(1-R_\lambda)]$ (6)

The equation of transfer through a non absorbing medium containing N particles per unit volume is written in terms of k'_{ext},

$d\ I_\lambda = -I_\lambda\ k'_{ext,\lambda}\ d\ell$ (7)

Equation (7) when integrated over a homogeneous path of depth L leads to

$I_\lambda(L) = I_\lambda(0)\ exp - \{[k_{abs} + k_{sca}(1-R_\lambda)]\ L\}$ (8)

The experimental optical depth calculated by Eq.(8) as $k'_{ext}\ L = Ln\ I_\lambda(0)/I_\lambda(L)$ can be related to k_{abs} by using Eq. (8), hence

$k'_{ext}\ L = k_{abs}\ L[1 + \dfrac{k_{sca}}{k_{abs}}(1-R_\lambda)]$ (9)

or $k'_{ext}\ L = k_{abs}\ L[1 + \dfrac{Q_{sca}}{Q_{abs}}(1-R_\lambda)]$ (10)

Equation (10) relates experimental attenuation to absorption. Hence we can define an absorption correction coefficient χ^a_λ which allows the absorption contribution to be separated from experimental attenuation, hence

$k_{abs}\ L = k'_{ext}\ L\ \chi^a_\lambda$ (11)

in which $\chi^a_\lambda = \dfrac{1}{[1+\dfrac{Q_{sca}}{Q_{abs}}(1-R_\lambda)]}$ (12)

Note that when R_λ varies from 0 to 1, χ^a_λ varies from $1-\omega$ to 1. χ^a_λ can be calculated theoretically from the Mie theory since it is independent of particle concentration and medium depth provided the assumption of single scattering holds.

3. NUMERICAL CALCULATION
3.1. The Mie theory for spheres

The Lorenz (12) (1890) and Mie (13) (1908) theory but commonly called the "Mie Theory" enables us by the resolution of Maxwell's equations with appropriate boundary conditions, to describe the interaction of an electromagnetic plane wave traveling in a non-absorbing medium with a homogeneous isotropic non-magnetic sphere of complex refractive index $m = n - ik$ and a diameter $2a$.

In Kerker (14) formalism, the scattered intensity distribution $I_{sca}(\theta, \Phi)$ is written as follows

$$I_{sca}(\theta, \Phi) = \frac{\lambda^2}{4\pi^2 r^2} [i_1 \sin^2 \Phi + i_2 \cos^2 \Phi] \tag{13}$$

The computation of the scattered intensities i_1, i_2 and the efficiency factors Q_{sca} and Q_{abs} is performed by using the so-called SUPERMIDI Code. This Code uses the Lentz algorithm (15) for Bessel functions calculations with neither particle size limitation nor complex refractive index. Full details of the structure of the code and its performance are given elsewhere (16,17).

3.2. Detection code

The code used to calculate the collected scattered power is an over simplification of the SPEMIE program developed by Cherdron et al (18).

The circular detector of radius R_d is located at a distance D from the center of a cartesian coordinate system X_i intrinsic to the incident wave (Fig. 1). The detector area is subdivided into $N_\rho \times N_\beta$ elements $\Delta\Sigma_{jk}$,

$$\Delta\Sigma_{jk} = (\frac{2\pi\rho_j}{N_\beta}) (\frac{R_d}{N_\rho})$$

ρ_j and β_k are the polar coordinates of $\Delta\Sigma_{jk}$ in the coordinate system ξ_i intrinsic to the detector.

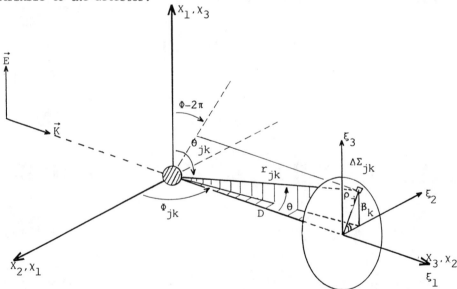

Fig. 1 : Coordinate systems X_i, χ_i and ξ_i used in the detection code.

The location of $\Delta\Sigma_{jk}$ in the χ_i coordinate system is given by the spherical coordinates r_{jk}, θ_{jk}, Φ_{jk}. The spherical coordinates r, θ, Φ required for the scattered intensity calculation are related to r_{jk}, θ_{jk}, Φ_{jk} by simple transformation relations.

$I_{sca}(\theta, \Phi)$ is calculated for each surface element and the collected scattered power P_R is given by :

$$P_R = \sum_j \sum_k I_{sca}(\theta,\Phi) \, \Delta \Sigma_{jk}$$

the total scattered power is calculated as

$$P_{sca} = I_o \, C_{sca}$$

Finally R_λ and x_λ^a are determined by using Eq. 5 and 12 respectively.

3.3. Numerical results and discussion

The sensivity of the code to the complex refractive index was first tested for three values of m

$$m_1 = 1.6 - 0.3i \quad (ref. \ 19)$$
$$m_2 = 2 \quad - 0.6i \quad (ref. \ 20)$$
$$m_3 = 2 \quad - \quad i \quad (ref. \ 19)$$

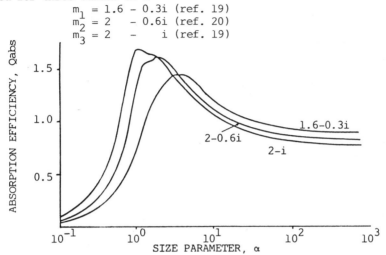

Fig. 2 : Absorption efficiency versus size parameter for three complex refractive indexes.

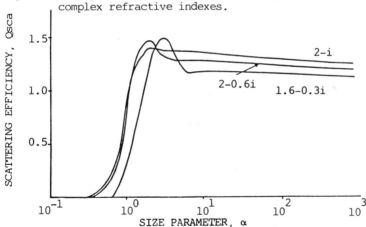

Fig. 3 : Scattering efficiency versus size parameter for three complex refractive indexes.

Significant variation is observed for Q_{abs} and Q_{sca} (Fig. 2 and 3), especially in the region of $\alpha \approx 1$. x_λ^a is less sensitive to the value of the complex index since the ratio Q_{sca}/Q_{abs} does not vary much. The value of m_1 corresponding to the coal similar to the one used in our experiments was

retained and variations with respect to wavelength as reported by Foster
[19] were considered,

$$m = 1.6 - 0.3i \qquad\qquad \lambda = 0.5 - 6\mu m$$
$$n = 1.6 + 1.5\ (\lambda\text{-}6)$$
$$k = 0.3 + 0.075(\lambda\text{-}6) \quad \Big\} \quad \lambda = 6 \ - 10\mu m$$

Figure 4 is ploted the normalized collected scattered power R_λ for a total collection angle of 3 deg. and for various particle diameters : 1, 10,
20, 40, 60, 80, 100 and 120μm. For d=1μm, R_λ is practically always equal
to zero. It increases rapidly with the increasing diameter. This is due
to the variation of scattering from an isotropic to a forward directed mode. R_λ
reaches rapidly a value of 0.85 for d > 40μm at 0.5μm. For a fixed diame-
ter, R_λ decreases with increasing wavelength since α decreases. χ_λ^a (Fig.5)
shows the same variation with respect to particle diameter and wavelength.
χ_λ^a varies from 0.85 at 0.5μm to 1-ω which lies between 0.45 and 0.53 for
the different diameters. The curve corresponding to d=1μm shows an inverse
variation, this is not in disagreement with the corresponding values of
R_λ ($R_\lambda \simeq 0$). In fact, at high wavelengths, the size parameter tends to 0.3
which is the begining of the soot region where scattering becomes insignificant, hence the χ_λ^a curve corresponds to the variation of 1-ω.

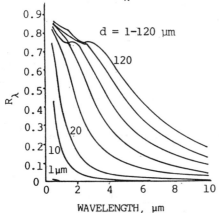 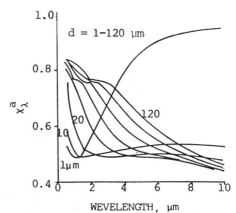

Fig. 4 : Normalized collected
scattered power, R_λ
versus wavelength.
Total collection
angle : 3 deg.

Fig. 5 : Absorption correction
coefficient χ_λ^a. Total
collection angle : 3 deg.

In Fig. 6 and 7 R_λ and χ_λ^a are plotted for a total angle of 12 deg. which is
generally taken as a limit angle over which more than 90 percent of the
scattered power is collected [21]. This is true for wavelengths less than
4μm and particle diameters greater than 40μm. For greater wavelengths, numerical results show that scattering has to be taken into account.
Variation of R_λ versus size parameter for apertures up to 20 deg. is
given Fig. 8. Fig. 9 is plotted the ratio (1-ω)/χ_λ^a showing an inverse
evolution versus aperture to that of R_λ . On the same figure is plotted the
single particle albedo showing the region where scattering becomes significant.
The numerical results given previously are calculated assuming a polarized incident plane wave. In fact, an unpolarized wave will be more realistic since infrared attenuation measurement are generally performed with un-

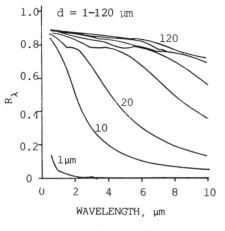

Fig. 6 : Normalized collected
scattered power, R_λ.
Total collection
angle : 12 deg.

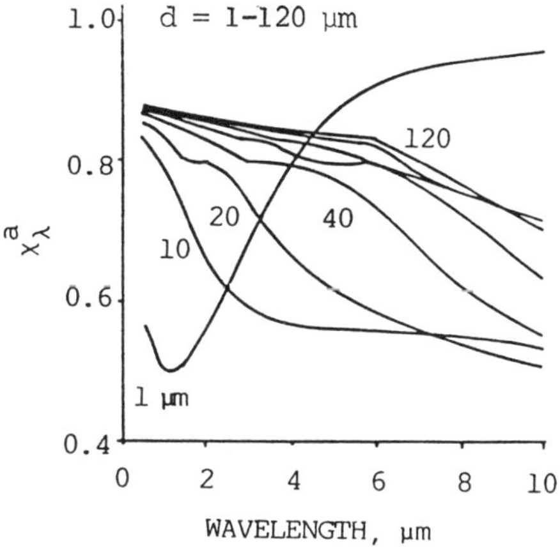

Fig. 7 : Absorption correction
coefficient χ_λ^a. Total
collection angle : 12 deg.

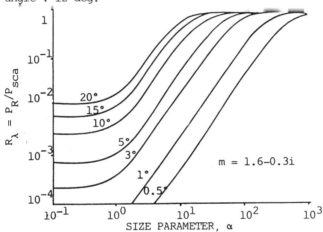

Fig. 8 : Normalized collected scattered power for different
collection angles.

polarized reference beam. In this case the expression of the scattered intensity distribution becomes[2]:

$$I_{sca}(\theta) = \frac{\lambda^2}{8\pi^2 r^2} (i_1 + i_2) \qquad (14)$$

Nevertheless, for the forward lobe and at narrow angles, calculations show that i_1 and i_2 are practically identical. Then equation (13) is reduced to equation (14) both independent of Φ.

4. EXPERIMENTAL
4.1. Apparatus
Full details of the experimental apparatus are given elsewhere (22).
Briefly , coal flames are generated in a vertical electric furnace of

384

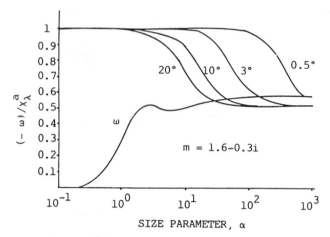

Fig. 9 : 1-ω to χ_λ^a ratio versus size parameter for different
collection angles. Included, variation of single
particle albedo ω.

5cm i.d and 1 meter long. A uniform wall temperature of 1400°K is main-
tened all along the experiments. Coal injection (4.17 g/min) is achieved
by the use of a mechanical injector. Oxygen mole fraction in the carrier
primary air (71/min dry air) was varied from 0.48 to 1. Hot argon generated by plasma guns and used as a secondary gas (2x71/min) provides rapid
thermalisation of the coal-gas mixture at the top of the furnace. Measurements were carried out at 15 cm from the injection point. The mean gas
temperature at this level, measured by a thermocouple, in the absence of coal,
is 1200°K. The residence time of the particles before reaching the
measurement point is about 150 ms.

Coal samples were prepared by aerodynamic sieving. Particles sizing
by laser diffraction gives a volumetric mean diameter (V.M.D.) of 93, 70,
50 and 34μm for nominal sieves dimensions of respectively 80-100, 63-80,
40-63 and 20-40μm. Analysis of the coal is given in table I.

TABLE I. Coal analysis.

Proximate analysis (as received)	Weight percent	Ultimate analysis (dry)	Weight percent
Moisture	2,5	C	79
Volatile matter	38,45	H	5,34
ASH	6	N	1
		ASH	6,1

Figure.10 is a chart of the infrared measurement set-up . The aperture of the collection optics is 3 deg. The He-Ne laser channel serves
for flame stability control.

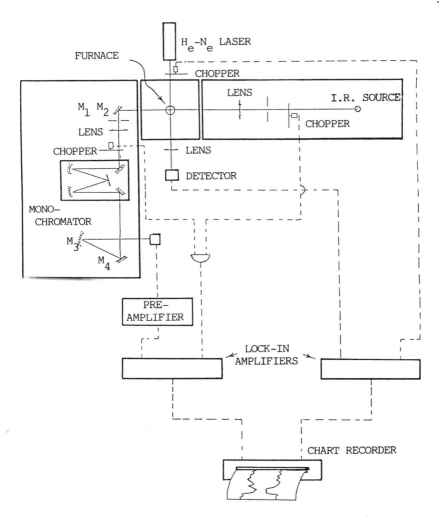

Fig. 10 : Top vue of the experimental set up showing
the optical path.

4.2. Data reduction

In general and in absence of multiple scattering (7), the emission signal emerging from an isothermal homogeneous slab of depth L containing a
suspension of absorbing, scattering and emitting particles is written,

$$I_\lambda(L) = \frac{k_{abs}}{k_{ext}} I_{b,\lambda} [1-\exp -(k_{ext} L)] \qquad (15)$$

Equation (15) corresponds to a total extinction situation where $I_{b,\lambda}$
is the black body intensity, hence k_{abs}/k_{ext} is identical to $(1-\omega)$. When
a fraction of the scattered power is collected by the detector, following
Eq.(11) we can write

$$I_\lambda(L) = \chi_\lambda^a [1-\exp-(k'_{ext} L)] I_{b,\lambda} \qquad (16)$$

When soot particles are present in the medium, Eq.(15) holds with

$$k_{abs} = k_{\lambda,soot} + \chi_\lambda^a k'_{ext}$$

and
$$k_{ext} = k_{\lambda,soot} + k'_{ext}$$

Then Eq. 15 is reduced to

$$I_\lambda(L) = \frac{1 + \chi_\lambda^a k'_{ext}/k_{\lambda,soot}}{1 + k'_{ext}/k_{\lambda,soot}} [1 - \exp -(k_{ext}.L)] I_{b,\lambda} \qquad (17)$$

$k_{ext}.L$ is determined experimentally. To estimate the emissivity of such a suspension, both ratios $k_{abs}/k_{\lambda,soot}$ and $k'_{ext}/k_{\lambda,soot}$ must be known. This estimation is possible if at least one of the soot of large particles contribution to attenuation can be determined separately. Two "sooty" situations can be encountered in P.C. flames. Soot and condensed matter are present surrounding the particle in a combustion situation (23). Such situation can be nonisothermal and Eq.(18) does not hold since the latter requires equilibrium of soot and large particles. On the other hand, classical attenuation methods cannot be applied to individual burning particles. The other case is when soot is present in the bulk gas, which implies a lack of oxygen and hence an incomplete combustion so that the assumption of thermal equilibrium of soot, large particles and gas becomes realistic. In that case, separation between soot and large particles can be achieved by determining the monochromatic transmission spectrum of the flame in the range of 0.4-5µm. The continuous base line of such a spectrum is due both to soot and large particles. Assuming a soot absorption coefficient in the form of $k_o \lambda^{-\beta}$, the Ln-Ln plot of the spectrum in terms of optical depth versus wavelength will show a slope break when soot and large particles contributions become comparable. The extrapolation of the curve to the infrared enables us to estimate k'_{ext} by difference and hence the ratio $k'_{ext}/k_{\lambda,soot}$.

4.3. Temperature

Particle temperature measurements were carried out by using the absorption emission method in the near infrared at 4.12µm. k_{ext} L was first determined by measuring the reference signal $S_\lambda(0)$ and $S_\lambda(L)$ respectively in absence and in presence of particles. $k_{ext}L$ is directly given by Eq.(8). The emission signal of the particles S_λ^P is compared to the signal S_λ^R given by a calibrated reference source of a known emissivity and temperature, respectively ε_λ^R and T_R. Using Planck's law, the particles temperature T_P is written :

$$\frac{1}{T_P} = \frac{\lambda}{C_2} Ln \left\{1 + \frac{\varepsilon_\lambda^F S_\lambda^R}{\varepsilon_\lambda^R S_\lambda^P} [\exp(C_2/\lambda T_R)-1]\right\} \qquad (18)$$

where $C_2 = 14.400$ µm.$^\circ$k and λ in microns.

Two situations have been investigated for four size graded coal samples. At 33 percent oxygen, the flames showed a very bright emission. Monochromatic study of the optical depth showed a total absence of any soot like variation. The measured transmission for all four samples was greater than 0.88. Hence an optically thin situation has been assumed and the emissivity used for particle temperature calculations is the one given by Eq.(16).

Sooty situations are produced by generating flames with a lack of oxygen (16 percent oxygen), the mesasured monochromatic optical depth versus wavelength for the 80-100µm sample is plotted Fig. 11 with a linear scale and show a soot -like variation. The four flames are practically identical, only the 4.3µm CO_2 band is significant. The Ln-Ln plot of k_{ext} L (Fig.11) shows a slope break at about 1µm. This was attributed to the presence of particles.

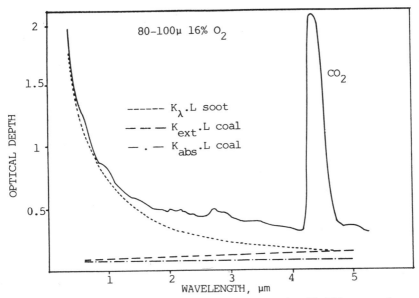

Fig. 11 : Monochromatic optical depth for the 80-100μm sample sooty flame.

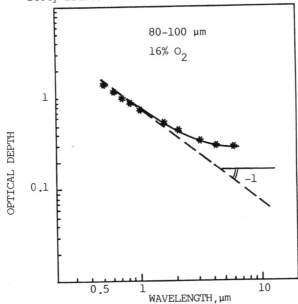

Fig. 12 : Ln-Ln plot of the monochromatic optical depth for the 80-100μm sample sooty flame.

The emissivity used in this case for particle temperature calculation is given by Eq. (17).

ature results are reported Tab. II and III. $k'_{ext}.L$ is the experi-
optical depth, T_1 is calculated temperature neglecting scattering
with an emissivity equal to $1-\exp-(k'_{aext}.L)$, T_2 and T_3 are temperatu-
taking scattering into account using χ_λ^a and $1-\omega$ respectively. T_v is the
visible temperature measured by the two-color method (at 0.65 and 0.9μm) for the
33 percent oxygen flames and by the emission absorption method (at 0.6μm)
neglecting scatter for the sooty flames. The two color pyrometer is presented elsewhere (22).

Taking the visible temperature as a reference, the values
of T_1 show a clear scattering effect and errors up to 30 percent are made
if scattering is not taken into account. For the 33 percent oxygen
flames (Tab. II), the corrected temperature with the use of χ_λ^a is reasonably adjusted to the visible one. The use of $1-\omega$ for correction (T_3) systematically gives higher values. The difference between T_2 and T_3 decreases
with decreasing particle diameter for which χ_λ^a tends to $1-\omega$. No significant difference is observed between the two types of correction for the 16
percent oxygen flames (Tab. III). The scattering effect on temperature is
attenuated by the presence of soot.

TABLE II. 33 percent oxygen flames, $\lambda = 4.12$μm (oxygen as a primary gas)

d, μm	80–100	63–80	40–63	20–40
$1-\omega$	0.44	0.45	0.46	0.47
χ_λ^a, 3 deg	0.6	0.56	0.52	0.49
$k'_{ext}.L$	0.05	0.08	0.1	0.12
T_1, °K	1720	1633	1530	1390
T_2, °K	2170	2120	2000	1866
T_3, °K	2560	2370	2146	1900
T_v, °K	2100	2100	2000	1850

TABLE III. 16 percent oxygen flames, $\lambda = 4.12$μm (48% oxygen in primary gas)

d, μm	80–100	63–80	40–63	20–40
$1-\omega$	0.44	0.45	0.46	0.47
χ_λ^a, 3 deg	0.6	0.56	0.52	0.49
β	1	1.17	1.15	1.1
$k_{soot}.L$	0.14	0.17	0.14	0.14
$k'_{ext}.L$	0.18	0.15	0.17	0.16
T_1, °K	1330	1410	1336	1312
T_2, °K	1480	1530	1495	1474
T_3, °K	1530	1580	1520	1482
T_v, °K	1530	1500	1450	1500

The values of χ_λ^a used for temperature correction correspond to the
volumetric mean diameter of the unburnt coal samples. Hence, it was assumed a particle combustion with a constant diameter, at least in the first
zone of the furnace, assumption not confirmed up to now . Consequently the
following antithesis can be put forward :

If a significant change in particle diameter did occur before passing in the measurement volume, then a nearly total extinction situation has to be applied for temperature calculations, which is practically the case of the 20-40μm particles. Hence T_3 will be closer to the real particle temperature than T_2. T_v, which is a two color temperature calculated assuming the particule has a grey body behaviour and theoretically unaffected by scatter, will in that case not as expected be representative of a particule temperature, especially in the case of the 80-100μm sample.

As Grosshandler showed in a recent paper (24), the two color temperature could represent a mean value of a large particle temperature and the temperature of a colder condensed matter surrounding the particle. The expression of the measured mean temperature as formulated by Grosshandler is

$$\overline{T} = T_P \left| 1 + \frac{1.0273 \, \varkappa}{0.04218 + \varkappa} \left(\frac{T_S^5}{T_P^5} - 1 \right) \right|^{0.25} \qquad (19)$$

where the ratio of the particle radius to the condensed matter zone radius (supposed spherical) is equal to 0.2 and assuming a single particle emissivity of 1. T_S and $\varkappa$ are respectively the temperature and the optical depth of the condensed matter.

In the case of the 80-100μm sample, assuming T_P equal to 2500 K $(=T_3)$ and $\overline{T}$ equal to 2100 K $(=T_v)$ and resolving Eq. (20), the most realistic solution is

$$\varkappa = 0.1, \qquad T_S = 1950 \text{ K}$$

The value of $\varkappa$ is improbable since the total optical depth of the medium including scatter does not exceed 0.05. Consequently, this result tends to support the validity of temperature correction using χ_λ.

5. CONCLUSION

Scattering corrections coefficients for attenuation measurements were presented and seem to be realistic when applied to particle temperature calculations by the emission-absorption method with intermediate solid angles of collection. The results for optically thin situations can be summarized as follows :

1) Neglecting scattering in temperature measurements at infrared wavelengths gives errors up to 30 percent.

2) For small particles i.e. mean diameter less than 30μm and for small apertures i.e. less than 3 deg., a total extinction situation can be assumed and correction using 1-ω is accurate.

3) For large particles and intermediate apertures, temperature correction using χ_λ^a seems to be more realistic than the use of 1-ω. Values of χ_λ^a depend on the wavelength at which measurements are carried out.

Acknowledgments : We would like to thank Professeur GOUESBET and Dr G. GREHAN from the U.A. CNRS 230, University of ROUEN, for their advice on Mie calculations.

_er P.J., Comb. and Flame, 7, 277 (1963).

_er J.M. and Howarth C.R., 12th Symposium (International) on Combustion _. 1205, The Combustion Institute (1969).

. Stull V.R. and Plass G.N., Emissivity of Dispersed Carbon Particles, Journal of the Optical Society of America, 50, 2, 121 (1960).

4. Lee S.C. and Tien C.L., 18th Symposium (International) on Combustion. The Combustion Institute, 1159-1166 (1981).

5. Vervisch P. and Coppalle A., Fire Flame Radiations, Combustion and Flame, 10, 2180 (1983).

6. Weill M., Flament P. and Gouesbet G., Appl. Opt., 22, 16, 2407 (1983).

7. Bayvel L.P. and Jones A.R., Electromagnetic Scattering and its Applications. Applied Sciences Publishers LTD (1981).

8. Bohren C.F. and Hauffman D.R., Absorption and Scattering of Light by Small Particles. Joh, Wiley and Sons (1983).

9. Hottel H.C. and Sarofim A.F., Radiative Transfert. Mc Grawhill, New-York (1967).

10. Hodkinson J.R., Particle Sizing by Means of the Forward Scattering Lobe, Appl. Opt., 5, 5, 239 (1966).

11. Maron S.H., J. Coll. Sci. 18, 107 and 391 (1963).

12. Lorenz L., Vidensk, Selsk, Skr, 6, 6, 1-62 (1890).

13. Mie G., Ann. der Phys., 25, 377 (1908).

14. Kerker M., The Scattering of Light and Other Electromagnetic Radiations Academic Press, New-York (1969).

15. Lentz W.J., Appl. Opt., 15, 3, 668-671 (1976).

16. Grehan G. and Gouesbet G., The Computer Program SUPERMIDI for Mie Theory Calculations, without practical size nor refractive index limitations. U.A. CNRS 230, Univ. Rouen, Report TTI/GG/79/03/20 (1979).

17. Ungut A., Grehan G. and Gouesbet G., Comparison between Geometrical Optic and Lorenz-Mie Theory, Appl. Opt., 20, 17, 2911-2918, sept.(1981)

18. Cherdron W., Durst F. and Richter G., Computer programs to predict the properties of scattered laser radiation, SFB 80, Ausbreitung und Transportvorgänge in Strömungen, Universität Karlsruhe SFB 80/TM/121, Januar (1978).

19. Foster P.J. and Howarth R.C., Optical constant of Carbons and Coal in the Infrared, Carbon, 6, 719-729, Pergamon Press (1968).

20. Mc Cartney J.T., Yasinsky J.B. and Ergun S., Optical Constants of Coals, Fuels 44, 349-354 (1965).

21. Grosshandler W.L. and Monteiro S.L.P., Attenuation of Thermal Radiation by Pulverized Coal and Char, J. Heat Transf., 104, 587-593 (1982).

22. Habib Z.G. and Vervisch P., Visible and Infrared Temperature Measurements in Pulverized Coal Flame, to be published.

23. Mc Lean W.J. Hardesty D.R. and Pohl J.H., 18th Symposium (International) on Combustion. The Combustion Institute, 1239-1248 (1981).

24. Grosshandler W.L., The effect of Soot on Pyrometric Measurements of coal particles temperatures, Comb. and Flame, 55, 59 (1984).

DISCUSSION

J.H. Pohl

What temperature error is caused by reglecting scattering in making two color temperature measurements in the usual wavelengths of 0.6 to 0.8 μm ?

Z. Habib

At visible wavelengths and for a fixed particle diameter, the effect of scatter do not vary much. So one could still assume that the expressions of the monochromatic emissivities of the particle suspension at both wavelengths are equal and hence the two color temperature is unaffected by scattering.

A. Sarofim

One would expect your results to be insensitive to the value of the complex component of the refractive index. Have you carried out a sensitivity analysis and if so can you comment on your results ?

Z. Habib

For the three values of the complex index tested in this paper, the theoritical results vary within five percent. In fact, the normalized collected scattered power is practically unsensitive to the value of the complex component of the refractive index since the scatterring pattern is not significantly affected by this parameter. Still, as the values of the absorption and scattering efficiencies are sensitive to this parameter the effect of the complex component of the refractive index will mostly be on the absorption correction coefficient. For a value of k greater than 0.3 the effect can be neglected. For ash for wich k is in the order of 0.05 which leads to large values of Q sca and low values of Q abs, the calcultations show that the absorption corection coefficient is comparable to the case of coal in the visible range because of the high value of the collected scattered power, but falls very rapidly in the infrared to reach very low values i.e. 0.2 or 0.3 because of the very low values of 1-ω.

TRANSFER TO THE MODELLING

TRANSFER OF FUNDAMENTAL RESULTS TO THE MODELLING OF BURNERS AND BOILERS

F.C. LOCKWOOD

IMPERIAL COLLEGE OF SCIENCE AND TECHNOLOGY, LONDON SW7 2BX, ENGLAND.

ABSTRACT
   The present status of the mathematical modelling of coal combustors is
outlined. The subject area is very large and can in no sense be comprehen-
sively treated in a single paper. For example the extensive and difficult
subject of turbulence modelling is a source of considerable uncertainty in
combustor prediction but not even the cold flow validation evidence of many
years of research on this topic can be fairly condensed into a paper which
addresses the topic of predicting combustors fired with that most fickle
of fuels: coal. An effort is made herein to give at least some perspective
on the capabilities and problem areas of current computer prediction
techniques. Industrial pulverised coal-fired combustors constitute the
central theme. Coal-liquid mixtures, coal-fired gas turbines and diesel
engines, gasifiers, and fluidised beds are excluded topics.

INTRODUCTION
   It has been demonstrated that engineering gas-fired combustors may be[1-3]
modelled mathematically with a precision acceptable to design engineers .
The production and destruction of soot and censopheres in oil-fired equipment and the simulation of the atomisation process pose substantial additional modelling problems but some progress has been made[4]. Pulverised-
coal flames present the greatest modelling challenge. Char particle
ignition and burn out, volatiles release and burning, and the char and ash
absorption and scattering of thermal radiation are some of the further processes requiring simulation. The current level of confidence in the modelling of coal-fired combustors must be considered to be significantly less
than that for gas-fired ones.
   This paper is intended to give an overview of coal combustor modelling
of practical equipment to an audience lacking in specialised experience
in the field. It has four primary objectives: to indicate industrial
combustion design priorities; to outline the nature of coal combustor
prediction procedures; to indicate the nature of the difficulties facing
the modellers; and to evaluate current predictive capabilities through
example applications.

GENERAL COMBUSTOR MODELLING DIFFICULTIES
Numerical problems
   These have two main sources. The more important one is the geometrical
complexity which typifies most engineering combustors. The other is a
consequence of the hybrid donor cell differencing schemes which character-
ise most of the current numerical solution procedures.
   Not only is the geometric form of the combustion chamber frequently
unsuited to convenient representation by conventional co-ordinate systems,

but usually there are large differences in dimensional scales between the burners and the chamber volume. Figure 1, taken from [1], shows a typical gas-fired industry glass melting furnace (panel (a)).

The near-burner field is handled by a subcalculation which, because in this case recirculation is absent from both the fuel and co-flowing air streams in this region, is parabolic. The form of the air inlet port is accommodated by an orthogonal curvilinear grid, and the refractory roof by an expanding and contradicting grid for which the effects of the relatively small non-orthogonality are ignored (panel (b)). Continuity of the magnitudes and fluxes of all variables is ensured along the frontier between the parabolic near burner calculations and separate and elliptic ones for the combustion chamber bulk volume. Predictions were obtained for the combustor velocities, temperatures, species concentrations, and local heat transfer rates to the glass. Panel (c) of Fig. 1 shows example predicted contours of heat transfer to the glass for the representative domain of calculation.

No data suitable for validation purposes were available for this furnace, a common situation for industry combustors, but the predictions of many conditions confirmed expectations based on operating experience. Of course the submodels for gas-firing have been extensively validated against laboratory data with encouraging results (see [5] for example).

Another gas-fired industry combustor is depicted in Fig. 2, panel (a): a refinery process heated. The combustor box is fired by a single burner and again we see the awkward difference in scale between the burner and bulk combustor volume.

The finite-difference grid is also shown on the figure. A curvilinear orthogonal co-ordinate system is employed to transgress from the axisymetric burner calculations to the rectilinear form of the bulk volume. In this case a swirl burner with a quarl is utilised so that the near burner subcalculations are necessarily elliptic. Also, radial injection of the gas fuel from six evenly spaced holes in the primary pipe necessitates calculations over a 60° sector and the application of cyclic boundary conditions.

Some predictions of velocity, temperature and mixture fraction are shown in panel (b) of Fig. 2 in cross-sectional planes downstream of the burner. The high three-dimensionality of the flow is evident. Panel (c) shows some predictions of wall heat transfer compared with a few available data points. The agreement is fairly reasonable. The wall heat transfer rates are some 40 kW/m^2, typical of gas-fired combustors of this size. It may also be noted that thermal radiation is the dominant heat transfer mode. Further details of this application of mathematical modelling may be found in [3].

The preceding two examples make clear that the specification of a finite-difference grid for industry combustors can be tedious. The practice of performing hybrid calculations in which near burner and bulk combustor calculations are matched at suitable interfaces invariably results in protracted run times. Impressive developments in specialised flexible or adoptive grid studies have been reported, e.g. [6]; there is a clear need to integrate the fruits of this work in industry combustor design codes.

The commonly employed first order donor cell differencing scheme, although generally reliable and robust, is susceptible to significant inaccuracy due to the effects of numerical (false) diffusion [7]. The problem is particularly severe when streamlines cross grid lines obliquely, a frequently encountered situation in industry combustors in which complex networks of recirculating flows are typical. It may well be that the inaccuracies reported in many furnace prediction validation studies have been as much due

to numerical diffusion as to the imperfections in the physical modelling
that the studies were designed to uncover[8].

Higher order difference schemes can be used to eliminate the false
diffusion problem but all of them give rise to 'undershoots and overshoots'
of the field variables prior to convergence of the iterative solution procedure. For combusting flows this can prove particularly troublesome as
negative concentrations, temperatures and densities may cause the procedure
to fail. There is no doubt that these difficulties will succumb to sufficient numerical research but thus far too little has been done in this
regard. And the numerical analyst is often reluctant to tackle the
disorderly and capricious difficulties which plague engineering applica-
applications.

Physical problems

The time and space scales of the smallest vrotices of the turbulence
spectrum are too small (order 10^{-4} s and 0.01 mm respectively) for the
governing Navier Stokes equations to be solved for engineering circumstances in time dependent fashion by any existing or foreseeable computer.
Instead time averaged forms of these equations are solved and the effects
of the turbulence on the solved for time-averaged variables is accommodated
through a model of turbulence. The turbulence modelling field has been the
centre of a very large amount of research activity, but in spite of this
no truly reliable and universal model has emerged. Mainly for reasons
of economy and convenience most combustor mathematical models employ a
two parameter scheme in which the turbulence is characterised by its turbulence kinetic energy, k, and the dissipation rate of that energy, ε, both
these quantities being solved for in their own equations. For highly
swirled burner flows Weber et al[8] recommended a full Reynolds stress tensor
closure in which modelled equations are solved for all six Reynolds
stresses. This is clearly a substantially more costly practice than the
use of the k-ε closure and the limited improvement in precision may not
always be justified.

In sum, combustion validation studies to date suggest that the accuracy
of the now rather dated turbulence models employed for combustor prediction
purposes is probably sufficient to meet most engineering needs. One may
be thankful for at least this degree of good fortune since, in spite of
much effort, economical turbulence models of significantly improved performance have thus far eluded researchers.

GENERAL CONSIDERATIONS ON APPLICATIONS TO COAL COMBUSTORS

Inclusion of the particulate phase

The accommodation of the solid phase in pulverised-coal (PF) fired combustor prediction represents a major complication over gas-fired
applications. Two approaches to the problem of handling the particulate
material are employed. In the one,[10] 'eulerian' equations similar in form
to those describing the gas phase are solved, one for each size group in
a discretised particle size range. The other and probably the more common
approach is based on the solution of lagrangian equations for representative particle positions and conditions, see[11] for example.

In the lagrangian technique the particle position is found from the
solution of its equation of motion:

$$m_p \frac{d\vec{V}_p}{dt} = C_D (\vec{V}_g - \vec{V}_p) [\vec{V}_g - \vec{V}_p] \rho_g \frac{A_p}{2} + m_p \vec{g} \tag{1}$$

where $m_p \equiv$ particle mass, $t \equiv$ time, V_p and $V_g \equiv$ particle and gas velocities,

$\rho_g \equiv$ gas density, $A_p \equiv$ particle cross-sectional area, and $g \equiv$ gravitational acceleration. The drag coefficient C_D of the geometrically irregular coal particle is almost always taken as that for an equivalent sphere, and uncertainty about what to do usually results in no correction being made for the effect of gas release due to devolatilisation and heterogeneous burning. These modelling deficiencies are of concern only in the near burner field, where V_p and V_g differ significantly, but the engineering importance of burner flame stability and of near burner pollutent reactions demands that further research attention be paid to them.

The particle/turbulence interaction is another complex phenomenon which has so far only been partially addressed. The particles will be dispersed primarily by the large scale turbulence motions. A frequently employed modelling approach is to presume that the fluctuating component of V_g is randomly directed with a magnitude probability described by a gaussian function, the variance of which is $2k^{\frac{1}{2}}/3$. Truelove[12] has adopted such a model with success in the simulation of the near field of a swirl burner, and Boyd and Kent[13] have used the same approach to predict the dispersion of particles in the bulk combustor volume. It needs to be stressed, however, that stochastic particle dispersion models are expensive of computer time because rather large sample sizes are required to achieve statistical independence. Also neither this type, nor any other type, of dispersion model has been subjected to systematic validation.

The energy balance equation for a particle in lagrangian form may be written:

$$C_p \frac{dT_p}{dt} = Q_p - \frac{L}{m_p} \frac{dm_p}{dt} \tag{2}$$

where $C_p \equiv$ particle specific heat, $T_p \equiv$ particle temperature, $Q_p \equiv$ specific particle heat gain, $L \equiv$ latent heat of devolatilisation. Discussion at this meeting has revealed controversy over the choice of value for L but there is some consensus that its role is negligible. Q_p is composed of: the heat gain by convection, Q_c; the heat gain due to char burning, Q_b; and the heat gain by thermal radiation, Q_r. Expressions analogous to and open to the same criticisms as those used for C_D are employed for Q_c. The energy release due to char burnout may be shared by the particle and adjacent gas. Numerical experiments show that predictions are little dependent on the distribution function and in practice the whole release is usually assigned to Q_b.

Equations (1) and (2) may be solved by an economical recurrence relation[14]. The eulerian gas phase and lagrangian solid phase calculations are combined in a hybrid solution procedure with the interactions being handled through appropriate source terms appended to the gas phase governing equations[14,15].

Modelling of devolatilisation

Gasification processes excepted, it is the general experience of the author's group that the release of volatiles in a PF combustor, once the particles have attained a threshold temperature, is so rapid that a precise prescription for the kinetic rate is unnecessary. Indeed, it is frequently sufficient simply to assume that the devolatilisation rate is constant above a threshold temperature of about $600^{\circ}K$[16]:

$$\frac{dx_v}{dt} = B \tag{3}$$

where $x_v \equiv$ fraction by weight of the total volatiles evolved, and B is a constant having a value of about 30 [14] for bituminous coals.

A much more complex formulation has been proposed by Anthony et al[17] which presumes that the devolatilisation kinetics may be represented by multiple and parallel reactions. The mass fraction of volatiles evolved at time t is given by:

$$V = \frac{V^*}{\sigma \sqrt{2\pi}} \int_0^\infty [1 - \exp(- \int_0^t K_0 \exp(- E/RT)dt] \times \exp[-(E - E_0)^2/2\sigma^2] \, dE \quad (4)$$

where $V^* \equiv$ mass fraction of volatiles evolved when $t \to \infty$, $K_0 \equiv$ a single rate constant for all of the multiple reactions, and E_0 and σ are the mean and standard deviation for a presumed Gaussian distribution of activation energies for the multiple reactions. This formulation has been employed with success in predicting test furnace performance of several Canadian bituminous coals[14,18], but the improved predictive accuracy over that achieved by the very simple eqn.(3) is not always sufficient to justify the substantially reduced computational efficiency of the eqn. (4) proposal. A single reaction model described by one rate constant and one activation energy may well constitute of compromise solution of considerable universality.

It is well known that the value of V^* realised under rapid particle heating in a PF combustor exceeds the volatile content, VM, determined under the slow heating rates of the standardised proximate analysis. The ratio V^*/VM is typically 1.5 for high volatile coals and as high as 2.5 for low volatile ones[19]. Validation studies[14,18,19,20] have revealed that the predictions are sensitive to the value of V^* which, because experimental values are frequently unavailable, must with disadvantage be estimated by numerical experimentation. It should be noted that the effect of high temperature volatile release is embraced by Wall et al[19] through the implementating of a two competing reaction scheme for the devolatilisation kinetics.

Volatiles combustion

The simulation of the combustion of the volatiles poses a major modelling problem which can be awarded only cursory discussion in the space available. Because the volatiles are released in a generally air deficient region at temperatures generally less than those required for ignition volatiles combustion in the immediate proximity of the particle surface of the isolated particle type cannot in general be envisaged. Rather the volatiles will premix with the surrounding air (and any products), most likely incompletely, so that we are forced to consider the most complex of all combustion phenomena: partially premixed turbulent reaction. It has to be admitted that state of the art models for this phenomenon are insufficiently developed and tested for engineering purposes. In the face of this some coal combustor modellers, e.g. [21], have preferred to assume that the volatiles burn in an unpremixed or diffusion flame for which reasonable theoretical formulations exist. Others, e.g. [14,20], take refuge in a suitably simplified description of partially premixed turbulent combustion which relates the reaction rate to the time scale of the large scale turbulence motions and the time-average of the controlling species concentration.

Char particle combustion

The heterogeneous combustion rate of a char particle depends on: the rate of arrival of oxygen at the particle surface or the external diffusion; the rate and extent of penetration of oxygen into the pore structure of the particle or the internal diffusion; and the chemical kinetics of the char material. Obviously the last two of these three effects depend on the nature of the char, that is on the coal type from which it is derived and, to a lesser extent one hopes, on its combustor history. The term char reactivity is frequently rather loosely used to refer to char particle reaction rate as governed by the combination of internal diffusion and heterogeneous kinetics effects. This subject area has been well discussed by Smith[22].

For engineering purposes model simplification is again warranted, especially in view of the shortage of reactivity data in relation to the multitude of char types encountered in practice. Smith[22,23] recommends describing the char reactivity by a first order reaction of rate constant:

$$k_c \equiv A \exp(- E_a/RT) \tag{5}$$

The apparent activation energy, E_a, is assigned a value of about 20 kcal/gm mole which is found to apply fairly well to a range of coals for the typical combustor condition where the effects of pore diffusion and char chemistry are both influential (regime II)[22]. The frequency factor, A, is based for convenience only on the external particle surface area alone, rather than the total of its internal plus external surface areas. The effect of coal type is accommodated by the single parameter A. In general its value decreases with increasing rank[19,20]. Analogous behaviour would require its value to diminish with increasing burnout, but such 'tail-end tailoring' has so far been ignored by modellers.

The preceeding arguments permit the reaction rate to be expressed as[22]:

$$R_c = k_c m_{O2,0} = k_d(m_{O2,\infty} - m_{O2,0}) = m_{O2,\infty}/(1/k_d + 1/k_c) \tag{6}$$

where $m_{O2,0}$ and $m_{O2,\infty}$ are the mass fractions of oxygen at the particle surface and in the surroundings, and $k_d \equiv$ is the external diffusion coefficient the value of which for engineering purposes may be sufficiently well determined[22].

Char particle swelling

The coal particles may swell during devolatilisation and diminish in diameter during burnout. The evidence relating this behaviour to coal type and combustion conditions is sparce. Since the A factor is conventionally related to unit external surface area, inconsistency in its usage between studies may occur due to different workers employing different particle size variation laws. It is probably not too unrealistic to presume constant diameter particle histories and for the most part this is why the author's group has adopted this convention.

Thermal radiation transfer

The role of thermal radiation in coal-fired combustors is substantial because they are physically large and because coal flames are highly absorbing. Indeed, thermal radiation is the dominating process in the very large combustors of power stations to the extent that the wall heat transfer distributions in them are strongly dependent on details of the radiation calculations and little dependent on those of the flow[18]. The variation of radiation intensity in an absorbing and scattering medium such as a coal flame is described by:

$$\underbrace{\frac{dI}{ds}}_{\text{change}} = \underbrace{k_g(\frac{\sigma T_g^4}{\pi} - I)}_{\substack{\text{gas} \\ \text{emission absorption}}} + \underbrace{k_p(\frac{\sigma T_p^4}{\pi} - I)}_{\substack{\text{particle} \\ \text{emission absorption}}}$$

$$- k_s I \quad + \quad \frac{k_s}{4\pi} \int_{4\pi} p(\underset{\rightarrow}{\Omega} \cdot \underset{\rightarrow}{\Omega'}) I(\underset{\rightarrow}{\Omega'}) d\Omega' \tag{7}$$

out   in scattering
scattering

where $I \equiv$ radiation intensity in the direction of $\underset{\rightarrow}{\Omega}$, $s \equiv$ distance in the $\underset{\rightarrow}{\Omega}$ direction, $\sigma \equiv$ Stefan-Boltzman constant, k_g and $k_p \equiv$ gas (including soot) and particle absorption coefficients, $k_s \equiv$ particle scattering coefficient, and $p(\underset{\rightarrow}{\Omega} \cdot \underset{\rightarrow}{\Omega'})$ is the probability of radiation travelling in other directions $\underset{\rightarrow}{\Omega'}$ being scattered into the $\underset{\rightarrow}{\Omega}$ direction.

The gas ∂r non-luminous absorption coefficient is highly non-grey[24], coal particles are highly absorbing and non-scattering[25], while the ash particles scatter strongly and isotropically[26]. Clouds of highly absorbing soot particles are also present in coal combustors. Fortunately recommendations are available which for the most part permit modellers to specify the mass of required radiation property information in a convenient manner for engineering purposes[14,24,25,26,27]. Because coal flames are almost always close to the optically think limit a precise specification of the radiation properties is frequently unnecessary.

The primary obstacle hindering the calculation of the thermal radiation transfer is the solution of eqn. (7). Many existing treatments, such as that described in [24] for example, involve the numerical solution of the three-dimensional and integro partial-differential equation which results when eqn. (7) is generalised for all spatial directions. Such methods, however elegant, are inevitably mathematically complex and they lack geometric flexibility because one is bound by the constraints of conventional co-ordinate systems when writing and solving the governing integro differential equation. These disadvantages are serious ones to the engineer who on the whole would prefer to avoid analysis based on complex mathematics and whose combustor most often does not fit neatly into the framework of conventional co-ordinate systems.

The 'discrete transfer' method which has been proposed in [28] avoids these difficulties. The essential feature of this method is that, rather than write eqn. (7) in three-dimensional space prior to its solution, eqn. (7) is solved directly for discrete solid angle elements within which the intensity is presumed uniform. The solution takes the form of an economical recurrence relation of the form:

$$I_{n+1} = \frac{E^*}{\pi}(1 - e^{-\delta s^*}) + I_n e^{-\delta s^*} \tag{8}$$

where the subscripts n+1 and n designate successive locations in the direction $\underset{\rightarrow}{\Omega}$ separated by $\delta s^* \equiv (k_g + k_p + k_s)\delta s$, $E^* \equiv (\sigma/\pi)[k_g T^4 + k_p T_p^4 + (k_s/4)\int_{4\pi} I(\underset{\rightarrow}{\Omega'})p(\Omega \cdot \Omega')d\Omega']/(k_g + k_p + k_s)$ and E^* is taken as uniform over δs^*. The solution of eqn. (8) is determined successively within each discretisation of the hemisphere about selected wall locations commencing from the known initial conditions at those locations and terminating when an opposing wall is encountered. In this manner the view factors are worked out implicitly and geometric irregularities pose no problem since the situations of surfaces merely define the initiation and termination location for application of the recurrence formula.

APPLICATIONS TO REAL COMBUSTORS
Some background
Gibson and Morgan[29] must take credit for the original application of a generalised computer based solution technique to a coal combustor. As

with many pioneer studies its importance was insufficiently recognised at
the time and as the whole subject of computational fluid mechanics was then
so novel the authors were disadvantaged in interpreting and exploiting
their results because knowledge about the validity of the many assumptions
involved was very small. Richter[30] in a subsequent significant study made
predictions of some anthracite flames fired in the IFRF furnace with a
level of precision which was remarkable for the time. Somewhat later still
Lockwood et. al[11] found that it was possible with available modelling knowledge to predict to an encouraging degree the result of the IFRF bituminous
coal, non-swirling, long flame trials.

The status of the coal combustor prediction field up to about 979 is
summarised in the book of Smoot and Pratt[31].

Swirl burner fired cylindrical furnace

Some predictions of swirl burner bituminous coal firing are reported
in [14,18,20] albeit for rather small swirl numbers. The furnace is the 0.8 m
diameter cylindrical one located at the Canadian Combustion Research
Laboratory (CCRL), the fuels were Canadian bituminous and low volatile
coals. Panel (a) of Fig. 3 shows the furnace dimensions and the computational grid.

Panel (b) of Fig. 3 shows predictions for a bituminous coal compared
with data for the distribution of the incident radiation energy flux impinging on the cylindrical walls. The parameter is the pre-exponential factor
A in the char burnout law, eqn. (5). The predictions are insensitive to A
for values greater than about 0.43 because the char particle reaction is
controlled by external diffusion in this range. The most that can be said
from this example of numerical experimentation is that the true value of
A for this coal type is no smaller than this value. Panel (c) shows prediction of char burnout for kinetic rate recommendations for several coals
plus the numerically optimised values of A and E for the test coal. Unfortunately one data point only is available! that for the exit section, and
again this fact severely limits the extent to which recommendations for the
values of A and E can be made. This graph is included here to emphasize
an important current need, that for the distributions of combustor char
burnout data. Panel (d) compares predicted and measured gas temperatures.
The two sets of data points give evidence of the typical difficulty of
obtaining truely axisymmetric flow in cylindrical combustors. The quality
of the predicted results is encouraging. However, it should be noted that
the most upstream profile is not very near the burner. It may well be that
prediction of the near burner field will prove much more difficult. Measurement in this region is particularly problematic and data remain scarce.

Burner flame stability

The important phenomenon of burner flame stability depends on the
characteristics of the near burner field. While measurement there may be
difficult the existence or not of a stable flame is readily observed and
may be correlated with the burner input and geometrical conditions. Much
experimentation of this kind has been performed on the vertical cylindrical furnace at Imperial College, see[32] for example. A representative experimental result is shown in panel (a) of Fig. 4 where the burner stability
performance of the Imperial College cylindrical furnace for fixed burner
geometry, excess air and firing rate is correlated in terms of the burner
swirl number, S_s, and the ratio of the primary to secondary axial momentum
ratio, R_m. Too low values of S_s or too high values of R_m lead to unstable
operation, that is to flames which are lifted rather than lit back and
anchored at the burner face.

Example predicted flow patterns corresponding to unstable, panel (b), and stable, panel (c), burner operation are also shown in Fig. 4. The unstable example shows the bulk of the coal particles punching through a relatively weak near burner or internal recirculation zone (irz) due to their high momenta. The stable one shows the majority of coal particles being retained by a stronger irz. In the latter case stable combustion results because the long residence times of particles in the irz ensures volatiles release, strong gaseous phase combustion, and particle ignition there. The Imperial College stability data are well supported by parallel predictions of this kind. There seems little doubt that the understanding of burner flame stability and progress towards improved burner designs can be greatly assisted by the use of existing mathematical models.

Comment on cement kiln prediction

The cement manufacturing industry is the second biggest user of coal fuel. The cylindrical geometry of cement kilns and their (usually) geometrically simple burners has attracted the interest of modellers. An early study is reported in [33]. This has been followed in the author's group by various proprietary studies one of which will be reported in [34]. There is insufficient space to summarise these specialised industry studies here. It suffices to say that current predictive capabilities have aroused international industry interest.

Power station combustors

The power generation industry is by far the largest and most important consumer of coal. The application of mathematical prediction models to the flow, combustion and heat transfer in power station combustors lags somewhat behind corresponding cement kiln experience primarily due to the more complex geometrical problems consequent of multi-burner firing of three-dimensional chambers. Panel (a) of Fig. 5 shows the geometry of a typical front-wall, swirl burner fired unit. The United Kingdom Central Electricity Generating Board (CEGB) has obtained high Reynolds number inert flow velocity data for a 1/35 model of full scale model of this geometry. It is clear that detailed computations of such a combustor would be intricate and costly because of the very fine grids which would be necessary in the vicinity of each of the 16 burners. Current philosophy advocates that the near burner and bulk combustor flows be computed separately. In [35] the bulk flow in the combustor of Fig. 5(a) is predicted using the grid illustrated in panel (b) in which the burners are simply treated as sources of mass and momentum in the adjacent finite-difference cell.

Predictions of the velocity vectors in two x-z planes are shown in panel (c), one plane coincides with the location of a column of burners, the other is located between burner columns. The persistence of the burner jets in traversing the combustor is noticeable, more noteworthy perhaps is the fact that the velocities in the hopper region are not everywhere negligible as is sometimes assumed. Sample comparisons of predictions with the CEGB data are presented in panel (d) of Fig. 5. On the whole the agreement is excellent.

Wall heat transfer rates have been determined for a similar coal-fired combustor in another study [18] in which the experimentally measured flow was an input condition to the calculations and in which an assumed distribution of heat release due to combustion was prescribed. The performances of two bituminous coals were examined, one from the Saar region of Germany and the other a high ash fuel from the Australian state of New South Wales. Typical heat transfer contours are presented in Fig. 5(d). The level of heat transfer is very dependent on both the emissivity of and thermal conductivity of any wall ash deposits. The values of some 400 kW/m^2

observed in the figure for the back combustor wall may be considered the maximum to be found in any combustor operating under optimal conditions. Important conclusions from this study were the facts that the wall heat transfer distribution is relatively insensitive to the assumed velocity field, while its dependence on the prescription for the heat release distribution can be strong.

Work is progressing in the author's group on the goal of the complete prediction of power station combustor performance. In this regard some interim predictions of the coal particle trajectories are shown in panel (e) of Fig. 5. Boyd and Kent[13] in Australia commencing with the basic Imperial College flow and heat transfer computational tools have been able to attain the final goal rather faster than ourselves. In addition they have been in the fortunate position of being able to make comparisons of their pre- dicted heat transfer rates with ones obtained on a real combustor belonging to the New South Wales State Electricity Commission. The level of agree- ment is quite remarkable and their study represents somewhat of a landmark in the evolution of mathematical modelling as a coal-fired combustor design tool.

CONCLUDING REMARKS

The subject of the mathematical modelling of combustors is a complex and multi-disciplinary one, this is especially true when the fuel is coal. The current status of coal combustor prediction methods is such that one can confidently advise interested industry parties that in spite of limitations, these methods all the same constitute a useful design tool provided that a sufficient commitment is made to acquire the necessary expertise to use the tool effectively. The current and international mushrooming of interest in combustor mathematical models is evidence that industry is, after some- what over a decade of diffidence, at last recognising their potential.

The further development of coal combustor mathematical models would be greatly hastened if enhanced cooperation in the common goal between experts in the various disciplines, and across the board between scientist and engineers, could be promoted, in human terms a tall order. Priority research areas are readily identifiable. Numerical solution schemes for the gas phase flow equations of improved geometric flexibility and which are robust, economic and free of the effects of numerical diffu- sion are required. Modern adaptive grid techniques have been slow to find their way into the engineering combustor prediction field. The requirement for robustness is a particularly demanding one for reacting flows since those numerical schemes which eliminate false diffusion invari- ably display 'over- and under-shoots' during the iterative evolution of the solution which can cause solution failure where combustion is concerned.

The total quantity of volatiles released by the char under combustion conditions has emerged as important in numerical sensitivity studies. More experimental research is needed to relate this quantity to coal type as characterised by its proximate, ultimate, and maceral compositions and to its in-flame temperature history. More empirical information on char re- activities as a function of coal type is needed, and especially on the dependency of reactivity on burnout if the flame tail is to be well simulated.

Current predictive techniques are based on volatiles combustion modell- ing which is rather too rough and ready. A model of the general case of turbulent partially premixed combustion is required which is above all economical and which preferably recognises the importance of carbon monoxide as an intermediate product of combustion. Readers familiar with this area will recognise this as a daunting request.

The importance of the role of the particle/turbulence interaction is as
yet unsettled. It is likely that this complex process need not be modelled with great precision but the validity of all existing models remains
untested and the commonly used stochastic one is insufficiently economic.

Fuel NO_x appears to play the dominant role in pulverised coal combustors.
Existing formulation for its formation and destruction have been too
little applied by the solid-fuel combustor prediction fraternity. It is
likely that some useful trends could be predicted now to the extent that
features of burner aerodynamics leading to lower NO_x emissions could be
better understood[35,36].

The preceding list of requirements and needs is not complete but in the
author's opinion it is a fair summary of the more obvious priorities. The
most urgent need of all is not on the predictive side. Rather, it is the
need for more experimental combustor data, particularly in the near burner
field, against which model formulations can be validated.

REFERENCES

1. Gosman AD, Lockwood FC, Megahed IEA and Shah NG: The Prediction of the
   Flow, Reaction and Heat Transfer in the Combustion of a Glass Furnace.
   AIAA Paper 80-0016, 1980, J. Energy, Vol. 6, No. 6, p. 353, 1982.
2. Carvalho MDGMDS and Lockwood FC: Mathematical Simulation of an End-Port
   Regenerative Glass Furnace. Proceedings of the Institution of Mechanical
   Engineers, Vol. 199, No. C2, 1985.
3. Abbas AS, Lockwood FC and Salooja AP: The Prediction of the Combustion
   and Heat Transfer Performance of an Experimental Refinery. Combustion
   and Flame 58: 91-101, 1984.
4. Abbas AS and Lockwood FC: Note of the Prediction of Soot Concentrations
   in Turbulent Diffusion Flames. Journal of the Institute of Energy,
   vol. 58, pp. 112-115, Sept. 1985.
5. Gosman AD, Lockwood FC and Salooja AP: The Prediction of Cylindrical
   Furnaces Gaseous Fueled with Premixed and Diffusion Burners. Proceedings
   of 17th Symposium (International) on Combustion, p. 747-760, 1978.
6. Thompson JF: Numerical Solution of Flow Problems Using Body Fitted
   Coordinate Systems. Comp. Fluid Dynamics. Edited by W. Kollmann,
   Hemisphere Publishing Co., 1980.
7. Patankar SV: Numerical Heat Transfer and Fluid Flow. Series in
   Computational Methods in Mechanics and Thermal Sciences. McGraw Hill
   Book Co. 1980.
8. Weber R, Boyson F, Swithenbank J and Roberts PA: Computations of Near
   Field Aerodynamics of Swirling Expanding Flows. 21st Symposium
   (International) on Combustion, Munich, 1986.
9. Leschziner MA: Practical Evaluation of Three Finite-Difference Schemes
   for the Computation of Steady State Recirculating Flows. Comp. Meth.
   Appl. Mech. Eng., Vol. 23, p. 293, 1980.
10. Spalding DB: The Numerical Methods for Multiphase Flows. Modelling and
    Solution Techniques for Multiphase Flow. Commission of the European
    Communities, ISPRA, MMF/85/2, 1985.
11. Lockwood FC, Salooja AP and Syed SA: A Prediction Method for Coal-
    Fired Furnaces. Combustion and Flame, Vol. 38, No. 1, pp. 1-15, 1980.
12. Truelove JS: Prediction of Near-Burner Flow and Combustion in Swirling
    Pulverised Coal Flames. 21st Symposium (International) on Combustion,
    Munich 1986.
13. Boyd RK and Kent JH: Three-Dimensional Furnace Computer Modelling.
    21st Symposium (International) on Combustion, Munich, 1986.

14. Lockwood FC, Rizvi, SMA, Lee GK and Whaley H: Coal Combustion Model Validation Using Cylindrical Furnace Data. 20th International Combustion Symposium, August 1984.
15. Migdal D. and Agosta VD: A Source flow Model for Continuous Gas-Particle Flow. J. Appl. Mech., Vol. 34, p. 860, 1967.
16. Baum MM and Street PJ: Predicting the Combustion Behaviour of Coal Particles. Comb. Sci. and Technology, Vol. 3, pp. 231-243, 1971.
17. Anthony DB, Howard JB, Hottel NC and Meissner HP: Rapid Devolatisation of Pulverised Coal. 15th Symposium (International) on Combustion, The Combustion Institute, p. 1303, 1975.
18. Lockwood FC, Rizvi SMA and Shah NG: Comparative Predictive Experience of Coal Firing. Proc. Instn. Mech. Engrs., Vol. 200, No. C2, pp. 79-87, 1986.
19. Wall TF, Phelan WS and Bortz S: The Prediction and Scaling of Burn-Out in Swirled Pulverised Coal Flames; An Analysis and Mathematical Modelling of the AP11, AP12 and AP13 Trials. International Flame Research Foundation, Doc. nr. F388/a/3, March, 1986.
20. Lockwood FC and Rizvi SMA: Prediction of Combustion Performance of Low Volatile Coal. Report to Canada Centre for Mineral and Energy Technology, July, 1986.
21. Smith PJ, Fletcher TH and Smoot LD: Model for Pulverised Coal-Fired Reactors. 18th Symposium (International) on Combustion, The Combustion Institute, p. 1285, 1981.
22. Smith IW: The Combustion Rates of Coal Chars: A Review. 19th Symposium (International) on Combustion, The Combustion Institute, p. 1045, 1982.
23. Smith IW: Private Communication, 1986.
24. Hottel HC and Sarofim AF: Radiative Transfer. McGraw Hill Book Co., 1967.
25. Gibb J and Joyner PL: Liddell Boiler Reheater Metal Temperature: Gas Side Heat Transfer Analysis. Report No. R/M/NI029, CEGB, Marchwood Engineering Lab., England (1979).
26. Gupter RP, Wall TF and Truelove JS: Radiative Scatter by Fly Ash in Pulverised-Coal-Fired Furnaces: Applications of the Monte Carlo Method to Anisotropic Scatter. Int. J. Heat Mass Transfer, Vol. 26, No. 11, pp. 1649-1660, 1983.
27. Truelove JS: A Mixed Grey Gas Model for Flame Radiation. Report No. AERE-R8494, UK Atomic Energy Authority, AERE, Harwell, 1976.
28. Lockwood FC and Shah NG: A New Radiation Solution Method for Incorporation in General Combustion Prediction Procedures. Proceedings of 18th Symposium (International) on Combustion, pp. 1405-1414, 1981.
29. Gibson MM and Morgan BB: J. Inst. Fuel, pp. 517-523, Dec. 1970.
30. Richter W and Quack R: A Mathematical Model for a Low Volatile P.F. Flame. Heat Transfer in Flames, Afgan and Beer (Ed.) pp. 95-110, Scripta Book Co., Washington, 1974.
31. Smoot LD and Pratt DT (Eds.): Pulverised Coal Combustion and Gasification. Plenum Press, New York, 1979.
32. Godoy S, Hirji K, Ismail M and Lockwood FC: Stability Limits of Pulverised Coal Firing. J. Inst. Energy, 59, p. 38, 1986.
33. Lockwood FC and Salooja AP: The Prediction of the Combustion and Heat Transfer from a PF Flame in a Cement Kiln. International Flame Research Foundation, Doc. nr. F 21/ca/43, Oct. 1979.

34. Lockwood FC, Lowes TM and Rizvi SMA: Mathematical Modelling of Cement Kiln Combustion and Heat Transfer. Submitted for 1st European Dry Fine Coal Conference, Institute of Energy, 1987.
35. Abbas AS and Lockwood FC: Prediction of Power Station Combustors. 21st Symposium (International) on Combustion, Munich, 1986.
36. Hill SC, Smoot LD and Smith PJ: 20th Symposium (International) on Combustion, The Combustion Institute, p. 1391, 1984.
37. Hirji KA, Lockwood FC, Offergeld H and Seitz CW: Measurement and Prediction of NO_x Emissions in a Coal-Fired Combustor. Imperial College, Mech. Dept. Rpt. FS/85/47, Dec. 1985.

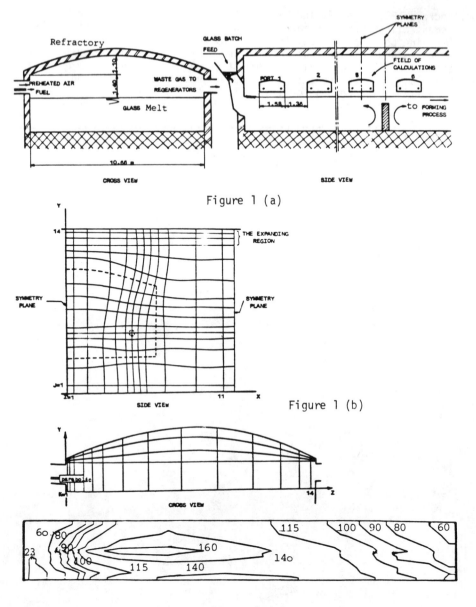

Figure 1 (a)

Figure 1 (b)

Figure 1 (c)

Figure 1. Prediction of a Glass Furnace(Ref 1)
    1(a)Furnace Geometry
    1(b)Finite difference grid
    1(c)Contours of Predicted Heat Transfer to glass kWm^{-2}

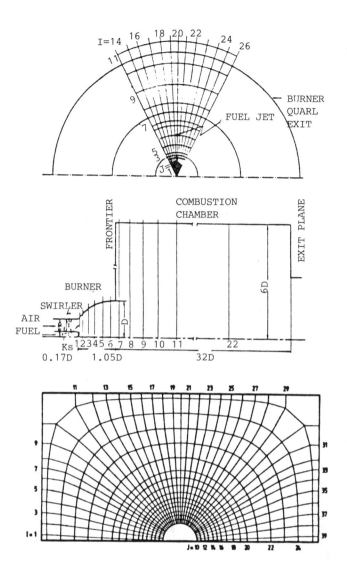

Figure 2 (a)

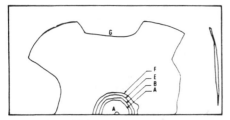

Contours of Axial Velocity, z/D = 1.15 (A = 60.2, B = 30.0, C = 15.0, D = 10.0, E = 5.0, F = 1.0, G = 0.0, H = −1.0 m/sec).

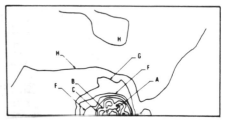

Contours of Temperature, z/D = 1.15 (A = 2000, B = 1600, C = 1300, D = 1100, E = 980, F = 960, G = 940, H = 920°K).

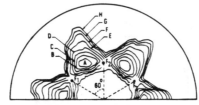

Contours of Mixture Fraction, z/D = 1.15 (A = 0.12, B = 0.11, C = 0.10, D = 0.09, E = 0.08, F = 0.07, G = 0.065, H = 0.06).

Figure 2 (b)

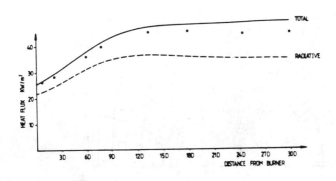

Figure 2 (c)

Figure 2. Prediction of a Refinery Heater (Ref 3)
Figure 2(a): Furnace geometry and finite difference grid
        2(b) Contours in cross sectional planes, D=quarl
              diameter.
        2(c) Predicted heat transfer to walls.
              (Points are data).

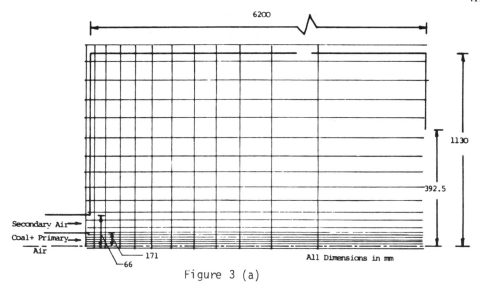

Figure 3 (a)

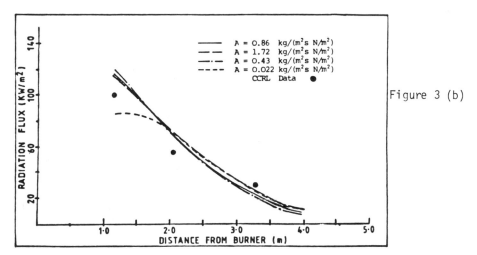

Figure 3 (b)

412

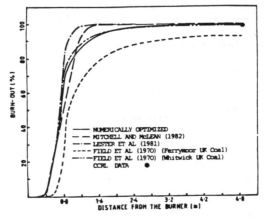

Figure 3 (c)

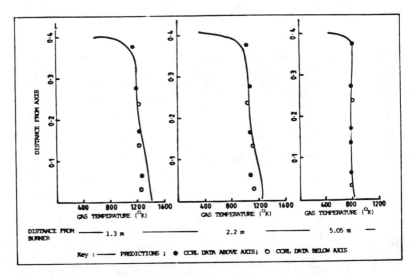

Figure 3 (d)

Figure 3. Prediction of Cylindrical Furnace of Canadian Combustion
          Research Laboratory (Refs. 14,18 and 20)
          3(a)Geometry and Grid
          3(b)Effect of Frequency Factor, A, on incident Wall Heat Flux
             for Bituminous Coal
          3(c)Char Burnout for Different Laws (Ref 14).
          3(d)Gas Temperature Predictions

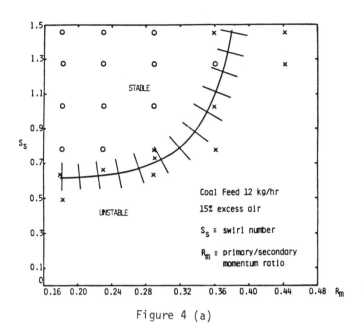

Figure 4 (a)

PREDICTION OF 1C COAL FURNACE   **UNSTABLE FLAME CONDITION**   Fig 4(b)
FEED RATE *12 Kg / hr , Rm 0.36 , Ss 1.0*

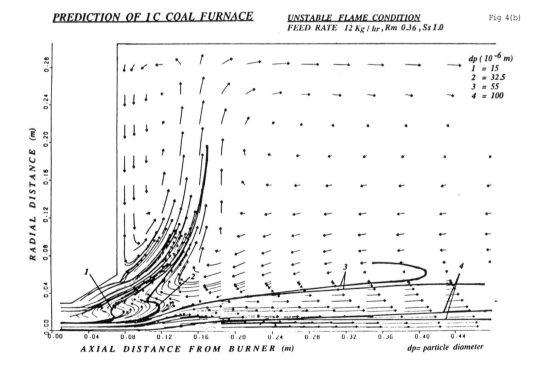

dp= particle diameter

414

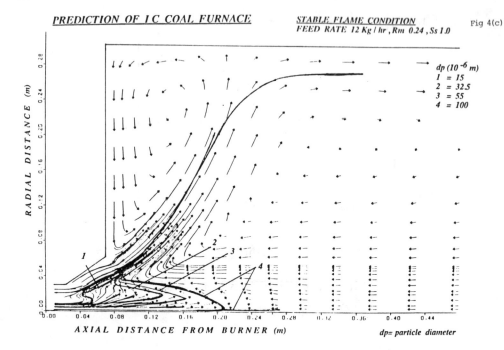

Figure 4. Prediction of Burner Flame Stability.
    4(a)Experimental Stability Curve (Ref 32)
    4(b)Predicted Near Burner Flow for unstable operation
    4(c)Predicted Near Burner Flow for stable operation

415

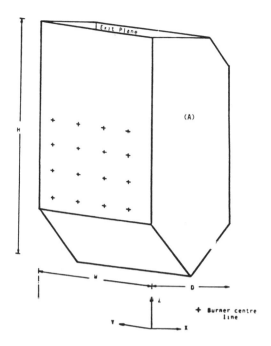

Fig. 5(a)

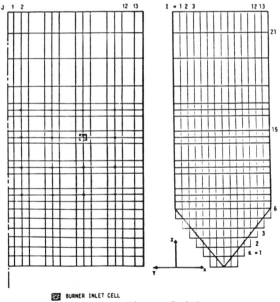

BURNER INLET CELL

Figure 5 (b)

416

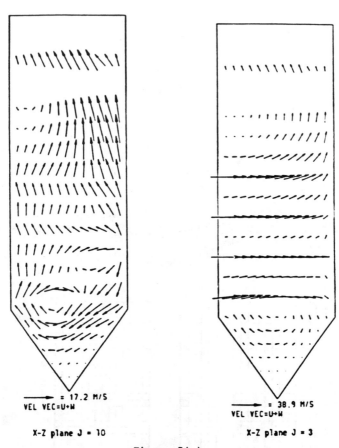

———▶ = 17.2 M/S
VEL VEC=U+W

X-Z plane J = 10

———▶ = 38.9 M/S
VEL VEC=U+W

X-Z plane J = 3

Figure 5(c)

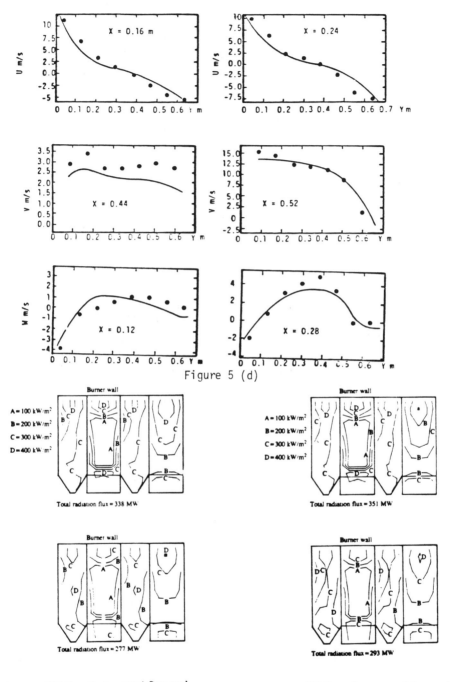

Figure 5 (d)

Figure 5 (e)

1100 K walls, top panel Saar coal, bottom panel NSW coal

733 K walls, top panel Saar coal, bottom panel NSW coal

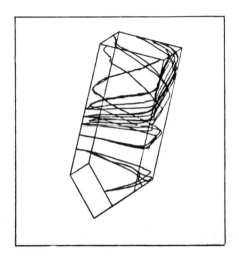

Figure 5 (f)

Figure 5. Prediction of Power Station Combustors
   5(a)Geometry of a Typical Front Wall Fired Unit (Ref 35)
   5(b)Predictive Grid
   5(c)Predicted Velocity Vectors in Vertical Planes
   5(d)Comparisons of Predicted Velocities with CEGB Data
   5(e)Predicted Wall Heat Transfer (Ref 18)
   5(f)Example Prediction of Coal Particle Trajectories

DISCUSSION

W. Zinser

In your presentation you have shown a computational grid for a front-wall fired furnace. Typical industrial coal burners in these furnaces are stabilized by strong swirl which affects probably the burner flow field as a whole. Having this in view, do you see a way to predict the turbulent flow field in these furnaces at reasonable expenses ?

F. Lockwood

Fortunately the local heat transfer are fairly insensitive to the bulk chamber flow field the essential features of which are not too sensitive to details of the near field flow. The only current economic calculation possibility is to make bulk chambers computations of the kind I have described to predict the wall heat transfer and to make separate fine grid computations for near field features such as flame stability. The two could ultimately be combined in a hybrid prediction procedure.

P.R. Solomon

Did I understand correctly that your simulations of flames are sensitive to your choice of devolatilization rates ? Whithin your model what rates agree with observations and which don't ?

F. Lockwood

On the whole we find an insensitivity of the predictions to the devolatilization rate specification. The temperature rise in the near burner field is generally so rapid that the volatiles evolve in a time scale which is short relative to other relevant time scale in this vicinity. It must be admitted, however, that there are few near field data and it is rather too early in prediction procedure history to justify a categoric statement. We have used the Anthony and Howard recommendation with success for bituminous coals. It has proved less satisfactory for the low volatile coals which are the subject of a current study. We hope to publish our findings on these coals early in the new year. Prof. Wall's recent study of low volatile coal prediction of the IFRF should also be consulted.

M. Morgan

Can you compare the state of development of aerodynamic, heat transfer, coal chemistry submodels ?

F. Lockwood

The available turbulence models, a fundamental component of all prediction procedures, are less than perfect. Their

420

defects are at least reasonably well established and not so great as to preclude the possibility of useful engineering predictions. The existing coal chemistry submodels seem capable of providing reasonable predictions, but it is still early days in the computer simulation of coal-fired combustors, and there is a lack of combustor data especially in the near burner field.

E. Saatdjian

Would you expect a big difference between theoretical and calculated particle temperatures ? The reason I ask this question is that in the heat transfer to the particle there are many phenomena that are very hard to model : the soot cloud around the particle, the diminished transfer during devolatilization, the "debatable" value of the heat of devolatilization, the fact that not all particles "see" the wall.

F. Lockwood

Our predictions so far suggest a maximum temperature difference between the particles and the neighbouring gas of 150-200°C.

K.H. Van Heek

Do the kinetic parameters used by you for modelling the pyrolysis match the kind of coal burned in the Canadian boiler for which the calculations were made ? If not, would it not be better (and easier) to describe the pyrolysis by mass balance (char, gases and tars) ?

F. Lockwood

We use Anthony and Howard's multiple parallel reactions devolatization recommendation for the Canadian bituminous coals even though this recommendation is derived from experiments on Pittsburgh coal. It appears to work adequatly well for our purposes but again I must stress that we have no near field data.

R.H. Essenhigh

For a particle of mass $m = \frac{4\pi}{3} a^3 \sigma$, the specific reaction rate, R , (mass loss from the total particle referred to unit surface area/time) is given by differentiating.

$$R_s = \frac{dm/dt}{4\pi a} = \sigma \, da/dt + (a/3)d\sigma/at = f \text{ (particle properties)}$$

surface   internal
regression   mass loss
rate   rate

When the particles burn internally and externally, the rate equation (above) is inexact : solution requires a relation between radius and density. The equation has to be integrated

along a particular path in σ-a space.
    What relation do you use between σ and a in your model-
ling ?

F. Lockwood

    Our usual practice is to assume that the particle diame-
ter remains constant during burn out. Of course this may not
always be acceptable but real combustor evidence is scanty.

R.H. Essenhigh

    Reinforcing Dr. Lockwood's point that his computational
results appeared to be somewhat insensitive to the pyrolysis
model used, in our one-dimensional, radiation stabilized
(laminar) flame, the modelling was successful for a bitummous
coal flame when the volatiles combustion was omitted entirely.
The reasons for this are complex but depend on the density/
radius relation invoked without which the system of equations
cannot be solved.

F. Lockwood

    Thank you for your reinforcement of our general observa-
tions. I agree that the reasons for the lack of dependency on
pyrolysis are complex. It may be that flows of practical interest may one day be identified for which the pyrolysis rate
emerges as a more important parameter.

T.R. Johnson

    1) This is a comment on the influence of fluid flow pat-
terns on heat transfer in furnaces. In earlier work I performed on zone modelling of the ljmuiden furnace, I investigated
the proportions of heat transport between zones. This showed
that the three major forms of heat transport, i.e. radiation,
chemical heat release and convection of sensible heat, were of
similar magnitude, particularly in the flame region. Thus the
fluid flow distribution will have a considerable effect on
distribution of temperature in the furnace.

    2) This comment concerns the relative influence of emis-
sivity and thermal conductivity of ash deposits on furnace
heat transfer. In the energy balance of an ash-covered wall,
both the emissivity and the thermal conductivity are included.
The emissivity defines the fraction of the incident radiation
which in absorbed by the wall, whereas the thermal conductivity (and thickness) governs the ash surface temperature and
hence the re-radiation of heat away from the surface. Thus
both of these ash parameters have a significant influence on
overall furnace heat transfer, and sensitivity analyses that I
have performed show them to have somewhat similar influence on
overall furnace heat transfer.

    3) You talked briefly of the turbulent diffusion of coal
particles in the particle-tracking part of your model. Others

422

have found it necessary to include turbulent particle diffusion in order to get realistic looking predictions of particle
paths. Do you include particle diffusion, and what influence
do you see it having on the realism of the particle path
predictions ?

F. Lockwood

1) All I can say is that it is our experience, and also
I believe in that of the CEGB, that the sensitivity of the
wall heat transfer distribution to details of the velocity
field is small. In other words the coupling between the
dominant radiation process and the aerodynamics is small. We
do find, however, that the influence of the heat release
distribution on the heat transfer distribution is strong.

2) Thank you for the comment with which I entirely
agree. I should say that for the exploratory heat transfer
calculations that we have performed we have generaly used ash
surface temperature and emissivity recommendations supplied by
the CEGB. For this reason we have so far avoided involving the
ash conductances in our calculations.

3) The matter of the turbulence/particle interaction is
very under researched. No validated model exists. We use a
simple gradient transport model which I have to confess does
not have a lot of physical justification. For these reasons it
is too early to make definitive statements about the effect of
turbulence particle dispersion on predictions. The large scale
turbulence in industry combustors will of course act to smear
ash deposition. The particles will respond less to the small
scales of the near burner field. However, if the particles
have insufficient momenta to penetrate the burner internal
recirculation zone, they can only do so by turbulence
dispersion and so the phenomenon can be relevant to flame
stabilization.

J.L. Roth

We believe that the prediction and mastering of the ignition point of the coal jet is also important in the case of
the blast furnace injection, because the ignition within the
tuyere (length ~ 500 mm, coal residence times ~ 5 to 10 ms)
can induce :
- on the one hand, accelerated tuyere wear/slagging/fouling,
  leading to economical and metallurgical losses.
- on the other hand, accelerated kinetics of coal gasification :
with the help of G. Prado, we showed that coal particles over
50 μm cannot be burned out within the tuyere + raceway trajectory (residence times less than 30 ms), and that their further
gasification (in a CO_2 - atmosphere, residence time about 1
sec.) can be strongly influenced by their initial heating
rate, and thus by the ignition distance.

In our industrial tests, we observed coal ignition distan-

ces variing from 300 mm to over 1 m with the coal type and
rate, with blast temperature and aerodynamic conditions.

Even if we cannot speak of a classical problem of flame
stabilization, there is an industrial compromise to be chosen,
and some installators of coal injection utilities sell their
know-how on this point (see Kobe Steel).

F. Lockwood

Thank you for this interesting information as to the relevance of ignition in blast furnaces.

R.H. Essenhigh

1) My contacts with utilities lead me to conclude that carbon burn-out is considered a problem mainly by those on the
"sidelines".Research in T.V.A. has given it some emphasis, but
the operators generally feel it is good enough for most
purposes and at this time it is a minor economic loss compared
with other factors.
2) I cannot believe that "flame stability" will ever be a
problem in the blast furnace fuel-injection. I had occasion
to discuss the fuel injection problem on three consecutive
weeks last year with representatives from Nippon Steel, IRSID,
and BHP. Independently of each other they agreed that ignition
was no problem at all : not in 1000 or 1700°C air. The concern
that did come up was the ignition time and the fraction of
ignition time as a fonction of residence time in the race way.
The Nippon Steel investigations were particularly clear. If
there was too little time left for burn-out the reaction could
be quenched and post-burned char would pass right through the
blast furnace burden. I think it is necessary to make a clear
distribution between flame stabilization and ignition distance. Sometimes they are closely coupled and sometimes not.

F. Lockwood

I do not disagree with your first comment although operators would I suppose be glad to win small improvements in
combustion efficiency in the absence of parallel consequences
of increased equipment costs.
Your second comment should be read together with the preceeding one of Dr. Roth. I do not think there is disagreement.
I would remark that flame stabilization is the result of
sufficient residence time in an industrial p.f. burner. In
this manner the blast furnace and conventional burner ignition
phenomena are not dissimilar.

INVESTIGATION OF THE INFLUENCE OF FURNACE GEOMETRY AND COAL PROPERTIES ON FURNACE PERFORMANCE WITH THE HELP OF RADIATION TRANSFER MODEL

W. THIELEN, H.P. Odenthal, W. Richter

L. & C. STEINMÜLLER GMBH, D-5270 GUMMERSBACH

1. PROBLEMS IN THE DESIGN OF FIRING SYSTEMS

The problems which can arise in the design and operation of firing systems are manifold.

As far as the design is concerned, in most cases the furnace output, the firing scheme, the mass flows of the fuels and their thermal characteristics are already dictated.

The main unknown quantities are the thermal flows towards the furnace walls and the temperature at furnace outlet, and these variables must be determined in order for design of the furnace to be possible at all. Up to now these variables have been determined by simple mathematical models and experience with a good degree of success.

The actual processes taking place in the individual furnace zones are, however, still a matter for speculation. This point is important for the firing system designer, who must know whether stable ignition can be expected under the desired operating conditions.

The heat flux distribution must be accurately known in order to avoid local thermal overloading of the furnace wall and fouling buildup.

Fouling of the heating surfaces in furnaces as well as emission of pollutants is considerably influenced by the velocity field, turbulent mixing and trajectories of the solid particles and their condition in respect of degree of burnout and temperature. To exercise any control over these processes the velocity and temperature fields must be known, as well as the mechanism governing combustion reactions in the coal particles.

2. BASIC PRINCIPLES IN THE CALCULATION OF FURNACE USING ENERGY BALANCE PROGRAMS

A whole series of complex transport processes takes place within a furnace, each of which affects all the others to a greater or lesser degree. These processes can be divided in principle into three basic groups, viz.
- the convective transport processes, which are heavily influenced by flow turbulence, particularly in the flame area,
- thermal transport by radiation, which is determined by the optical properties of the gases and solid particles, as well as by temperature distribution
and
- the chemical reactions, which are influenced by the degree of turbulence, e. g. the combustion of volatile constituents

in the flame, which is affected by mixing, or burnout of a coal particle, which is a function of reaction kinetics and diffusion.

The complexity of the physical and chemical processes in industrial furnaces requires more or less drastic simplification of the physical description. In the following a solution method consisting of a combination of a mathematical model and model-cum-experiment procedure is being presented. Fig. 1 shows the basic approach for solving the problem. The furnace is divided into a number of separate balance zones. It is a good idea to assign one or more zones - depending on the degree of resolution required - to each individual burner. The energy balance must then be solved separately for each balance zone i. e. the energy flows across the balance envelope and the energy of the heat sources derived from the transformation of the chemical energy in the fuel must be worked out.

On the r. h. side of Fig. 1 can be seen in principle how such a balance is drawn up for the volume component. The heat liberated by transformation of the chemical energy in the fuel is expressed as the heat source $\dot{Q}_S$. Thermal transport by radiation is effected by emission ($\dot{Q}_e$) and absorption ($\dot{Q}_a$). The convective term $\dot{Q}_k$ embraces energy transport by sensible heat and the term $\dot{Q}_d$ transport by diffusion and thermal conductivity.

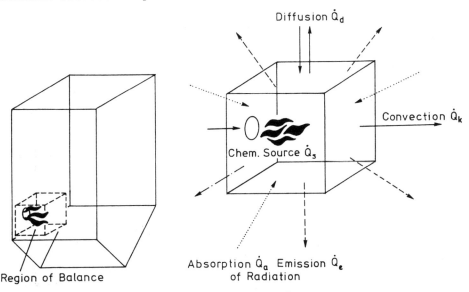

Diffusion $\dot{Q}_d$

Convection $\dot{Q}_k$

Chem. Source $\dot{Q}_s$

Absorption $\dot{Q}_a$ Emission $\dot{Q}_e$
of Radiation

Region of Balance

$$\dot{Q}_e - \dot{Q}_a + \dot{Q}_k + \dot{Q}_d - \dot{Q}_s = 0$$

FIGURE 1. Balance of energy in a volume element

The problem can be solved in three separate stages:
1. Working out the reaction kinetics for fuel input in the individual zones

426

2. Ascertaining the flow field in the furnace and calculating the mass flows across the balance zone boundaries
3. Calculating thermal transport via radiation, the temperature fields and fuel transformation by means of a suitable radiation exchange model which takes into account the
flow field and reaction kinetics.

2.1 Reaction Kinetics Model for Coal Combustion

The process of combustion in a coal particle can be divided conceptually into two main parts: the combustion of the
volatiles and the burnout of the degasified coke particle.

The coal particle which is due to undergo combustion is
heated up on entry into the combustion chamber partly by
transfer of heat from the hot flue gases and partly by its
own rapidly commencing reaction. The particle is degasified
as a result, i. e. the volatiles are expelled. The rate of
this degasification depends upon the rate at which the particle heats up, the size of the particle and the composition
of the coal itself. The exact composition of the volatiles
again depends on the progress of the heating-up reaction,
since intermediate reactions of various types can occur here.

For coal particles under 100 microns, which form the
greater part of the p. f. content in coal firing systems, and
at temperatures above 1000 °C the time of devolatilisation is
below 100 ms.

It is assumed for this purpose that devolatilisation takes
place within a negligible time, i. e. that the volatiles and
the degasified coke particle are already separated at the
point where combustion begins. Combustion of the two components then takes place side by side, though not completely
independent of each other.

The reaction rate in the combustion of the volatiles is so
great that, for industrial firing systems, it is of negligible effect in comparison with the influence of the mixing
times, which are determined by flow turbulence in the flame.
This means in effect that combustion of the volatiles is affected solely by the mixing effect and is thus essentially
dependent on the type of burner and the manner in which it is
operated.

Generally speaking, combustion of the volatiles can be expressed as an exponential function.

Immediately on heating up of the coal particle, burnout of
carbon at the particle surface commences. The burnout mechanism of the coke particle can be extremely complicated and
depends partly on the structure of the particle and its condition as a result of the events during heating-up and devolatilisation, partly on the thermal and chemical environment.

But already the use of a very simple burnout model (Fig. 2
and [1]) gives good results. It is assumed that the coke particle is surrounded by a stagnant gas layer several particle
diameters in extent. Oxygen diffuses through this boundary
layer and reacts at the particle surface to form carbon dioxide or carbon monoxide, which then diffuse out into the free
gas flow.

Given that it is mainly carbon monoxide which diffuses

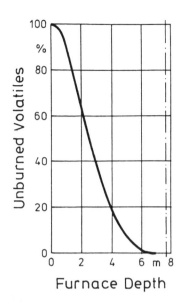

$$\dot{r} = \frac{p_{O_2}}{(1/k_D) + (1/k_f)} \quad [kg/m^2\,s]$$

Diffusion controlled
$$k_D = \frac{4\,M_C\,D_0\,(T_m/T_0)^{1.75}}{d_p\,R\,T_m}$$

$$D_0 = 3.49 \cdot 10^{-4}\,m^2/s \quad at \quad T_0 = 1600\,K$$

Kinetics controlled $\quad k_f = k_0\,\exp\,(-E/R\,T_s)$

$T_m\,(T_s)$ Temp. of bound. layer (particle surf.)

k_0 Frequency factor $[kg\,Char/m^2\,s\,bar\,O_2]$

Furnace Depth

Combustion of Volatiles   Combustion of Char

FIGURE 2. Burnout model

back into the free gas flow, the surface reaction rate can be described in terms of molecular diffusion, a first-order surface reaction at the particle itself and the partial pressure of oxygen in the free gas stream.

Values for the activation energy E and the frequency factor k_0 depend heavily on the coal type and can be calculated approximately from [2] or experiments.

During burnout of the coke particle, temperatures considerably in excess of the surrounding gas temperature may develop at its surface. These temperatures must necessarily be calculated and taken into account in any consideration of the energy balance derived from the energy flows to and from the coal particle.

In considering real combustion processes, the whole spectrum of coal grain sizes must be taken into account. For this purpose the coal particles are subdivided into individual coal grain classes and the burnout behaviour of each class considered in isolation. This procedure is described in detail in [3].

2.2 Flow field
For calculation of the mass flows across the balance zone boundaries, the knowledge of the flow field is necessary.

Strictly speaking, it is only possible to contemplate the flow field in conjunction with the process of combustion and gas temperature distribution i. e. as a general rule, only an

alternating reiterative series of calculations for the velocity and temperature fields will produce a correct answer.

In furnaces where the flue gas flows exhibit a high degree
of turbulence it is however possible, by observing certain
modelling rules, to determine the flow field separately in an
aerodynamic model by taking a series of isothermal flow measurements. The mass flows across the balance zone borders
required for calculation of the radiation exchange can be obtained from the isothermal measurements by integration across
the balance zone envelopes.

A very good reason for applying this procedure is that the
development of computer programs for calculating 3-dimen-
sional flow fields in large furnaces with many burners is,
even today, not so advanced that the application of such programs is necessarily easier or economically sounder. By contrast, a fund of long-term experience in the modelling of
flow processes can be drawn upon.

Further, experience has shown that the practice of taking
isothermal flow measurements in furnace models is capable of
revealing and thus allowing the correction of flow anomalies
which would otherwise occur in the fullsize furnace.

Fig. 3 and 4 show now for example the flow fields in the
burner cross section of a furnace with swirl burners and a
full-wall tangential firing system.

These model measurements have been carried out with a
triple-split-film probe [4], [5] which allows a fully com-
puter-controlled measurement and automatic evaluation.

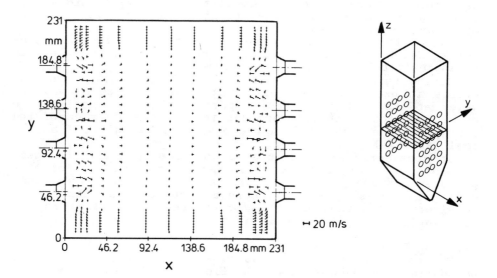

FIGURE 3. Furnace with swirl burners

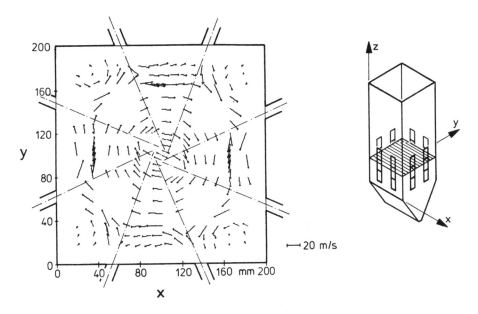

FIGURE 4. Full-wall tangential firing system

2.3 Calculation of Radiation Heat Transfer

Calculation of the temperature distribution, the radiation exchange, the fuel transformation and the gas composition in the various furnace zones is then finally carried out using a 3-dimensional radiation exchange mathematical model. This model, a zone procedure, operates on a so-called semi-stochastic method which represents a further development of the Monte Carlo method [3].

The model permits calculation of radiation transport to the furnace walls, the temperature distribution and gas composition, together with coal burnout calculations for coal firing systems.

By inclusion of a factor for depth of fouling and the thermal conductivity of the fouling layer on the walls, the influence of fouling on furnace behaviour can be investigated.

In this model the furnace is devided into individual zones. Starting from each zone, a discrete number of rays with controlled random points of origin radiate out and are followed along their paths through the furnace until they to all intents and purposes lose their identity through absorption in the various zones.

In contrast to the point of origin of the ray and its direction, other factors such as the absorption along the path of the ray are treated in a deterministic manner. In this way the unavoidable statistical errors of the Monte Carlo method are largely obviated.

3. APPLICATION OF THE MODELS TO LARGE UTILITY BOILERS

Boiler performance is influenced by many items, but one of the most important is the coal itself. Investigations within the presented model have shown, that the coal properties may not be considered isolated but in conjunction with the furnace design [6].

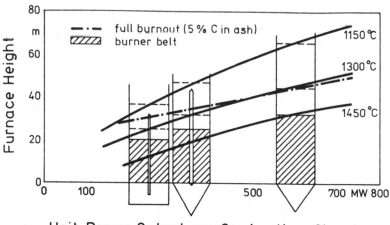

FIGURE 5. Characteristic temperatures and burnout points for different furnace sizes

A good survey is given by Fig. 5 where some characteristic temperatures for different furnace sizes are shown. As can be seen, temperatures in the burner belt increase with the furnace size. If coals with low ash melting points are fired, slagging on the furnace walls has to be taken into account in larger furnaces more than in smaller ones. But if some fundamental design rules are observed the problems can be handled.

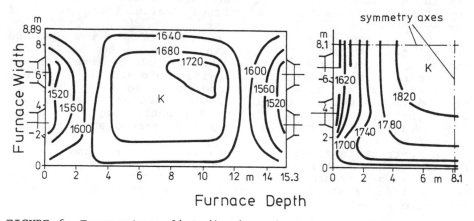

FIGURE 6. Temperature distributions in a burner plane

Fig. 6 shows the calculated temperature distribution in a burner plane of two swirl burner fired furnaces with different sizes but identical thermal cross section loads. The temperatures in the centre of the larger furnace are about 100 K higher than in the smaller one. But the choice of a large distance between burner and wall helps to decrease the temperatures in the near wall region so that the difference is here only about 60 K.

Furthermore the calculation of the trajectories and burnout of coal particles in the burner region shows that impingement of molten ash particles onto the furnace walls can even be reduced by optimising the burner arrangement.

In the radiation transfer model anisotropic scattering at fly ash particles can be considered.

The radiation transport equation can be written as:

$$\frac{dI}{ds} = \underbrace{- (k_a + k_s) \, I}_{(A)} + \underbrace{k_a \, I_b}_{(B)} + \underbrace{\frac{k_s}{4 \, \pi} \cdot \int_0^{4\pi} P \, (\Phi^*) \; I \, (\Omega^*) \; d \; \Omega^*}_{(C)}$$

(A) is the sum of absorbed and scattered radiation intensity, (B) the growth of radiation intensity by emission at the location s and (C) the growth of radiation intensity by scattering from outside.

As can be derived from this equation, scatter is able to reduce the radiation transport if the optical depths are high enough. This reduction can be even higher than the increasing of emissivity dependent on ash load, particle size, size distribution and particle material [7].

The impact of this anisotropic scattering is depicted in Fig. 7. Heat flux density and temperature distribution were calculated in a full wall tangential fired furnace of a 600 MW$_{el}$ boiler with and without inclusion of scatter. As can be seen, scatter increases the temperatures inside the furnace and decreases the heat transfer to the walls. As a consequence of this, heat transfer is shifted towards the convective heat exchangers behind the furnace chamber.

The comparison of several calculations with furnace measurements indicates, that
- scattering has to be taken into account if the ash content is high, optical pathes are long and ash material is optical transparent
- it seems at present time that the influence of ash absorption is nearly compensated by scatter if scatter has to be taken into account.

The strong influence of scatter on heat transfer especially in larger furnaces and a lag of knowledge about the influence of the ash composition suggest more systematic investigations of the effect of scatter in large p.f. furnaces.

In further studies the simple reaction model which is used for the description of the char burnout was examined [8].

Coal burnout in big furnaces is strongly influenced by
- coal particle size and size distribution which are determined by the fineness of grinding

432

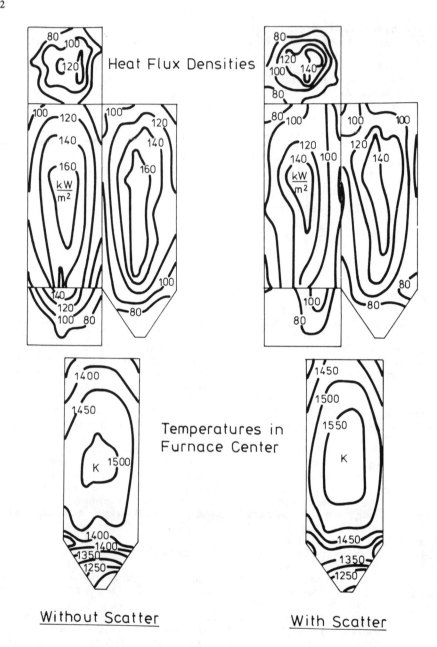

FIGURE 7. Impact of anisotropic scattering on net heat flux
density and temperature distribution in a fullwall
tangential-fired furnace

- coal composition and reactivity which roughly can be derived from the content of volatile matter
- flue gas temperature which is determined by furnace design, fouling and air ratio
- residence time in the furnace
- oxygen partial pressure which varies with the air ratio or if flue gas is recirculated.

The influence of particle size and content of volatile matter in coal on burnout is well known from a lot of measurements. There exists an empirical rule which relates the required fineness of grinding to the proximate of volatile matter content to obtain a definite content of unburned carbon in ash.

Such an empirical relation is shown in Fig. 8 for a value of 6 % of unburned carbon in ash and valid for a 700 MW$_{el}$ boiler.

The calculations 1, 2 and 3 fit the 6 % curve if k_O = 400 kg char/m^{2}s bar O_2 and E = 89890 kJ/kmol.

By varying the content of volatile matter for case 2 and 3 we obtain two other constant carbon burnout curves (2 % and 13 %) which allow the extrapolation of experimental data.

Calculations with different coal particle size distributions and measurements show that the unburned carbon in ash originates from the big coal particles, so that the residue on the 200 μm mesh is a better criterion for grinding quality with respect to burnout than the residue on 90 μm mesh.

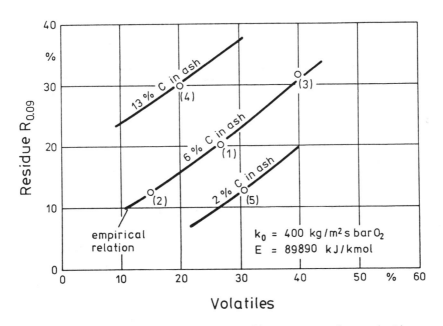

FIGURE 8. Relationship between fineness of grinding and volatile content for constant carbon in ash (1750 MW$_{th}$ furnace)

434

If coal mixtures with different reactivities/contents of
volatile matter are fired, a separate particle size distribution and reaction equation has to be assigned to each component.

A very interesting result is in that case that the burnout of the high reactive coal is intensified at the beginning
because the low reactive coal doesn't consume so much oxygen
and that the burnout of the low reactive coal is decelerated
because the overall oxygen concentration is reduced by the
accelerated burnout of the high reactive particles.

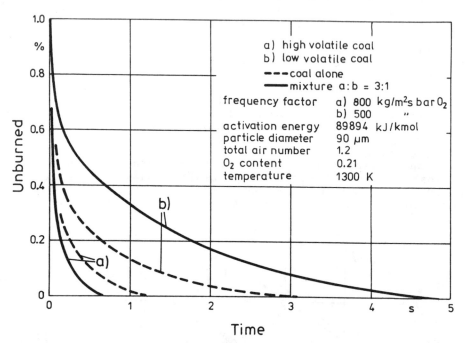

FIGURE 9. Burnout behaviours of different coals

Fig. 9 shows the results of a burnout calculation with
high and low reactive coal alone and high and low reactive
coal in a mixture.

The influence of coal mixture, temperature and oxygen content of combustion air has been investigated and compared
with measurements in a 1600 MW_{th} opposed fired boiler combustion chamber, see Fig. 10.

Though the absolute values of predicted burnout differ
from the measurement results due to insufficient knowledge of
physical model parameters and due to the simplifications of
the combustion process the general behaviour is represented
very well.

Burnout depends strongly on the coal composition, on furnace temperature which in this case is determined by the
fouling of furnace walls and oxygen content in combustion
air. Residence time has to be taken into account too, but

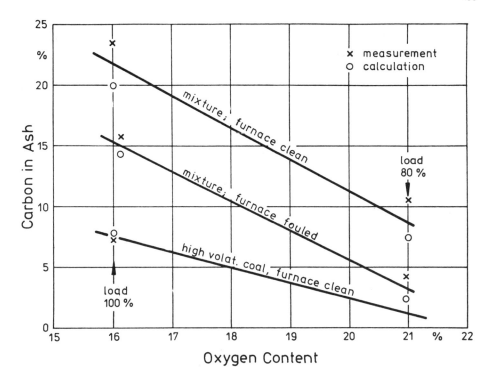

FIGURE 10. Influence of oxygen content in combustion air,
          coal composition and furnace condition on unburned
          carbon in ash

it is of minor significance in comparison to coal composition, temperature and oxygen content.

4. CONCLUSIONS
    With the help of a 3-dimensional heat transfer model,
aerodynamic model measurements and a simple burnout model it
is possible to predict the overall furnace performance, heat
flux, temperature distribution and carbon burnout.
    The influence of furnace geometry and fuel properties has
been investigated with the model and compared with measurements in large p.f. fired utility boilers.
    Absolute values of predicted variables may still be some-
what uncertain but reliable predictions of the impact of relative changes of geometrical, operational and fuel-related
parameters on thermal performance of a furnace are possible.

436

5. LITERATURE

[1] Field/Gill/Morgan/Hawksley: Combustion of Pulverized Coal.
BCURA 1967, edition 1974, Edition Institute of Fuel

[2] Essenhigh, R.H.: Fundamentals of Coal Combustion.
In M.A. Elliot, ed. Chemistry of Coal Utilization, Second
Supplementary Volume
John Wiley & Sons, New York 1981, pp. 1153 - 1312

[3] Richter, W.: Mathematische Modelle technischer Flammen,
Grundlagen und Anwendungen für achssymmetrische Systeme
Dissertation Stuttgart 1978

[4] Hinze, J.O.: Turbulence.
Mc-Graw-Hill 1959

[5] Jørgensen, F.E.: Characteristics and Calibration of a
Triple-Split Probe for Reversing Flows
DISA Information No. 27 (1982), pp. 15 - 22

[6] Wall, T.F., Lowe, A., Wibberley, L.J. and Mc.C.Stewart:
Mineral Matter in Coal and the Thermal Performance of
Large Boilers.
Prog. Energy Combust. Sci., Vol. 5, pp. 1 - 29

[7] Thielen, W.: Practical utilisation of calculation models
for furnaces taking the 700 MW unit Mehrum as an example.
VGB-Kraftwerkstechnik 66 (1986) 6, pp. 515 - 520

[8] Richter, W., Payne, R., Thielen, W.: Application of
advanced computer models for performance analysis of p.f.
and CWM-fired industrial furnaces and boiler combustion
chambers.

APPLICATION OF MATHEMATICAL FLAME MODELLING TO NO_x EMISSIONS FROM COAL FLAMES

W. ZINSER, U. SCHNELL

Institut für Verfahrenstechnik und Dampfkesselwesen
University of Stuttgart, FRG

1. INTRODUCTION

Formation and reduction of nitrogen oxide from fuel-bound nitrogen in pulverized coal flames strongly depends on combustion conditions. From experimental investigations it is known that important parameters envolved are residence times, temperature distributions and the availability of reducing flame zones. While these effects can be directly studied in laboratory scale experiments, problems arise in practical applications where these parameters are generally unknown. Mathematical flame models offer a possibility to predict the turbulent flow field and local flame conditions at least to some engineering accuracy.

This paper describes an attempt to exploit these features of mathematical flame models for the prediction of NO formation and reduction in coal flames. The investigation starts from a model for single enclosed pulverized coal flames developed at the University of Stuttgart. After a short discussion of basic modelling assumptions, the predicted flame pattern is used to introduce a simplified scheme for the kinetic mechanisms governing NO emissions. This scheme has been proposed by deSoete (1) based on laboratory investigations. Turbulence modelling of temperature fluctuations revealed amplitudes of up to 300 K in the primary flame zone, which is again the region dominating NO emissions. As the relevant reaction rates show high activation energies, it is proposed to model these effects by way of a probability density function for the gas phase temperature.

Prediction examples indicate that the model is able to give a realistic overall picture of NO formation in pulverized coal flames. Finally, the paper discusses some implications of the approach and its use for practical investigations.

2. THE PULVERIZED COAL COMBUSTION MODEL

2.1 Introductory Remarks

Mathematical flame models for the prediction of pulverized coal flames have been proposed by various authors. Fundamental studies at the British Coal Utilization Research Association (2) led as early as 1970 to the publication of a simple fluid-dynamical model for axi-symmetric anthracite flames by Gibson and Morgan (3). Richter and Quack (4) extended this approach by including turbulence modelling in coal combustion and were able to test their model successfully against experimental data. Following the recognition that fluid mechanics of the gaseous and of the particulate phase play an important role in coal combustion - see for example the discussion in the review of Beer et al. (5) - various models have been proposed to model the turbulent particle phase. A Lagrangian description of the particulates offers advantages in the characterization of particle slip and temperature-time history of single particles. This approach has been followed by various workers (6), (7), (8), while in the present work a Eulerian description has been preferred which offers advantages in depicting turbulent particle dispersion (9).

Mathematical flame models for three-dimensional furnace predictions have become available recently (10), (11), and can be expected to offer a clear potential for transferring mathematical coal combustion modelling to utility furnace applications. For a model of nitrogen oxide formation it is save to assume that the processes involved do not affect the heat release or the local density of the flow field. Thus, flame predictions can be performed using a simplified heat release model. Based on the flame pattern obtained in this way, the modelling of nitrogen oxide can be performed independently. This is described in section 3.

2.2 Fluid Dynamical Modelling

Figure 1 gives a short sketch of flame processes which are of interest in mathematical flame modelling.

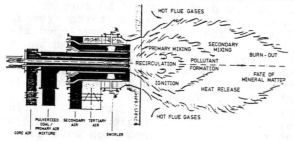

FIGURE 1. Fields of interest in flame modelling

The turbulent flow field is influenced by the combustion process only by way of local flow density. This allows to treat the turbulent transport processes as independent of the combustion model for a given density distribution. Computationally, as the flow field will influence the flame pattern, this approach will require to re-adjust any predicted flow field to the predicted local flame density until convergency is achieved.

Fluid-mechanically, pulverized coal flames are turbulent particle-laden reacting flows. As the volumetric fraction of the particulate phase is typically less than 0.001 and most particles are crushed to diameters less than 100 μm, the particle cloud is approximated as a continuum and the mean particle velocity is assumed to be approximately equal to the gas phase velocity. The latter assumption excludes flames with strong velocity gradients, but it can be expected to hold in long coal flames as found in many practical applications.

Variable	Description	Units	$Pr_{t,\phi}$	Source term $(\overline{\rho_m} \tilde{S}_\phi)$
$\tilde{u}_i$	Velocity	$\frac{m}{s}$	1.0	$\frac{\partial}{\partial x_j}\left[\overline{\rho_m}\{\nu_t+(1-\tilde{m}_p)\nu\}\{\frac{\partial \tilde{u}_j}{\partial x_i} - \frac{2}{3}\delta_{ij}\frac{\partial \tilde{u}_l}{\partial x_l}\}\right]$ $- \frac{2}{3}\frac{\partial}{\partial x_j}\overline{\rho_m}\tilde{k} - \frac{\partial \overline{p}}{\partial x_i} + \overline{\rho_m}g_i$
$\tilde{k}$	Kinetic energy of turbulence	$\frac{kJ}{kg}$	1.0	$\left[\overline{\rho_m}\nu_t\left(\frac{\partial \tilde{u}_i}{\partial x_j} + \frac{\partial \tilde{u}_j}{\partial x_i}\right) - \frac{2}{3}\delta_{ij}\left(\overline{\rho}\tilde{k}+\overline{\rho_m}\nu_t\frac{\partial \tilde{u}_l}{\partial x_l}\right)\right]$ $\cdot \frac{\partial \tilde{u}_i}{\partial x_j} - \frac{\nu_t}{\overline{\rho_m}}(\frac{\partial \overline{p}}{\partial x_i} \cdot \frac{\partial \overline{p}}{\partial x_i})$
$\tilde{\varepsilon}$	Dissipation rate of turbulent kinetic energy	$\frac{kJ}{kg\ s}$	1.3	$1.44\cdot\left[\overline{\rho_m}\nu_t\left(\frac{\partial \tilde{u}_i}{\partial x_j} + \frac{\partial \tilde{u}_j}{\partial x_i}\right) - \frac{2}{3}\delta_{ij}\left(\overline{\rho_m}\tilde{k}\right.\right.$ $\left.\left.+ \overline{\rho_m}\nu_t\frac{\partial \tilde{u}_l}{\partial x_l}\right)\right]\frac{\partial \tilde{u}_i}{\partial x_j} \cdot \frac{\tilde{\varepsilon}}{\tilde{k}} - 1.92\cdot\overline{\rho_m}\frac{\tilde{\varepsilon}^2}{\tilde{k}}$
$\tilde{f}$	Mixture fraction	-	0.7	-
$\tilde{h}$	Enthalpy of mixture	$\frac{kJ}{kg}$	0.7	$-4\ \sigma\left[K_{a,P}\ T_P^4 + K_{a,G}\ T_G^4\right] + \dot{q}'''_{abs}$

Additional nomenclature: x_i m spatial coordinate, ρ_m kg/m³ density of mixture, $\tilde{m}_p$ kg/kg mass fraction of particulates, ν m²/s molecular kinematic viscosity, ν_t m²/s turbulent kinematic viscosity, $\overline{p}$ N/m² static pressure, $\overline{\rho_m}g_i$ N body force, δ_{ij} Kronecker symbol, σ 5.67·10⁻¹¹ kW/m²K⁴ Stefan-Boltzmann constant, K_a 1/m volumetric absorption coefficient for particulates and gas phase, T K temperature, $\dot{q}'''_{abs}$ kW/m³ heat absorpted from radiation.

Table 1. Modelled transport equations

If one defines a mean density $\overline{\rho_m}$ as mass of mixture per unit flow volume, a Favre-averaged transport equation for a general variable $\tilde{\phi}$ may be cast in the following form (9):

$$\frac{\partial}{\partial x_j}(\overline{\rho_m}\ \tilde{u}_j\ \tilde{\phi}) = \frac{\partial}{\partial x_j}\left[\overline{\rho_m}\{\frac{\nu_t}{Pr_{t,\phi}} + (1-\tilde{m}_p)\frac{\nu}{Pr_\phi}\}\frac{\partial \tilde{\phi}}{\partial x_j}\right] + \overline{\rho_m}\ \tilde{S}_\phi \quad (1)$$

The eddy viscosity of the mixture is derived from turbulence modelling to be given by

$$\nu_t = 0.09 \cdot \{(1-\tilde{m}_p) + \tilde{m}_p\left(1 + \frac{t_P}{t_L}\right)^{-1}\}(1-\tilde{m}_p)^{1/2} \cdot \frac{\tilde{k}^2}{\tilde{\varepsilon}} \quad (2)$$

440

$\tilde{m}_P$ denotes the mean local mass fraction of particulates. The
relation between particle relaxation time and a Lagrangian
time scale of eddy motion, t_P/t_L, is approximated by

$$\frac{t_P}{t_L} = 0.478 \frac{d_p^2 \, \rho_S \, \tilde{\varepsilon}}{\mu \, \tilde{k}} \tag{3}$$

d_P and ρ_S denote diameter and material density of the
particles, while μ is the dynamic viscosity of the gas phase.
If the local particulate loading $\tilde{m}_P$ vanishes, the standard
$\tilde{k}-\tilde{\varepsilon}$ model for gas flow is recovered.

Details of turbulence modelling applied here are given in
ref. (9). Table 1 gives an overview of transport equations
solved for the fluid-dynamical description of the flame.
Local gas phase temperatures T_G can be deduced from h, while
the temperature of the particulates T_P is calculated from
thermal equilibrium at the surface. The radiation model which
appears in the source term of $\tilde{h}$ is a four-flux model taken
from Richter (12).

2.3 Combustion Modelling
    Modelling of heat release and of major species reactions is
done in a rather simplified way. A discussion on the validity
of the underlying assumptions can be found in (2), (13), (14).
Figure 2 gives a scheme of the assumed global reaction paths.

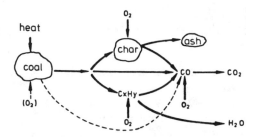

FIGURE 2. Simplified scheme of coal combustion

Coal is considered to devolatilize in a single step to form
pyrolysis products which are represented by CO, H_2O and a general hydrocarbon C_xH_y. In the presence of oxygen, the hydrocarbon C_xH_y is instantaneously oxidized to yield H_2O and CO.
Char oxidation is dependent on the oxygen concentration at the
surface, which again is a function of particle size and
pyrolysis rate. The product of char oxidation is assumed to be
CO, which will be oxidized to CO_2 in the presence of water
vapor H_2O and be rate controlled by eddy mixing (15).
    Thus, in the formulation of the burn-out model, the following
wing species are balanced: Volatile Coal (VC) which is released
as volatiles during pyrolysis, Char (C), and the gaseous
species C_xH_y, O_2, H_2O, CO, and CO_2. Particle size effects are
included by considering various size classes for the solids.
Ash is treated as an inert compount. The modelled global
reaction scheme is given in table 2.

Pyrolysis:

$$|\nu_{1,VC}| \; VC \xrightarrow{(1)} \nu_{1,C_xH_y} \; C_xH_y + \nu_{1,CO} \; CO + \nu_{1,H_2O} \; H_2O$$

Char oxidation:

$$|\nu_{2,C}| \; C + |\nu_{2,O_2}| \; O_2 \xrightarrow{(2)} \nu_{2,CO} \; CO$$

Gas-phase reactions:

$$|\nu_{3,C_xH_y}| \; C_xH_y + |\nu_{3,O_2}| \; O_2 \xrightarrow{(3)} \nu_{3,CO} \; CO + \nu_{3,H_2O} \; H_2O$$

$$|\nu_{4,CO}| \; CO + |\nu_{4,O_2}| \; O_2 \xrightarrow{(4)} \nu_{4,CO_2} \; CO_2$$

Relations for mass-based stochiometric coefficients:

$$\sum_j \nu_{i,j} = 0 \qquad \sum_{j=RHS} \nu_{i,j} = +1 \qquad \sum_{j=LHS} \nu_{i,j} = -1$$

VC denotes "volatile fraction of coal", C denotes "char"

Table 2. Modelled global reaction scheme

2.4 Results of Flame Predictions

A comparison of prediction results with measurements for the present model has been attempted in (9), (10), (15). For the present purposes, the experimental work of Michel and Payne (18) seems especially suited as both field values of velocity, temperature, major species concentrations, and nitrogen oxide concentrations are reported. In this experiment, pulverized high volatile Saar coal was burned in a parallel-flow burner at a thermal load of roughly 2 MW. Details of the measurements in 6 attached and detached flames are given in the original report (18).

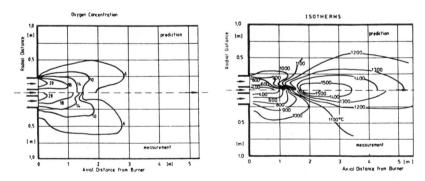

FIGURE 3. Measured and predicted distributions in a detached coal flame

In predicting these flames, problems arouse with a realistic calculation of the flame ignition length. Although the char burning model described in table 3 gave improved predictions of char ignition conditions if compared to more common approaches previously used (15), a strong influence of the coal flow conditions inside the burner could be observed. This matter requires further studies. Still, for the present purposes, the prediction results were considered to provide a reasonably sound basis for pollutant formation studies. Figure 3 shows a comparison of predicted (upper half) and measured (lower half) profiles of temperature and oxygen in a detached coal flame from (18).

Variable	Description	Unit	$Pr_{t,\phi}$	Source term $\overline{\rho}_m \tilde{S}_\phi$
$\tilde{m}_{VC}$	Volatile fraction of coal	kg/kg	0.7	$-1.5 \cdot 10^5 \, \overline{\rho}_m \, \tilde{m}_{VC} \exp(-8900./T_P)$
$\tilde{m}_C$	Char fraction	kg/kg	0.7	$-1248. \cdot \dfrac{\overline{\rho}_m}{\rho_S d_P} \cdot \tilde{m}_C \cdot \tilde{m}_{O_2,S} \cdot \exp(-9553./T_P)$ Oxygen concentration at particle surface S: $\tilde{m}_{O_2,S} = \chi + (\tilde{m}_{O_2} - \chi) \exp\left[-\dfrac{\chi \cdot \tilde{m}_{O_2,S} \, d_P}{2 \, \rho_G D} \right]$ Mass flux ratio in particle boundary layer: $\chi = \dot{m}_{O_2}/\dot{m}_{tot}$ Effect of burn-out B: $d_P = (1-B)^\alpha \, d_{P,0} \quad , \quad \rho_S = (1-B)^{1-3\alpha} \, \rho_{S,0}$
$\tilde{m}_{CO_2}$	CO_2	kg/kg	0.7	$-\overline{\rho}_m \cdot \tilde{\varepsilon}/\tilde{k} \cdot \min\,(6.3\,\tilde{m}_{CO}, \; 11.\,\tilde{m}_{O_2})$

Additional nomenclature: T_P K Particle surface temperature, ρ_S kg/m^3 material density of char, d_P m particle diameter, $\dot{m}_{O_2}, \dot{m}_{tot}$ kg/m^{2}s mass flux of oxygen and all species in particle boundary layer, $\rho_G D$ kg/m·s diffusion coefficient in particle boundary layer, B kg/kg local burn-out, α - empirical parameter between 0 and 1/3 to account for shrinking.

Table 3. Modelled transport equations for coal combustion

Following a proposal by Richter and Saha (16), the assumption of equal turbulent Schmidt/Prandtl-Numbers for all species allows to rearrange species concentrations in table 2 in such a way that combined variables ξ_i are obtained which are free of source-terms and are linearly related to the mixture-fraction $\tilde{f}$ from table 1. A description of the Shvab-Zeldovich formulation may be found in (17).

In this case, a typical combined variable may be given by

$$\xi_1 = \frac{\nu_{4,CO_2}}{\nu_{4,O_2}} \frac{\nu_{3,O_2}}{\nu_{3,C_xH_y}} \tilde{m}_{C_xH_y} - \frac{\nu_{4,CO_2}}{\nu_{4,O_2}} \tilde{m}_{O_2} + \tilde{m}_{CO_2} + \frac{\nu_{4,CO_2}}{\nu_{4,O_2}} \frac{\nu_{2,O_2}}{\nu_{2,C}} \tilde{m}_C$$

$$- \frac{\nu_{4,CO_2}}{\nu_{4,O_2}} \frac{\nu_{3,O_2}}{\nu_{3,C_xH_y}} \frac{\nu_{1,C_xH_y}}{\nu_{1,VC}} \tilde{m}_{VC}$$

(4)

From this variable, the modelling assumption that C_xH_y and O_2 do not coexist at the same time at any flame location in a mean sense can be introduced by postulating $\tilde{m}_{C_xH_y} \cdot \tilde{m}_{O_2} = 0$.

Table 3 gives a review of the balanced mass concentrations to close the coal combustion model.

3. NITROGEN OXIDE FORMATION AND DESTRUCTION
3.1 The Reaction Scheme

Nitrogen oxide emissions from pulverized coal flames are dominated by conversion of fuel-bound nitrogen. Detailed reviews about this subject have been provided by various authors, see for example (13), (19), (20), (21).

In the present work, an attempt is made to include the global features of NO-formation mechanisms into the flame model for volatile coals. This approach is expected to allow a prediction of NO formation as a function of local residence time, temperature, equivalence ratio and turbulent mixing in the flame.

During the course of pulverized coal combustion, fuel-bound nitrogen is released in two separate stages. Under flame conditions, a major part of the nitrogen evolves during pyrolysis. Pohl and Sarofim (22) report that the relative portion of nitrogen released in this way is strongly increasing with pyrolysis temperature. In the second stage, residual nitrogen which is retained in the char may be oxidized during char burn-out. Song et al. (23) have found in laboratory furnace experiments that oxidation of residual fuel nitrogen and char is not selective, but that at high temperature a secondary pyrolysis can be observed of tightly bound fuel-nitrogen. Still, under some circumstances the conversion rate of nitrogen to NO might be even less for char-bound nitrogen than for volatile nitrogen (22).

In modelling these effects we follow a proposal by Hill et al. (24) and assume that the rates of nitrogen release during pyrolysis and during char oxidation are proportional to the equivalent rates of carbon release (see table 3). The relative amount of fuel-nitrogen evolving during pyrolysis is guessed as an empirical parameter of this simple model. From ref. (22) it can be deduced that this parameter strongly exceeds the volatile content from proximate analysis and may reach up to 90 % at high flame temperature.

The fate of fuel-nitrogen after release from the coal is modelled following a proposal by deSoete (1), this is again adopted from ref. (24). DeSoete concludes from work in doted premixed flames that HCN occurs as major intermediate species and is formed very rapidly after nitrogen release from the coal. HCN may react further in two ways. In the presence of oxygen, HCN will be oxidized to form NO. As a competing reaction, HCN can be considered to reduce NO to N_2. In reality, the latter reaction will involve NH_i species (21). For the

present purpose of global modelling, these intermediate steps
are considered as fast in the presence of OH-radicals.

In modelling the fate of residual nitrogen released from the
char, we have adopted an argument from (22) and assumed that
inside the particle boundary layer oxygen is depleted. From
this we conclude that fuel-nitrogen is released as HCN at the
rate of char burning and can be treated in the same way as
volatile components. Fig. 4 shows a sketch of the modelled
reaction scheme.

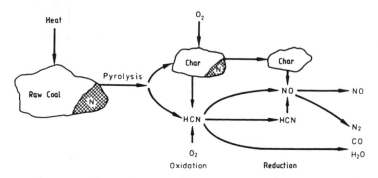

FIGURE 4. Modelled scheme of fuel-NO reactions

Variable	Unit	Source term $(\overline{\rho}_m \tilde{S}_\phi)$
HCN	kg/kg	$2.3 \cdot 10^5 \cdot \overline{\rho}_m \cdot \tilde{m}_{VC} \cdot m_{N,O} \cdot \exp(-8900./T_p)$ $- 10^{10} \, \overline{\rho}_m \, \tilde{m}_{HCN} \cdot \tilde{x}_{O_2}^b \cdot \int P(T) \cdot \exp(-34000/T) \, dT$ $- 3 \cdot 10^{12} \, \overline{\rho}_m \, \tilde{m}_{HCN} \cdot \tilde{m}_{NO} \cdot \int P(T) \cdot \exp(-30000/T) \, dT \cdot \dfrac{\overline{M}_G}{M_{NO} (1-\tilde{m}_P)}$ $+ 1203.4 \cdot \dfrac{\overline{\rho}_m}{\rho_S \, d_P} \cdot \tilde{m}_C \cdot \tilde{m}_{O_2,S} \cdot \dfrac{m_{N,C} \cdot m_{N,O}}{1 - m_{VM}} \cdot \exp(-9553./T_p)$
NO	kg/kg	$10^{10} \, \overline{\rho}_m \cdot \tilde{m}_{HCN} \, \tilde{x}_{O_2}^b \cdot \int P(T) \cdot \exp(-34000/T) \, dT \cdot M_{NO}/M_{HCN}$ $- 3 \cdot 10^{12} \, \overline{\rho}_m \cdot \tilde{m}_{HCN} \cdot \tilde{m}_{NO} \cdot \int P(T) \cdot \exp(-30000/T) \, dT \cdot \dfrac{\overline{M}_G}{M_{HCN} (1-\tilde{m}_P)}$ $+ \{4.86 \cdot 10^{21} \cdot \overline{\rho}_m \cdot \tilde{m}_{O_2} \cdot T \cdot \exp(-41150/T) - 2.4 \cdot 10^{20} \cdot \overline{\rho}_m \, \tilde{m}_{O_2} \cdot \tilde{m}_{NO} \cdot \tilde{m}_{NO}\}$ $/ \{1.6 \cdot 10^{10} \cdot \tilde{m}_{NO} + 6.4 \cdot 10^6 \, \tilde{m}_{O_2} \cdot T \cdot \exp(-3150./T)\}$
Additional nomenclature: $m_{N,O}$ kg/kg nitrogen content of raw coal, b fractional reaction order of oxygen as function of x_{O_2}, given by deSoete (1), $m_{N,C}$ kg/kg mass fraction of nitrogen retained in char, m_{VM} kg/kg volatile matter in raw coal (from proximate analysis), x_{O_2} kmol/kmol mole fraction of oxygen in the gas phase, M_i kg/kmol molar weight of species i (i = G: gas phase)		

Table 4. Modelled reaction rates of HCN and NO

Formation of thermal NO we have modelled following a simplified
scheme from Richter and Chae (25). Heterogeneous reduction of
NO in the presence of char may be significant in some situations, but the effect observed in computations using literature
rate data was less than 10 ppm NO in the cases investigated
here. The kinetic data used in the model are given in table 4.

3.2 Turbulence Interactions
Reaction rates in turbulent flows become strongly dependent
on turbulent fluctuations of local flame conditions if reaction
time constants are of the same order of magnitude or smaller
than mixing time constants of the turbulent field (26). Reactions with high activation energy show largely increased mean
reaction rates if they are subjected to significant temperature
fluctuations. This situation is typically found for the formation of nitrogen oxide in turbulent flames where radical formation plays an important role. In investigating the formation
of fuel nitrogen from coal flames, the influence of turbulence
on local reaction rates should not be confused with the apparent insensitivity of NO formation to changes in mean temperature. The latter effect may be attributed to a trade-off
between pyrolysis yield affecting primary stoichiometry and
nitrogen conversion (22).
Smoot and Smith (13) propose to model these turbulent
oscillations by local fluctuations of the mixture fraction.
In large coal flames, radiation becomes a dominant mechanism of
heat transfer and local mixing conditions are intricately
dependent on gasification mechanisms of the coal dust. Regarding the high activation energies of the reactions of the
nitric species specified in table 4, we assume that concentration fluctuations are not strongly correlated with temperature
oscillations in these flames. This effect can be described by
defining a probability density function P(T) for the local
temperature at any location in the flow. The mean value of the
Arrhenius-term of each reaction is then given by simple
convolution over P(T):

$$\overline{\exp(-T_A/T)} = \int_T P(T) \exp(-T_A/T) dT \tag{5}$$

In table 4 the mean reaction rates of the NO formation and reduction model are obtained in this way.
With this formulation the problem has been transferred to obtaining a probability density function for the local temperature. Here we have assumed that P(T) may be characterized by
two parameters, the mean temperature $\tilde{T}$ and the variance of
temperature $\tilde{T}''^2$ in a Favre averaged sense. Following Jones (27)
a good choice for a two-parametric probability distribution
of scalars in flames may be given by a ß-function. This is
shown in table 5.
Starting from mean values of temperature $\tilde{T}$ at any flame
location, the problem remains to specify local temperature
oscillations. This has been derived from modelling a transport
equation for the local variance of sensible enthalpy $\tilde{g}_h$, which
is connected with temperature fluctuations via the local
specific heat capacity c_p: $\tilde{g}_h = c_p \tilde{T}''^2$. The turbulence model

is an extension of the $\bar{k}-\tilde{\varepsilon}$ turbulence model, basic steps are described for example in (9). The model is able to take the effects of radiation and local reaction rates on turbulence fluctuations into account. Table 5 shows the modelled equations for the description of temperature turbulence.

Variable	Description	$Pr_{t,\phi}$	$\bar{\rho}_m \tilde{S}_\phi$
$\tilde{g}_h$	Variance of sensible enthalpy	0.7	$2.7\,\bar{\rho}_m c_p^2 \nu_t \left(\frac{\partial \tilde{T}}{\partial x_i}\right)^2 - 1.787\,\bar{\rho}_m \tilde{g}_h\ \tilde{\varepsilon}/\bar{k}$ $- 32\,\sigma\,K_a \frac{1}{c_p} \tilde{T}^3\,\tilde{g}_h - 2\,\bar{\rho}_m \frac{\tilde{g}_h}{\tilde{T}^2}\frac{1}{c_p} \cdot$ $\Sigma\,(\dot{m}_{k,0}''' \cdot \Delta h_k \cdot T_k)$

β-function P(T)

$$\tilde{T}''^2 = \frac{\tilde{g}_h}{c_p^2}\ ,\ \text{normalized}\ \tilde{t}''^2 = \frac{\tilde{T}''^2}{(T_{ad}-T_{inlet})^2}$$

$$\text{normalized variable}\ t = \frac{T-T_{inlet}}{T_{ad}-T_{inlet}}$$

$$P(T) = \frac{t^{p-1}\,(1-t)^{q-1}}{\int_0^1 t^{p-1}\,(1-t)^{q-1}\,dt}$$

$$p = \tilde{t}\left[\frac{\tilde{t}(1-\tilde{t})}{\tilde{t}''^2} - 1\right]\quad q = p \cdot (1-\tilde{t})$$

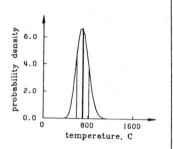

Typical shape of β-function

Additional nomenclature: c_p kJ/kg K specific heat capacity, $\dot{m}_{k,0}'''$ kg/kg s mean reaction rate of reaction k without fluctuations, Δh_k kJ/kg heat of reaction; reaction k, T_k K activation temperature, T_{ad} K adiabatic temperature

Table 5. Modelling of temperature fluctuations

3.3 Prediction results

The flame model has been applied to a variety of coal flames from the "long coal flame trials" of Michel and Payne (18). Presently, we shall discuss only very few results.

Figure 5 shows a sketch of the measured NO concentrations, plotted over a cross-section of the furnace in a three-dimensional view. From this graph one can clearly identify the peak of NO after devolatilization of the coal and the reduction of NO in the secondary flame region. On the RHS of fig. 5, the modelled variance of temperature fluctuations is shown.

Maximum values of oscillations are observed to occur directly
after the ignition region where large temperature gradients
have been developped. In the post-flame regions, these
fluctuations are largely damped by mixing and radiation.

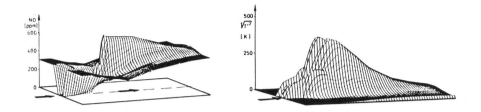

FIGURE 5. Measured NO distribution and predicted temperature
fluctuations in the coal flame

From these predictions we conclude that large temperature fluctuations are to be expected in the flame zone where primary reduction of NO occurs. The effect of this is shown in figure 6.
On the LHS of figure 6, prediction results on the flame axis
and for two radial cross sections of the flame are compared
ignoring turbulence interaction. Obviously, reaction rates are
largely underpredicted. It is interesting to note, that this
effect is much stronger in predicting peak values of NO inside
the flame than in predicting mean outlet values.
   On the RHS of figure 6, predictions are shown including
turbulence effects as described in table 5. In this case, the
trends of the NO distribution at the jet axis can be considered
as predicted quite correctly.

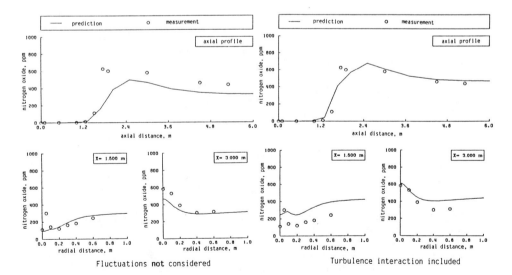

FIGURE 6. Prediction results compared with measurements

4. CONCLUSIONS

Although many questions concerning NO emissions from coal flames are not sufficiently understood, the simplified mathematical model appears able to predict at least some trends and levels of NO formation rather realistically. At present, the model has only been applied to non-swirling flames of high volatile coals. To increase confidence in the capability of this model, a wide range of various flame types should be investigated.

However, there are similarities to flame conditions to be found in tangentially or wall-fired furnaces where non-swirling burners are used. Future work will show whether a model of the presented type included in furnace prediction codes can perform realistic calculations of NO-emissions of full-scale pulverized-coal fired furnaces.

ACKNOWLEDGEMENT

The present work was supported by the German Ministry of Research and Technology (BMFT) and the Federal Government of Baden-Württemberg with the TECFLAM project. This is gratefully acknowledged.

REFERENCES

1. DeSoete GG: Overall Reaction Rates of NO and N_2 Formation from Fuel Nitrogen. 15th Symp. (Int.) Combustion, The Combustion Institute, Pittsburgh, pp. 1093-1102, 1975.
2. Field MA, Gill DW, Morgan BB, Hawksley PGW: Combustion of Pulverized Coal. The British Coal Utilization Research Association, Leatherhead, 1967.
3. Gibson MM, Morgan BB: Mathematical Model of Combustion of Solid Particles in a Turbulent Stream with Recirculation. J. Inst. Fuel, Vol. 43, pp. 517-523, 1970.
4. Richter W, Quack R: A Mathematical Model of a Low-Volatile Pulverized Fuel Flame. In: Afgan NH, Beer JM(ed): Heat Transfer in Flames. Scripta Book Company, Washington, pp. 95-110, 1974.
5. Beer JM, Chomiak J, Smoot LD: Fluid Mechanics of Coal Combustion: A Review. Prog. Energy Combust. Sci, Vol. 10, pp. 177-208, 1984.
6. Lockwood FC, Salooja AP, Syed SA: A Prediction Method for Coal Fired Furnaces. Combustion Flame 38, pp. 1-15, 1980.
7. Smith PJ, Fletcher TH, Smoot LD: Model for Pulverized Coal-Fired Reactors. 18th Symp. (Int.) Combustion, The Combustion Institute, Pittsburgh, pp. 1285-1293, 1981.
8. Boysan F, Weber R, Swithenbank J, Lawn CJ: Modelling Coal-Fired Cyclone Combustors. Combustion Flame 63, pp. 73-86, 1986.
9. Zinser W: Zur Entwicklung mathematischer Modelle für die Verfeuerung technischer Brennstoffe. Fortschr.-Ber. VDI Reihe 6 Nr. 171, VDI-Verlag, Düsseldorf, 1985.

10. Görner K, Zinser W: Prediction of Three-Dimensional Flows in Utility Boiler Furnaces and Comparision with Experiments. ASME 107th Winter Annual Meeting, Symposium on Calculations of Turbulent Reactive Flows, Anaheim CA, 1986 (to be published).
11. Abbas AS, Lockwood FC: Prediction of a Corner-Fired Utility Boiler. 21st Symp. (Int.) Combustion, paper 13, The Combustion Institute, Pittsburgh, 1986 (to be published).
12. Richter W: Mathematische Modelle technischer Flammen. PhD tesis, University of Stuttgart, 1978.
13. Smoot LD, Smith PJ: Coal Combustion and Gasification. Plenum Press, New York, 1985.
14. Elliot MA(ed): Chemistry of Coal Utilization. Second Supplementary Vol. John Wiley and Sons, New York, 1981.
15. Zinser W: Untersuchung technischer Kohlenstaubflammen durch mathematische Modellbildung. VDI-Berichte Nr.574, VDI-Verlag, Düsseldorf, pp. 595-616, 1985.
16. Richter W, Saha RK: The Prediction of Low and High Volatile Pulverized Fuel Flames. Vth Int. Symp. Combustion Processes, Krakow, 1977.
17. Williams FA: Combustion Theory, Addison-Wesley, Reading, 1965.
18. Michel JB, Payne R: Detailed Measurement of Long Pulverized Coal Flames for the Characterization of Pollutant Formation. Int. Flame Research Found., Doc.No. F09/a/23, 1980.
19. Wendt JOL: Fundamental Coal Combustion Mechanisms and Pollutant Formation in Furnaces. Prog. Energy Combust. Sci., Vol. 6, pp. 201-222, 1980.
20. Pohl JH, Chen SL, Heap MP, Pershing DW: Correlation of NO_x-Emissions with Basic Physical and Chemical Characteristics of Coal. Joint Symp. on Stationary Combustion NO_x Control, Paper 36, Palo Alto, 1982.
21. Schulz W: Experimentelle Untersuchung der Bildung von Stickoxiden bei der Kohlenstaubverbrennung. PhD-thesis, University of Bochum, 1985.
22. Pohl JH, Sarofim AF: Devolatilization and Oxidation of Coal Nitrogen. 16th Symp. (Int.) Combustion, The Combustion Institute, Pittsburgh, pp.491-501, 1976.
23. Song YH, Beér JM, Sarofim AF: Oxidation and Devolatilization of Nitrogen in Coal Char. Combustion Science and Technology, Vol. 28, pp.177-183, 1982.
24. Hill SC, Smoot LD, Smith PJ: Prediction of Nitrogen Oxide Formation in Turbulent Coal Flames. 20th Symp. (Int.) Combustion, The Combustion Institute, Pittsburgh, pp. 111-120, 1984.
25. Richter W, Chae JO: Die Vorausberechnung von Erdgasflammen einschließlich eines einfachen Modells der NO-Bildung. International Flame Research Foundation, 4th Members Conference, 1976.
26. Libby PA, Williams FA(ed): Turbulent Reacting Flows. Springer-Verlag, New York, 1980.
27. Jones WP: Models for Turbulent Flows with Variable Density and Combustion. In: Kollmann W(ed): Prediction Methods for Turbulent Flows. Hemisphere Publ. Corp., New York, 1980.

DISCUSSION

G. Soete

In a turbulent (coal) flame you have fluctuations of temperature, but also of local equivalence ratio. The latter may
lead to increase or decrease of the NO formation rate (remind:
fuel-NO is strongly oxygen content related!). Why didn't you
take also into account the statistical, local, equivalence
fluctuations?

W Zinser

Temperature and concentration fluctuations are probably
correlated to some degree in these flames and this would require the use of joint probability density functions for all
variables involved. In this case, we have assumed that the
high activation energy of the reactions involved in NO formation and reduction lead to a situation, where the influence of
temperature fluctuations dominates that of concentration
fluctuations. To check numerically upon this assumption, we
have assumed alternatively a joint pdf of the square wave
oscillation type for oxygen and temperature in such a way that
a temperature rise was correlated to a decrease in oxygen
concentration. This situation was considered as a worst case
in reducing the effect of turbulent fluctuations. In the
predictions, this assumption showed to decrease peak values of
NO, but in no way as much as neglecting temperature fluctuations would do. In order to avoid the need to predict this
joint pdf in a general way, we felt it more realistic in the
present case to neglect this correlation and to model temperature fluctuations in radiating and reacting flame situations.

E.M. Suuberg

How sensitive is the model to the fraction of nitrogen released with the volatiles? Presumably an increased value could
raise the peak NO value to the measured value. Is there a problem with modelling the outlet NO level if a higher fraction
of volatile nitrogen is assumed?

W Zinser

The model assumes that nitrogen retained within the char
during devolatilisation is released together with the carbon
in the char burning region. As only very few NO reducing species are available in this region (neglecting heterogeneous
reduction of NO), NO formed in this way will mainly be emitted. This leads to the effect that an increase of nitrogen
content in the residual char increases flue gas concentrations
of NO, but reduces at the same time the observed peak in the
pyrolysis zone. The measured distribution of NO in the flame
shows a very clear peak after pyrolysis, and the ratio between
peak values and exit values was indeed used to guess the
nitrogen retained in the char. As devolatilization is probably
rising up to 1900 K in the flame under investigation, a value

of 10 - 20 % seems not unreasonable.

Measurements of Michel and Payne (18) indicate a variation of this value for various types of coal flames with different heating rates.

To answer your question, the trends predicted by the model are indeed as you are expecting. The modelling problem was to find a consistent formulation which predicts both peak values and outlet values of NO realistically. Application of the model to predict various coal flames showed that NO predictions were mainly sensitive to the quality of the predicted temperature field.

P.R. Solomon

Can you compare your results where you assume 15 - 20 % of N is retained in the char with those of Smoot where the char N is assumed to be proportional to the char carbon.

W. Zinser

The model proposed by Smoot and coworkers (13), (24) is in many respects similar to the one presently proposed. Major differences are the Eulerian description of the burning coal cloud and the treatment of temperature fluctuations with respect to mean gas phase reaction rates. As mentioned in the answer to the question by Suuberg, a trade-off in the NO outlet level has been observed between temperature fluctuations and nitrogen retention in the char. Further considerations should be based on field measurements of NO, analysis of the coal under high temperature pyrolysis conditions, and, possibly, in-situ measurements of temperature and concentration fluctuations.

PRACTICAL USE OF COAL COMBUSTION RESEARCH

P.T. Roberts   C. Morley

Shell Research Ltd.
Thornton Research Centre
P.O. Box 1
Chester CH1 3SH, U.K.

0. SUMMARY

Laboratory measurements of coal rapid pyrolysis char yield and char reactivity, together with a simple model of pulverized coal combustion, have been used to predict coal combustion efficiency in utility boilers. Several bituminous coals were tested and found to burn out to different extents under boiler conditions. Burnout performance did not correlate well with proximate volatile matter content.

Burnout predictions agreed with measured power station data for three coals. Model calculations showed that furnace residence time distribution and excess air are parameters strongly affecting burnout. Grind fineness (% w < 75 micron) has a potentially large affect on combustion efficiency but the extent to which this is realized in practice depends on the mill classification characteristics.

1. INTRODUCTION

Traditional coal quality indices do not necessarily provide a reliable measure of the combustion and ash behaviour of pulverized coal. There is a need to develop practical tests which provide this information quickly, at reasonable cost and which use small coal samples to allow characterization of bore-cores and spot cargoes as well as contract supplies. Because combustion behaviour depends on furnace chamber conditions as well as on coal properties, test procedures must take account of the differences in combustion environment that exist in boilers of different design. This paper describes the development of a method for assessing coal burnout performance in utility scale boilers.

Coal burnout performance is commercially important because the presence of carbon in ash lowers the overall boiler efficiency and can affect electrostatic precipitator performance. In certain cases the suitability of fly ash as a cement component also depends on its carbon content. Carbon in ash comprises partially burnt out char particles formed by coal pyrolysis. The degree of burnout achieved in a boiler depends on the amount of char formed by pyrolysis under flame conditions and the burning rate of the char particles under the boiler operating conditions of particle size, temperature, oxygen concentration and residence time distributions. In this work account is taken of each of these factors, however, for practical reasons, the separate contributions to burnout of

the different maceral components of coal (4) have not been
identified and averaged properties of whole coal are used
instead. Fuel quality parameters are measured in two
laboratory experiments, coal pyrolysis and char oxidation, and
a mathematical model is used to calculate burnout under
operating conditions relevant to large boilers.

2. COAL PYROLYSIS

The coal pyrolysis apparatus is intended to reproduce the
heating conditions experienced by coal particles injected into
a hot furnace. It comprises a vertical flat flame burner
fueled with a stoichiometric hydrogen oxygen mixture. Nitrogen
is added as a diluent to give a combustion product temperature
of 1675 K. Trace amounts of ethylene are added to render the
flame visible. The combustion products are confined by a
quartz tube to prevent ingress of air. Particles of pulverized
coal, metered at approximately 3 g/h, by a screw feeder are
entrained in the combustible mixture and delivered to the
centre of the burner. The residence time of particles in the
hot gases is 34 ms. Sampling at shorter times showed that most
of the weight loss occurs in the first 20 ms of heating as
found by Kobayashi (1) working at a similar temperature of
1740 K. Char particles are collected via a cooled probe onto
a filter paper for char yield determination and in a three
stage cyclone separator for longer char preparation runs.

Coal is prepared by crushing a representative sample to
smaller than 212 microns in a laboratory hammer mill. The mill
product is then sieved at 125 microns and oversized material
reground with a pestle and mortar. This procedure minimizes
the production of very fine, < 10 micron, particles whilst
removing large particles which may not heat up to the
pyrolysis temperature in the available time. Char particles
much larger than 100 micron are also less suitable for high
temperature oxidation studies because the particle burning
rate is usually close to the external diffusion control limit
and hence estimation of the chemical reaction rate is
unreliable.

The particle size distributions of the coal and char are
measured using a commercial particle size analyser (Malvern
Instruments type 2200). This provides a volume size
distribution from a light scattering measurement. In some
cases measurement accuracy is affected by the presence of very
small particles in the samples which are therefore wet sieved
at 5 micron as a precautionary measure.

Volatile matter yields obtained by pyrolysis under the above
conditions are usually greater than those obtained by
proximate analysis as expected. Some results are shown in
Figure 1. Values of the Q-factor, (flame volatiles/proximate
volatiles) lie in the range 1.0-1.25, which compares
reasonably well with the data reviewed by Howard (2). Our
values of volatile yield may be underestimated due to the
inclusion of soot in the measured char yield. The proportion
of soot formed is hard to quantify as the parent coal samples
contain small particles and fines are also produced through
coal fragmentation during pyrolysis.

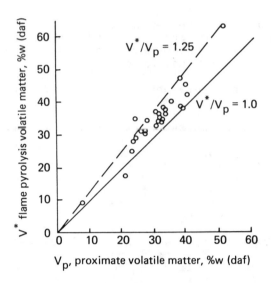

FIG. 1 — Volatile matter yield under rapid heating
conditions for a range of coals

It is estimated that soot may contribute up to ten percent of
the char yield for higher volatile coals.

Significant changes in particle size on flame pyrolysis were
rarely observed. Swelling was marked by a fractional increase
(or decrease) in particle size and swelling factors were
independent of particle size in the range tested. However none
of the coals were classed as strongly swelling. Fragmentation
was distinguished from swelling by significant differences in
the size distribution of coal and char. We suspect most coals
fragment to some small degree but this is confused by soot
formation. Only a couple of high volatile coals have shown
extensive break up. Larger particles appeared to be the most
likely to fragment. Unfortunately we have not been able to
derive an explicit description of the fragmentation process
and because of uncertainty over char particle size are unable
to predict burnout performance for these coals. This remains a
subject of study.

Char is prepared for oxidation studies by wet sieving into
narrow size fractions. A 50-63 micron sieve range is used as a
standard for coal assessment since smaller particles are hard
to characterize at other than high burnoff whilst larger
particles are liable to burn under external diffusion control
at high temperatures. Oxidation measurements have however been
made using particle sizes between 20 and 100 microns. Prior to
oxidation the ash content of each char fraction is measured
and the bulk particle density determined.

3. CHAR OXIDATION

The char oxidation apparatus is a vertically mounted
entrained flow reactor. Oxidation takes place inside a 38 mm
i.d. ceramic tube heated externally by four silicon carbide
elements. The temperature distribution along the tube is
nearly isothermal and internal tube temperatures can be up to
1750 K. In our studies the normal operating range used is 1320
- 1520 K. Particle temperatures are higher depending on char
properties and oxygen concentration.

Char particles are entrained in a small flow of nitrogen
from a fluidized bed feeder. The particle mass flow rate is
small, 0.3 g/h, to avoid perturbing conditions inside the
reactor. The particle stream passes through a preheater to
raise the particles to test temperature and is injected along
the axis of the reactor. The main gas flow through the ceramic
tube comprises a mixture of air and nitrogen. Ratios of these
components are chosen to give the required oxygen
concentration. With the small char feed rates used oxygen
concentration is uniform over the reactor length.

Particle residence times of between 0.15 and 0.7 s are
controlled using different gas throughputs. At all flow rates
the gases are in laminar flow and the particles remain on axis
so that their residence time is well defined. A greater
residence time range can be achieved by also changing the path
length between the char feed and collection points but
alignment of probes can be difficult and time consuming.

Char particles are collected on a filter using a
water-cooled probe into which is also injected cooling
nitrogen to ensure rapid quenching of char combustion. The
mass flow rate of particles passing through the reactor under
nitrogen is determined by weighing the sampling filter after a
3 minute interval. The gas composition is changed to give
oxidizing conditions and the weight collected on the filter
over a 6 minute interval measured. The initial nitrogen run is
repeated to ensure constancy of the flow rate. Burnoff is
determined from the mass collected under oxidizing conditions
relative to that collected under nitrogen with an appropriate
correction for ash. This procedure takes account of other
factors such as drying, further devolatilization or collection
losses which could give weight loss not attributable to
oxidation. The method works well in practice but we do find,
for a limited number of coals, that repeatability of the
burnoff measurement is poor. We suspect that density
differences within a char sieve size, leading to aerodynamic
separation in the fluid-bed feeder, are responsible for
nonuniformity of feeding. A problematic coal has recently been
identified as dividing cleanly into high and low density
fractions and also a separate ash fraction at these small
particle sizes.

3.1 Oxidation results

Char oxidation results comprise a set of burnoff versus time
measurements obtained normally at nominal furnace temperatures
of 1320, 1420 and 1520 K ,oxygen concentrations of 3, 5 and 10
%v., and spanning a burnout range of 10-90 %w. A mathematical

description of the char combustion process is needed in order
to derive a value for the chemical reaction rate of the coal
char. Empirically we have found that, for different sized
particles, the particle mass burning rate is first order in
oxygen concentration and proportional to the particle external
surface area. Char reactivity can therefore be characterized
by a surface reaction rate coefficient K, kg/(m^{2}s atmO$_2$), as
for "shrinking sphere" mode combustion. This is recognized to
be an inexact description. Char particles are known (3) to
burn internally and we have observed changes in particle
density and internal pore structure with increasing burnoff.
Char particles are also not homogeneous in composition being
made up of a blend of different char types reflecting the
different macerals in the coal. However, the assumption that
particles burn as impervious "shrinking" spheres
satisfactorily accounts for the particle size dependence of
the burning rate. One possible reason for this behaviour is
that the internal pore structure and hence internal surface
area of the char is a function of particle size (4) and not a
uniform material property as assumed in many analyses of
porous particle combustion. This point deserves further
investigation.

The range of surface oxidation rate coefficients, corrected
for external diffusion, is indicated in Figure 2, which is a
plot of char reactivity versus reciprocal particle
temperature. The difference between the latter and the furnace

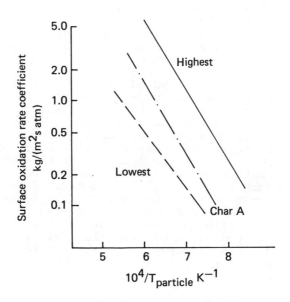

FIG. 2 — Range of char surface oxidation rate
coefficients measured in bituminous coal
studies

gas temperature is calculated using a simple heat balance and
assuming that the primary product of char oxidation is CO (5).
Between coals the surface oxidation rate coefficient can vary
by up to a factor of ten. Arrhenius plots for other chars
tested lie between the bounds shown in Figure 2. A large
proportion are distributed about the central line, char A,
having fairly similar reactivities at a temperature of 1400 K,
Figure 3(a), and differing in their activation energy,
Figure 3(b). Values are widely spread about an average of
approximately 115 kJ/mol which is consistent with the
particles burning between Zone 1 and Zone 2 conditions (6).

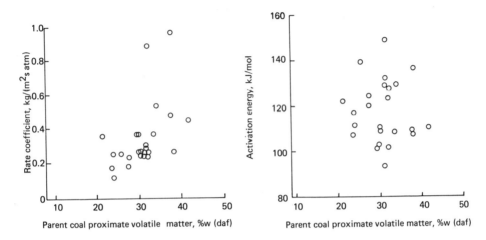

FIG. 3a — Surface oxidation rate coefficient at 1400 K   FIG. 3b — Activation energy of the char surface oxidation
for chars of different coals   rate coefficient for different coals

4. BURNOUT MODEL

The burnout model assumes that the flow pattern inside a
utility boiler can be considered to comprise a set of distinct
flow paths originating from each operating burner and
terminating at the furnace exit plane. It is assumed in the
first instance that these flow paths fill the furnace chamber
so that the cross-sectional area of each path is equal to the
furnace chamber cross-section divided by the number of
operating burners, and the furnace volume is the sum of the
different path lengths multiplied by their cross-sections.
Equal fuel and air distribution to each burner is assumed.

Each flow path is divided into two sections : a burner zone
extending typically 3 m from the burner and a char burnout
zone which occupies the remaining flow path. The burner zone
is associated with the main pf. flame where devolatilization
and volatile burning together with combustion of the smaller
char particles takes place. This is treated as a simple
well-stirred reactor. The larger char burnout zone is treated
as a simple plug-flow reactor along which temperature and
oxygen concentration decrease as burn out progresses.

Calculation of the appropriate temperature distribution is beyond the scope of this work. Instead, data on the temperature field inside boilers of different design were obtained from the Central Electricity Generating Board Marchwood Engineering Laboratories (MEL). The data were generated by a computer program which predicts both the temperature and wall heat flux distributions in large boiler plant (7). The original data were in the form of temperature-distance histories along burner flow paths as described above. Analysis showed that, along the plug-flow component, temperature distributions were self-similar if the distance along each plug-flow reactor was normalized with respect to the reactor length. The temperature distribution obtained in this way was almost linear in dimensionless plug-flow reactor length decreasing from a typical peak pf. flame temperature of $\sim$ 1900 K to the furnace exit plane temperature. Thus the temperature field can be characterized by a single parameter; the exit plane temperature.

Figure 4 shows burner path length frequency distributions for three boilers. They show a grouping of flow path lengths shorter than the average for the boiler (furnace volume/ furnace cross-section) with a wider distribution of longer paths. A single calculation based on the average path length gives nearly the same result as the average of the burnouts calculated for the different length burner paths and

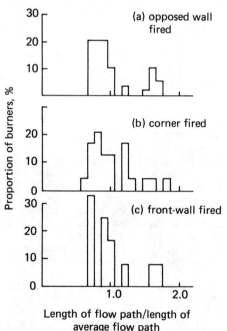

FIG. 4 — Burner flow path length distributions for three boilers

has proved satisfactory for assessing differences in burnout performance between coals.

The coal feed size distribution, specified as Rosin-Rammler, is divided into 15 equally spaced monosized classes. Volatile matter, determined by the flame pyrolysis experiments, is assumed to burn in the well-stirred reactor. The amount of char burning in the reactor is calculated from the char oxidation rate, the residence time distribution and the local oxygen concentration. The well-stirred reactor is assigned an average temperature of 1600 K. On entry to the plug-flow reactor the particle size distributions are resorted into monosized classes. This is necessary because, due to their different residence times, single sized particles entering the burner zone exit as a continuous size distribution. Burnout and oxygen concentration changes along the plug-flow reactor are calculated for the input temperature distribution.

5. BURNOUT PREDICTIONS

Figure 5 shows the predicted loss of unburned carbon for the three coals spanning our measured char reactivity range (Figure 2) in a 360 MWe opposed wall fired boiler at Castle Peak Power Station, Hong Kong. The input particle size distribution was typical of pf being Rosin-Rammler of index = 1, with 71%w of coal less than 75 micron (8). The temperature distribution was that calculated by MEL for a 330 MWe unit. An average pathlength was used and calculations made for several

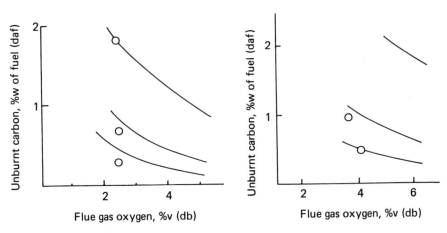

FIG. 5 — Carbon loss from a 360 MWe opposed fired boiler at Castle Peak Power Station, Hong Kong.

Points are data
Lines are model predictions

FIG. 6 — Carbon loss from a 260 MWe corner fired boiler

Points are data
Lines are model predictions

excess air values. No allowance was made for the possible
effect of excess air on furnace temperature.

Comparison of the calculated burnout performance and the
measured values of unburned carbon, shown as points in Figure
5, shows that real differences in the combustion performance
of the coals exist and that these have been correctly
identified. Method predictions are in good absolute agreement
with observation, better than might have been expected from so
simple an approach.

Figure 6 shows another comparison between method predictions
and power station data for the same coals burned in a 250 MWe
corner fired unit. The furnace mean residence time is shorter,
1.5 s, than that for the wall fired boiler, 2.4 s, and higher
excess air values are used in normal operation. MEL
calculations suggest that temperature distributions should be
similar. As before the assessment method correctly identifies
the difference in combustion efficiency between coals and
gives realistic values for carbon loss.

Figure 7 shows the range of burnout performance that would
be expected from different coals. Combustion conditions have
been kept constant in these calculations and are appropriate
to the 360 MWe wall fired boiler. The differences in carbon
loss shown in Figure 7 therefore derive solely from
differences in char yield and reactivity, the latter being the
most important. The most important feature is the lack of a
good correlation between burnout performance and coal

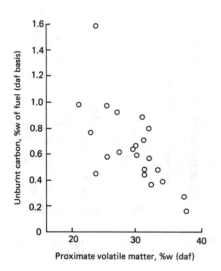

FIG. 7 — Variation in burnout performance due to
differences in coal properties. 360 MW$_e$
boiler, 15% excess air

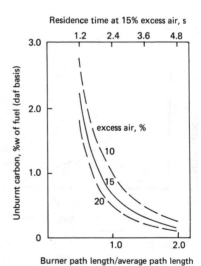

FIG. 8 — Dependence of carbon burnout on burner
path length (residence time) at three
values of excess air

proximate volatile matter content. Although it is true that
carbon loss is likely to be higher for a low volatile coal
significant variation exists in the combustion efficiency of
coals in the volatile matter range 25 - 35 %w (daf) used for
steam raising.

6. BOILER OPERATING PRACTICE
   The mathematical model allows the effect of certain boiler
operating practices on burnout to be investigated. Already
shown in Figure 5 is the effect of excess air. Unburnt carbon
is predicted to increase rapidly as excess air is reduced to
below 10 % (2.0%v flue gas oxygen). Increasing excess air
above 20% (4.0%v flue gas oxygen) gives only a slow
improvement in burnout which is broadly in line with
industrial experience.
   The contribution to total carbon loss from different
burners, if the path lengths are known, can be estimated from
Figure 8. This shows the change in carbon loss with burner
path length for a 360 MWe boiler. Path lengths have been given
relative to the average. Unburnt carbon concentration
increases very rapidly as path length decreases so that top
row burners may be expected to contribute significantly to
carbon loss. Similarly, poor furnace aerodynamics causing
under utilization of the furnace volume and an effective
reduction in path lengths would also result in significantly
increased carbon loss. Fuel-air distribution to the burners
can also be difficult to control in utility plant. If a
combination of short path length and low, < 10 % , excess air
occurs then boiler performance may be severely affected.
   The effect of particle size distribution on carbon burnout
is easily calculated but it is difficult to identify exact
values for the size distribution of coal produced by large
mills. Coal grind fineness is usually characterized by a
single parameter; the weight percent passing a 75 micron
sieve. The size distribution of larger particles typically
follows a Rosin Rammler distribution. For "standard" pf. the
slope of the R-R distribution is close to unity in value and
70-75 % w of material is sized smaller than 75 micron (8).
If it is assumed that all coal psd's are R-R with unit slope
then carbon burnout is a very strong function of 75 micron
grind fineness as shown by the hashed line in Figure 9. This
is however an unrealistic situation. Power station mill
product is influenced by the presence of the mill classifier
which limits the largest particle size sent to the burners. If
it is assumed that the coal distribution is R-R but that the
mill classifier fixes the largest particle size ,such that for
example 99.99%w of coal is always less than a certain value
d_{max}, then burnout is a weak function of 75 micron grind
fineness as shown by the solid line in Figure 9. The best,
albeit limited, information available to us on the variation
in coal particle size distribution with 75 micron grind
fineness supports the second hypothesis above. This leads to
the conclusion that only limited scope for modifying coal
combustion performance through finer grinding exists unless
the utility is able to alter the mill classifier settings.

462

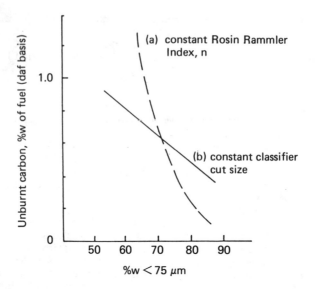

FIG. 9 — Effect of changing grind fineness on carbon
burnout for two mill operating modes

7. CONCLUSIONS

Measurements of coal combustion properties, char yield and char reactivity, have shown that significant differences exist between coals and that proximate volatile matter content is not a good guide to combustion performance. Char particle combustion can be described to a practical level of accuracy by assuming external surface reaction. Burnout predictions for large boilers made using this approach agreed well with measured data and most importantly correctly identified differences between coals. The calculation procedure also allows the effect of boiler operating conditions on burnout to be investigated so that burnout assessments can be made for different boilers and reasons for poor combustion efficiency investigated. Fuel air maldistribution, particularly to upper burners, emerges as a likely prime cause for high carbon in ash concentrations.

On a scientific level, more work is needed to understand the mechanism of char particle combustion. As well as data on burnout, which would ideally comprise instantaneous measurements of burning rate as a function of burnoff, size, temperature and oxygen concentration, better measurements of

463

pore structure are needed together with models of transport and combustion within the pores. These will need to take account of pore structure dependence on parent coal type and particle size, pyrolysis history and burnoff. In addition the above comments apply to the different maceral components within the coal. We also suspect that catalysis of the char oxidation reaction by ash components may be responsible for significant differences in char oxidation behaviour over and above pore structure effects. While it would be desirable to incorporate these into an advanced method of coal characterization such an approach is likely to be uneconomic for routine evaluation.

ACKNOWLEDGEMENTS

The authors wish to thank Dr. S. Cooper of Marchwood Engineering Laboratories, C.E.G.B, for information on boiler conditions and Mr. K. Stott, Station Superintendent, Castle Peak Power Station, Hong Kong, for boiler performance data.

REFERENCES
1. Kobayashi H., Howard J.B., Sarofim A.F. Coal Devolatilization at High Temperatures. 16th Symposium (Int.) on Combustion. The Combustion Institute. p411 (1975)
2. Howard J.B. Fundamentals of Coal Pyrolysis and Hydropyrolysis. Chemistry of Coal Utilization. 2nd Supplementary Volume. Wiley Interscience. (1981)
3. Essenhigh R.H. Fundamentals of Coal Combustion. Chemistry of Coal Utilization. 2nd Supplementary Volume. Wiley Interscience. (1981)
4. Morley C., Jones R.B. Entrained Flow Reactor and Image Analysis Study of Maceral Effects in Coal Char Oxidation. submitted to: 21st Symposium (Int.) on Combustion. Munich. (1986)
5. Ayling A.B., Smith I.W. Measured Temperature of Burning Pulverized Fuel Particles and the Nature of the Primary Reaction Product. Combustion and Flame 18 173-184 (1972)
6. Laurendeau N.M. Heterogeneous Kinetics of Coal Char Gasification and Combustion. Progress in Energy and Combustion Science. 4 221-270 (1978).
7. Cooper S. Private Communication. Marchwood Engineering Laboratories. (1983).
8. Steam its Generation and Use. Babcock and Wilcox Publications (1978).

464

DISCUSSION

R.H. Essenhigh

1) Support for the "flow tube" concept is supplied by cold-model experiments and analysis reported on the 14th combustion Symposium (Zeinalov et al.). Contours of mixing delay could be defined, and an "information flow path" identified flowing normally to the contours. This may be of interest.

2) A perfectly stirred reactor "flame ball" model was constructed to look at the extinction in a boiler (data from Waibel). The results showed the need for substantial char combustion to maintain the flames, and also indicated that the flames could be uncomfortably close to extinction is normal operation.

P. Roberts

Thank you for your support. I am also of the opinion that early flame development relies on char oxidation and that flame stability is largely determined by the reactivity of the coal char.

H. Jüntgen

Can we take the message from your modelling and practical experience that volatiles (under the conditions of heating rates in flames), reactivity of the residual char and its particle size are the only coal or char properties, which describe the behavior of different coals ?

P.T. Roberts

Yes, of these, char reactivity and particle size would seen to be the most important in determining combustion performance.

G. Flament

I would suspect that 34 ms might be a bit short for reaching 100 % devolatilization for large coal particles. Therefore the factors (V^x/V proximate) that you have derived from your measurements might be somewhat underestimated.

P.T. Roberts

The maximum coal particle size is restricted to 125μ m to ensure that the largest particles reach the pyrolysis temperature. The Q factor is slightly underestimated due to the inclusion of soot in the char yield.

T.F. Wall

I suggest that in furnaces flame temperatures will be

higher for high VM coals than for low. Therefore the tempera-
ture/time paths will depend on VM and a more pronounced effect
of burn-out with VM can be expected than your predictions
based on the same T-t path.

P.T. Roberts

The detailed temperature distribution in the burner zone
can be expected to depend on fuel quality. This we treat as a
well-stirred reactor with an average temperature and fuel
burn-out figures are found to be insensitive to changes in
this temperature. There will also be small changes in residence time with fuel quality changes for a boiler fired at
constant heat input. However these factors do not affect our
conclusions to any significant degree.

I.W. Smith

You showed 1st order rate coefficients. What range in p_{O2}
did you use to get this ?

P.T. Roberts

Burnout measurements are made at oxygen concentrations
of 3, 5 and 10 % v. It is possible that a fractional reaction
order would also describe the data. This order would not be as
small as 0.5 as TGA measurements at low temperature have
shown. (Entrained flow reactor and image analysis study of
maceral effects in coal char oxidation). C. Morley and R.B.
Jones, 21st Symposium (Int.) on Combustion, Münich (1986).

466

SYNTHESIS AND RECOMMENDATIONS FOR FUTURE WORK

INTRODUCTION

The three principal fossil energy resources on which the
world currently depends are oil, gas, and coal, accounting
between them for 89 percent of the 7.2 gigatonnes oil equivalent of primary energy consumed in 1984, with oil contributing
39% of the total, gas 20%, and coal 30%, the balance being
nuclear and hydroelectric [1]. The economically recoverable
reserves of oil and gas are however much smaller than that of
coal--64.8 gigatonnes for oil, 86.7 gigatonnes oil equivalent
for gas, and 2616 gigatonnes oil equivalent for coal[1]. The
current consumption pattern and energy reserves are clearly
not in balance, and the present heavy reliance on oil and gas
is not sustainable for very long. Although short-term trends
are difficult to predict, it is reasonable to anticipate that
much heavier reliance will need to be placed on energy from
coal, probably within ten to fifteen years, and much of the
coal will be consumed directly in coalburning equipment.

The technology for burning coal evolved slowly until the
last century. Early man burned coal on hearths. By the eigh-
teen century the major improvement made was to utilize a grate
to support the burning coals and to provide some underfire air
through the grate. Further refinements in the technology were
the introduction in the nineteenth and early twentieth centuries of moving grates and overfire air-jets. Pulverized coal
was introduced in 1961 and has become the method of choice for
large coalburning power stations, although a significant
market penetration is anticipated for fluidized-bed combustion
in both traditional bubbling beds and in circulating beds.
Advanced concepts for coal combustion involving cyclone
burners, direct-fired large-bore diesel engines, gas turbines,
and open-cycle MHD are at the research stage and are based
primarily on the use of pulverized and sometimes micronized
coals. The present workshop with its focus on pulverized coal
combustion is therefore addressing issues of importance in
both the large number of existing boilers and the next generation of advanced coal combustors.

Coal is less favored as fuel than gas and oil because as a
solid it is more difficult to burn in confined spaces, the ash
residue released during combustion reduces the availability of
furnaces through the fouling and slagging of heat transfer
surfaces, and the contaminants in coal lead to a wide variety
of environmental problems. Environmental problems related to
coal have been of concern for centuries, starting with an
early prohibition of the burning of Seacoal (bituminous coal)
in Soutwark, England early in the fourteenth century by King
Edward I, and continue to provide a major deterrent to the
greater use of coal. The pollutant of greatest importance has
changed with time. Soot, tars, and other hydrocarbon emissions
were undoubtedly the ingredients of the emissions from early
combustors that Sir John Evlyn in 1661 described as "an impure
and thick mist, accompanied with a fuliginous and filthy

vapor..."[2] . Coal generated deposits in chimneys were also the source of the compounds that led Sir Percival Pott in 1775[3] to establish a link between cancer in chimney sweeps and exposure to chemicals, now identifified as the polyclic aromatic hydrocarbons. More recently, the smoke and sulfur oxides generated by coal combustors were implicated in the large numbers of excess deaths resulting from respiratory impairment experienced during periods of atmospheric inversion in London, England during the 1950s, in the Meuse Valley, Belgium in 1930, and in Donorra, Pennsylvania in 1948[4]. The coal combustion products of current concern are SO_2 and NO_x because of their role as precursors to acid rain, and CO_2 and N_2O because of their contributions to potential global climatic effects in the twenty-first century. It is the challenge of coal combustion research to develop technologies to burn coal efficiently and cleanly.

Coal is a substance of great variability in chemical and physical composition, which undergoes a complex series of transformations during heating and oxidation. A critical component of coal research is to develop an understanding of how to measure and characterize the coal properties that influence combustor design and performance and to use that information to guide the development of improved technologies.

The presentations at the workshop reviewed the progress that had been made in understanding some of the many phenomena that occur during pulverized coal combustion and in identifying the gaps in the knowledge base. Three groups were formed to summarize the status of the field and to develop research recommendations. The working groups and their cochairmen were:

I. Devolatilization and Heterogeneous Combustion of Coal.
   Co-Chairmen: J.B. Howard and G. Prado.

II. Pollutant Formation and Destruction.
   Co-Chairmen: B.S. Haynes and G. De Soete.

III. Mathematical Modeling: Transfer to Industrial Applications. Co-Chairmen: F.C. Lockwood and G. Flament.

A final plenary session was held to review the recommendation of the Working Subgroups and to identify topics of research which cut across the responsabilites of the individual Subgroups (discussion leaders A.F. Sarofim and J. Lahaye).

I. DEVOLATILIZATION AND HETEROGENEOUS COMBUSTION OF COAL

J.B. Howard   Department of Chemical Engineering
               Massachusetts Institute of Technology
               Cambridge, MA 02139 USA
G. Prado   C.R.P.C.S.S.
               24, Avenue du President Kennedy
               Mulhouse, France.

The discussion was structured in the following five topics:

1) Definition of the relevant phenomena and problems

2) Status of the field

3) Research needs

4) Innovative experimental tools

5) Overall Recommendations.

1. Definition of the relevant phenomena and problems

Four topics were identified as important steps to describe:

1.1 Devolatilization, including kinetics and product yields

1.2 Heterogeneous combustion, with description of controlling phenomena: intrinsic kinetics, pure diffusion, adsorption-desorption, catalysis by inorganic species

1.3 Ignition delay - Mechanism under different conditions

1.4 Char structure - Its development during devolatilization - Its modification during char burnout and the evolution of active sites.

2. Status of the Field

Coal devolatilization and heterogeneous combustion are too complex to describe from first principles. We define quantitative descriptions of devolatilization and heterogeneous combustion as inputs for overall descriptions of coal combustion. We have had considerable success in developing simplified global models such as the multiple parallel reaction description of coal devolatilization, or the three-zone description of the O_2 - CO_2 - H_2O char reactions with order-of-magnitude correlation of the O_2- char reactivity with temperature.

There is much need to broaden the data base for these models to cover a range of coal types and char produced under different conditions. Also, there is need for more information on the associated physical processes and material properties.

3. Research Needs

The following topics need an extended research effort:

3.1 Kinetics of devolatilization, with special focus on
Flame stabilization
Trade-offs of staged combustion
Reactivity rate control (regimes of kinetics vs heating rate control)
Effect on carbon burnout

Role in pollutant formation
Heating values of coal volatiles
Better understanding of the mechanism, in relation to oxygen content, mineral matter, gas pressure, particle size, chemical structure of coal.

3.2 Ignition

Quantitative understanding of ignition delay
Role of devolatilization and heterogeneous reactions.

3.3 Extent of coal softening, swelling, fragmentation and agglomeration under combustion conditions.

3.4 Effects of coal type, maceral composition, and rank dependence of maceral behavior.

3.5 Char structure and regulation. There is a need of a data base for oxidation rate, product composition and char structure as functions of temperature and extent of gasification.

The development and measuring techniques for:

identification of active sites for $C-O_2$ reaction
identification of degree of graphitization of char should provide relationships between intrinsic reactivity of char and parameters above.

3.6 Order of magnitude correlations of char-CO_2 and char-H_2O reactions similar to that for char-O_2 reactions.

3.7 Identification of more meaningful surface area mesuring techniques for coal combustion conditions.

3.8 Mineral matter behavior under combustion conditions:

fly ash production
catalysis of char burnout
effects on ignition delay
ash deposition.

4. Innovative Experimental Tools

Two groups of methods need special development.

4.1 Optical diagnostics for in situ measurements in partical systems.

4.2 Single particle methods for simultaneous measurement of temperature, rate of weight loss, particle size, gas product composition (Drop tube furnace, suspended particle...).

5. Overall Recommendations

The group felt it to be extremely important to promote better interactional cooperation between scientists, with three
recommendations:

5.1 Establish a bank of coals, chars, available internatio-
    nally. Make these available to calibrate equipments and
    to conduct complementary experiments.

5.2 Promote exchange of scientists.

5.3 Find a mechanism for NATO collaboration with Australian
    Scientists.

II. POLLUTANT FORMATION AND DESTRUCTION

B.S. Haynes   Department of Chemical Engineering
                 University of Sydney
                 SYDNEY N.S.W. 2006 - Australia

G. De Soete   Institut Français du Pétrole
                 B. P. 311
                 92506 RUEIL-MALMAISON CEDEX - France

This subgroup concentrated its discussion on aspects of
pollutant formation and destruction of practical importance in
pulverised coal combustion. Inputs required for pf combustion
modeling were given special emphasis.

The bases of the group's discussions were the presentations by Prof. Sarofim, Dr. Pohl, and Dr. Jackson.

Soot and PAH

Soot and PAH are formed in substantial quantities in pf
combustion but conditions in the combustor are such that the
emission of these pollutants is not normally a problem. Within
the combustion chamber, particularly in the early (volatiles)
combustion zone the presence of soot may modify significantly
the local heat transfer distribution and the furnace temperature profiles.

On the basis that soot appears to be formed from the evolved tars, the group recommended that the tar --> soot conversion requires more detailed study, particularly under combustion conditions. Not only is this required for radiation
calculations, but also the conversion of tar to soot may
reduce heat release rates in the volatiles combustion zone as
soot is more resistant to burnout than are the tars. In this
light, the relative burnout rates of soot (particularly if it
is condensed into "tails") and tars also become important.

SO$_x$

At the high temperatures occurring in pf combustors, all the coal sulfur, both mineral and organic, is converted to SO$_2$. In this light, the sulfur distribution in the coal is not important. However, it should be pointed out that for low rank coals lower temperatures pertain and this question may need further study. Prof. Cypres drew attention to his studies of calcite/S interactions in fixed bed combustion. Prof. Jüntgen pointed out that future coal desulfurization technologies will be based on specific sulfur distributions.

Areas which the subgroup identified as requiring further work are those of heterogeneous SO$_3$ ans H$_2$SO$_4$ formation on fly ash and wall deposits. The condensation and reaction of these species on fly ash, especially on the high-area fume, also require attention. Inasmuch as sulfur, presumably as H$_2$SO$_4$, influences the precipitability of fly ashes, these issues are also important in ash collection.

While much work on sulfur capture with limestone has already been carried out, further investigations of temperature controle in the process is needed. The use of promoters to improve capture efficiencies is a promising development.

NO$_x$

Professor Sarofim dealt in detail with the relevant issues in this lecture. The subgroup considered the following areas to be of particular importance:

a) the distribution of fuel-N between the volatiles and char during devolatilization. The relationship between volatiles-N and the usually observed HCN and NH$_3$ must be established if simple model inputs, as [HCN] and [NH$_3$], are to be available.

b) the heterogeneous NO- and HCN-soot reactions may give favorable N$_2$ production rates. These processes could be important during devolatilization.

c) the homogeneous NO-hydrocarbon reaction(s) to form HCN are incompletely understood. Given that these are crucial to the success of staged combustion applications in reducing NO$_x$ formation, these reactions require further study.

d) the climatological implications of substantial N$_2$O emissions from pf, and other combustors require further investigations of the kinetics of this species. Fundamental flame kinetic studies are needed to include this species in the otherwise fairly well established gaseous fuel-N mechanism.

e) while the gas-phase fuel-N mechanism is in good shape, current modeling efforts do not take advantage of this. Some of the simple, quasi-global mechanisms should be compared with detailed benchmark calculations using more complete mecha-

472

nism_.

f) the formation of NO from char-N requires further attention. The heterogeneous reduction of NO on chars appears to be
too slow to be of importance in pf combustion.

g) the interaction between NO_x and SO_x may not be a major
effect but requires further quantification.

Mineral Matter

The influence of mineral matter on pf combustion is perva-
sive, and, from a practical standpoint, often of paramount
importance in determining boiler availability. The subgroup
considered the mineral matter transformations as giving rise
to three distinct inorganic classes: volatiles, comprising
completely volatilised inorganics such as Na, SO_3, and Cl
which condense only on cool surfaces; the submicron fume
formed by the vaporisation/condensation phenomena described by
Prof. Sarofim; and the residual fly ash left behind when the
char burns away.

The following issues were considered:

Radiation: Radiation scatter from the residual fly ash fraction influences radiative transfer in pf boilers. A knowledge
of the particle size distribution is needed to predict this
effect. Emission by ash particles is probably negligible.

Slagging and Fouling: The volatiles and the fume may be involved in "sticky layer" formation and boiler tube corrosion. The
bulk of deposits comes from the residual ash whose deposition
is a function of size and composition.

Precipitation: The prediction of precipitability is reasonably
reliable, although the Australian coals are more difficult to
predict. Mechanisms are not well understood, however. The
influence of H_2SO_4, whose formation and behavior are controlled to some extent by the ash volatiles and the fume requires
further investigation.

A specific recommendation of the subgroup followed from Dr.
Jackson's and Dr. Pohl's presentations: because the problems
in ash deposition and fouling involve such a vast array of
physical and chemical interactions, guidelines to further
fundamental risk should be obtained first from investigations
into phenomena occurring in full-scale units. Fundamental work
which could follow from such guidelines would be in the areas
of heterogeneous kinetics (ash/SO_2, HCl, etc.), the deposition
fluxes of ash components, the transport of species through
deposited ash layers, and the thermodynamics of molten ash
layers.

A number of new techniques are available which could provide valuable information in ash studies. These include surface
reflectometry, surface analysis by ESCA and Auger, synchrotron

radiation, LAMMA, laser vaporization techniques, and FTIR.

DISCUSSION

R. Cypres (Université Libre de Bruxelles)

It would be of interest to study interaction of mineral matter of the coal and SO_2 pollution. We published results of work on fixed bed combustion of an Italian coal (Sulcis) with 5% of S but which is rich in calcite ($CaCO_3$) in reduction conditions the H_2S formed is tied up as CaS. Under oxidation conditions CaS is oxidised to $CaSO_4$. This was published some years ago in Fuel but the conditions are not those of pulverized coal burners.

H. Jüntgen (Bergbau-Forschung GmbH)

1) Comment to NO formation mechanism by burning volatiles:

The highest percentage of N in volatiles is found in tar. There is a need to study the chemical structure of N containing products in tar and to study their special burning behavior in relation to NO_x formation. The combustion of model substances e.g. pyrrolidine and pyrrol, has shown, that different nitrogen containing compounds behave quite differently.

2) Comment to S distribution in coal:

I agree that S distribution is not relevant for burning behavior and SO_2 formation. However it is important in relation to future work of developing processes for S removal before burning. As to the research of development of biochemical processes it could be shown that for the removal of pyritic sulfur and for the removal of organic sulfur very different process conditions and different bacteria are needed.

K.H. van Heek (Bergbau-Forschung GmbH)

As to the pollutants originating from mineral matter in coal, researchers have also to deal with the need that the ashes have to be disposed or utilized. With respect to the possible impact to the environment or human health much more research is needed on composition, mineral state and long term fate of constituents of the solid residues. The results expected from such work could lead to a classification of ashes from a perspective of environmental impact. Further, they form a broader basis for a proper design of location for disposal and for public discussion about the acceptability of large coal fired power stations.

III. MATHEMATICAL MODELING: TRANSFER TO INDUSTRIAL
    APPLICATIONS

F.C. Lockwood   Department of Mechanical Engineering
                 Imperial College of Science and Technology
                 Exhibition Road
                 LONDON SW7 2BX - England

G. Flament   C.E.R.C.H.A.R.
                 Société des Ciments Français
                 Centre Industriel et Technique
                 Rue du Château
                 78931 GUERVILLE CEDEX France

It was decided to exclude fluidised beds, gasification processes and diesel engine applications from discussion as they
were considered peripheral to the primary theme of the
conference. The principal industry users of pulverised coal
were identified. These appear in the left hand column of Table
I. Broadly, the entries are in descending order of total coal
usage although it is recognised that consumptions among
industries will vary somewhat between countries. In general
the power generation industry is fare the greatest coal
consumer. Note that it was felt proper to include all forms of
steam raising plant within the dominant power generation
application.

The principal phenomena relevant to the coal burning industries were identified and these are listed horizontally along
the bottom of Table I. Numbers 1 to 3 have been entered in the
Table, as far as possible by members of the relevant industries. The number one means that the phenomenon is of prime
importance to that industry, two designates some importance,
while three indicates little relevance. The number one may
also be taken to imply a need for further model development in
the area. It was noted that the factors governing flame
stability in steam raising combustion differ somewhat from
those governing the ignition and possible extinction of coal
injected into the raceway of a blast furnace, and of course it
would be possible to identifiy other imprecisions in the
classification of Table I. Discussion on the importance of
ignition in blast furnace injection suggested that the ignition delay is important as this limits the time for combustion. For the extremely short time available this delay is
significant and for this reason it has been allocated priority
one. With the exception of the cement kiln application, flame
stability emerges as the topic of greatest general interest.
The gas turbine industry makes the strongest demands on
modellers.

The next task of this working group was to identify the
extent to which modern mathematical models are being applied
by the relevant industries.

Unfortunately there was time enough only to consider the power generation industry. The entries in Table II were made by a representative of a leading European boiler manufacturer. The table indicates both the actual use of mathematical modeling by this enterprise, for each of the physical phenomena identified in the previous Table, along with and the panel's view on the potential of existing mathematical models to simulate the phenomena. The panel was in general surprised and encouraged by the extent to which this particular manufacturer was relying on mathematical modeling. It was felt that competitors in other countries were probably less active in this respect, although interest was growing, and that in general the other industries probably lagged behind by greater and smaller amounts.

The panel then turned its attention to identifying gaps in the component submodels. The modelers expressed the wish for extensive and reliable recommendations for devolatilization and char burn out kinetics, especially in view of the discussion which these topics had generated at this workshop. They felt that the confidence in their predictions was in proportion to the certainty of these and the other submodels. Although the need for improved kinetic information was not denied the interest of the panel rather quickly focussed on the need to identify new laboratory experiments which would serve to bridge the gap between the classic drop tube and heated grid studies and the full scale combustor.

The new experiments would emphasize fluid mechanical effects in response to the general lack of velocity data reported by the modellers and because of unexpected discrepancies observed by researchers between ignition in quiescent surroundings and in flows. It was stressed by the industry representatives that the flows in their equipment are almost exclusively characterized by very high Reynolds numbers. The modellers stated that laminar effects could not be incorporated in existing turbulence models without recourse to additional and imperfect modeling inputs. It was concluded that the flow in any experiment designed to emphasize the practical effects of the fluid mechanics would have to be characterized by fully turbulent Reynolds numbers, a requirement which is not always convenient at the laboratory scale. As an example it was stated that for a simple jet mixing experiment the Reynolds number based on the injection pipe diameter and the velocity difference between that of the injected fluid and that of the surroundings would have to be in excess of 20,000. In a straightforward laboratory experiment this would typically imply an injector diameter of 10^{-2} m and a velocity difference of 10 m/s.

Two basic types of experiments were identified which may be termed A and B, the latter being a logical extension of the former. Both experiments would make use of the recently developed analytical techniques described at the conference. These now allow non-intrusive measurements of particle and gas temperatures, gas and particle velocities and soot and gas concentrations. The type A experiment would comprose a high velocity jet of cool, or slightly warmed, air (or possibly nitrogen) issuing into a non-trubulent, low velocity, co-flowing secondary stream of air heated to combustion temperatures. The primary stream would be seeded with individual particles, and later particle clouds, of coal which would devolatilize and ignite as a result of the turbulent mixing with the secondary stream. This experiment could also be used to establish the effect of turbulence on burning rates of pulverized coal. For practical reasons it might be necessary to contain the experiment by glass walls, but if this were to be done the walls should be film cooled to eliminate the unwanted complicating effect of thermal radiation transfer to the fuel. The walls would have to be sufficiently remote to avoid jet de-entrainment and consequent recirculation. It was further remarked that the flow geometry should be cylindrical rather than planar for several reasons, not the least being that existing turbulence models are more developed for this case.

The type B experiment would involve a suitably scaled down industry type swirl burner firing into a cylindrical chamber. It would be important to ensure the axial symmetry of the flow as the performance of model validation in three dimensional flows is extremely inefficient. Ideally local gas phase measurements of the mean and fluctuating velocities, temperatures and species are required as well as the determination of the particulate matter velocities, temperatures, number counts and elemental and proximate analyses. It was recognised that existing non-intrusive techniques would be unable to penetrate a useful distance into a PF flame ant that, regyetably, initial experiments would have to be performed on gas flames seeded with dilute quantities of coal or possibly inert particles. Velocity data were of particular interest to modellers at this time.

It was acknowledged that the modellers could be of considerable assistance in designing the new experiments, for example in ensuring the adequacy of the local turbulent Reynolds numbers in the type B facility, and that the mathematical models, when run in parallel with the experiments, would greatly assist the establishment of the important experimental parameters and their ranges.

Table I: Pulverised Coal Burning Industries in Order of Importance versus Relevant Physical Phenomena.

1 = phenomenon of major relevance
2 = some relevance
3 = no relevance

	Flame Stability	Heat Transfer	Ash Deposition	Carbon Burnout	Pollution and Emissions	Fuel Quality Tolerance	Turned Down Operation
1. Power Generation and Industrial Steam Raising	1	2	1	1	1	2	2
2. Cement Kilns	3	2	2	3	1	2	3
3. Blast Furnace Injection	1	3	2	1	3	2	3
4. Smelting	1	1	3	3	2	2	3
5. Process Heating	1	1	2	2	2	1	1
6. Gas Turbines	1	1	1	1	2	1	1

Table II. Relative Exploitation of Modeling by the Power
        Generation Industry.

        1 = successful use, or availability of good models.
        2 = no use, or nonavailability of good models.

Physical Phenomenon	Model Use	Model Availability
Flame Stability	2	2
Heat Transfer	1	1
Ash Deposition	2	2
Carbon Burnout	1	1
Pollutant Emission	2	1
Fuel Quality Tolerance	2	2
Turn down Operation	2	2

GENERAL RECOMMENDATIONS

A.F. Sarofim   Department of Chemical Engineering
                    Massachusetts Institute of Technology
                    Cambridge, MA 02139 USA
J. Lahaye   C.R.P.C.S.S.
                    24, Avenue du Président Kennedy
                    Mulhouse, France

The plenary sessions involved a review and discussion of the Working Group reports, summarized above. There were several recurring themes deserving emphasis that emerged. The framework for the discussion was provided by Fig. 1, provided by Prof. Robert Essenhigh (Ohio State University, USA), showing the progression of research from the fundamental sciences to practical applications. The processes occurring in a coal-fired combustion chamber involve complex interactions between homogeneous and heterogeneous chemical kinetics; material properties, particularly those that determine the pore structure and surface reactivity of the char produced during coal devolatilization; heat transfer; and aerodynamics. As an illustration of the progressive addition of layers of complexity in addressing a problem, consider the issue of flame stability. At a fundamental level we are concerned with the rate of evolution and composition of the volatile products; these in turn are influenced by the secondary reactions which can occur within a particle, reactions which are strongly influenced by the plasticity and/or pore structure of the coal particle during pyrolysis. Flame ignition requires consideration of transport phenomena, by diffusional processes and radiation in a one-dimensional flame, complicated by turbulence and recirculation in two-dimensional flames, by heat transfer to the walls in a pilot-scale boiler, and by burner-burner interactions in practical combustion systems.

Experimental design. The above illustration on flame stability points out the need for experimental apparatus of different scale and design to evaluate the processes of importance in coal flames. At the fundamental level there is a need to work with equipment in which the temperature is controlled by external energy inputs. Problems such as ignition where the flame is stabilized by the energy feedback from combustion products need to be studied in self-subtained flames. The desirability of having standard equipment being reached since there is much merit to the current trend to design equipment for particular goals. The importance of characterizing the conditions, particularly temperature, within the test apparatus was stressed. The need for involving modellers in the design of experiments, particularly in aerodynamically complex systems, was emphasized.

Advanced diagnostics. Much of the recent progress that has been made in coal science is attributable to advances in modern physical instrumentation. The inceased use of NMR, ESR, FTIR, GC-MS, GPC, HPLC... have provided insights on the relation of the structure of pyrolysis products and char to that

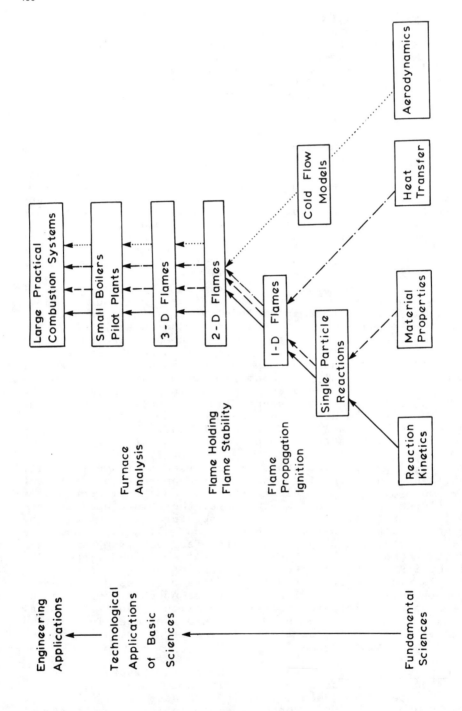

Fig 1. Hierarchical Approach to Experimental Studies of Coal Combustion (Essenhigh)

of the parent coal. ESCA, Auger, AA, PIXE, INAA, Mossbauer, high resolution electron-microscopes yield information of the physical and chemical state of the mineral matter in coals and chars sometimes down to a nanometer resolution. Lasers and solid state detectors permit the in-situ measurement of particle size, temperature, and velocity. New instrumental analysis, see Working Group II recommendations, continue to be developed and there is a need to apply them to coal since the current understanding of the chemical reactions and physical transformations occuring in the solid state is still at a formative state. There is a need for more refined methods of analysis of the pore size distribution in chars during reaction, with distinction of accessible and inacessible porosity and some measure of pore interconnectivity; SAXS combines with mercury porosity and BET surface areas give a partial measure of these parameters. Progress has been made on characterizing the viscosity and bubble distribution in plasticizing coals but much remains to be done.

Selection of coals and chars. The need to establish a seriesof reference coals and chars to be used by different coal science laboratories was recognized. At present coal banks are being established at a national level. International collaboration is desirable. An argument was also presented (D.R. Hardesty, Sandia, USA) for the use of model coals and chars that can be selected to systematically study the effect of changing a single parameter at a time. Examples of synthetic chars are those produced by the pyrolysis of different organic polymers with and without mineral additives. Such chars can be designed with different distributions of micro and macroporosity and with different degrees of dispersion of mineral additives from molecularly dispersed to micron-size inclusions. Model compounds can also be used to examine the effect of organic structure on combustion behavior, as has been done for example by the selection of pyrrolidine and pyrrol to examine the fate of organically-bound nitrogen during pyrolysis (H. Jüngten, see above).

International exchanges. The cross-fertilization of ideas that occurred at the Workshop underlined the value of exchanges of scientists between different coal science laboratories not only within NATO but also with other countries active in coal research, with Australia being singled out for particular consideration.

Technology transfer. From the presentations at the Workshop and from the long list of research topics identified by the Working Subgroups, it may be, and sometimes is, argued that there are so many gaps in our understanding of the properties and combustion behavior of coal as to impede any ability to apply the results to practical combustors. There was, however, heartening evidence at the Workshop of the application of the fundamental knowledge to the design and optimization of practical combustors. Models for radiative transfer in boilers are being used increasingly to predict the balance of the heat transfer to the radiant chamber of boilers, models of volatile yields and char oxidation kinetics to examine the extent of carbon burnout, and models of fuel nitrogen chemistry to guide the development of pollutant control strategies such as the reburning of nitrogen oxides. For some of the more complex problems, such as fouling and slagging, it was felt that research should be conducted simultaneously at both the fundamental level, to determine the factors that govern ash particle size distribution and the composition of vaporized ash, and on full-scale units to infer from the morphology and composition distribution of deposits a reasonable mechanism for deposition. Research is needed at all scales of operation in order to translate the fundamental science to industrial practice, following the roadmap of Fig. 1.

In summing up, one can find in the Workshop papers and Subgroup discussions evidence of both the significant progress that has been made in understanding many of the processes occurring in a coal-fired combustor and in translating that understanding to practice and of the major gaps remaining in our understanding. The improbability of an early resolution of all the issues related to the physics and chemistry of coal burning is well expressed by a quotation cited by Thomas[5]. In a short story "On the feasibility of a coal-driven power station" purportedly written in the year 4995 A.D., Otto Frisch concludes"... the kinetics of coal-oxygen reactions are much more complicated than fission reactions and not yet completely understood". The challenges for the coal research scientist and engineer are clear and the opportunities abundant.

REFERENCES

1. BP Statistical Review of World Energy, The British Petroleum Company, London, England, June, 1985.

2. Evlyn, J., Fumifuguim (1661), in "The Smoake of London, Two Prophecies (J.P. Lodge, Ed.), Maxwell Repring Company, Elmsford, New York, 1969.

3. Pott, P., London, Printed for L. Hawes, W. Clarke, and R. Collins (1775).

4. National Air Pollution Control Association "Air Quality Criteria for Sulfur Oxides", NAPCA Publication no. AP-50, H.E.W., Washington, D.C., 1970.

5. Thomas, J.M., Carbon, 8, 413-421 (1970).

LIST OF PARTICIPANTS TO THE WORKING GROUPS

I. Devolatilization and Heterogeneous Combustion of Coal

Co-chairmen: J.B. HOWARD - G. PRADO

Participants:

R. CYPRES - R.H. ESSENHIGH - M. HERTZBERG - J. JÜNTGEN
F. KAPTEIJN - R.E. MITCHELL - H.J. MÜHLEN - J.F. MULLER
N.P. ODENTHAL - J.R. RICHARD - I. SMITH - J. SMITH -
E.M. SUUBERG - J.M. VLEESKENS - A. WILLIAMS.

II. Pollutant Formation and Destruction

Co-chairmen: B.S. HAYNES and G. De SOETE

Participants:

F. BERETTA - A. GARO - B.S. HAYNES - T.R. JOHNSON
J.H. POHL - A.F. SAROFIM

III. Mathematical Modeling:
     Transfer to Industrial Applications

Co-chairmen: F.C. LOCKWOOD and G. FLAMENT

Participants:

J.M. BEER - M. DESPREZ - I. GULYURTLU - Z. HABIB
D.R. HARDESTY - J.D. HICKERSON - M. MORGAN - P. ROBERTS
J.L. ROTH - E. SAATDJIAN - P.R. SOLOMON - W. THIELEN
K.H. Van HEEK - T.F. WALL - W. ZINSER.

BANKS OF COALS

AUSTRALIA
FRANCE
THE NETHERLAND
U.S.A.

AUSTRALIA

ADDRESS :

CSIRO
Division of Fossil Fuels
Delhi Road, North Ryde, N.S.W.
Australia
Doctor I.W. Smith

SAMPLES AVAILABLE :

A wide range of Australian samples is available (from
low volatile bituminous to brown) in dried, pulverized
form.

COST :

Only freight cost.

FRANCE

ADDRESS :

CERCHAR
B. P. n° 2
60550 Verneuil-en-Halatte
Mrs Malechaux

SAMPLES AVAILABLE :

Four french coals are available :
semi-anthracite ; medium volatile bituminous ; high
volatile bituminous B ; high volatile bituminous C.
An anthracite will be added in 1987.

BASIC CHARACTERIZATION :

The following data are available :
- proximate analysis (moisture ; ash and volatile contents ; calorific value)
- ultimate analysis
- ash analysis (chemical and fusion analysis)
- mineralogical analysis
- swelling index.

COST :

Originally, the bank is for national use. If samples
are required, the Cerchar must be contacted.

THE NETHERLAND

ADDRESS :

Dutch Centre for Coal specimens, SBN
Postbox 151
6470 ED Eygelshoven
The Netherlands
Dr. Ir. K.A. Nater

SAMPLES AVAILABLE :

105 coal samples (01 August 1986) from the following
countries : Australia, Belgium, Canada, China,
Columbia, France, Germany, Hungaria, Indonesia,
Norway, Poland, South Africa, Soviet Union, United
Kingdom, United States.

All coals are stored in inert atmosphere.

BASIC CHARACTERIZATION :

On all coals the following analyses have been performed :
- proximate analysis
- heat of combustion
- element analysis (ultimate analysis)
- chlorine content
- sulphur forms and total sulphur
- composition of the ash
- maceral group analysis

- vitrinite reflections
- composition of the ash
- mineral composition (REM/EDS)

In several cases the trace elements have been determined.

RATE SCHEDULE : (August 1986)

10 kg samples

single sample	Hfl. 600,-
series of 10 samples form one coal	5 000,-
larger series	price on request

1 kg samples

single sample	150,-
series of 10 samples from one coal	1 000,-
larger series	price on request

WSS-samples

single sample 20 grams in glass vial	30,-
series	price on request

ST-samples

single sample 30 grams in glass vial	60,-
series	price on request

WSS = WHOLE SEAM SAMPLE
ST = SPECIAL TREATED SAMPLE
        washed at density 1.6
        screened -18 to 100 mesh
For specific information see also
the SBN sample catalogue and the
SBN flyer.

U. S. A.

ADDRESS :

The Penn State Coal Sample Bank
The Pennsylvania State University
513 Deike Building
University Park, PA 16 802, U.S.A.
Dr. C. Philip Dolsen

SAMPLES AVAILABLE :

Over 1300 coal samples from U.S.A.
The great majority of samples are raw coal (no clea-
ning or beneficiation procedure).
"Premium" coal samples are prepared with require special precautions in sampling, crushing and handling to
avoid "weathering" or "oxidation". All premium samples
have been sealed in an inert gas at the collection
site and retained in an inert atmosphere in all subsequent handling.

BASIC CHARACTERIZATION :

The following data are available :
- proximate analysis,
- ultimate analysis
- analysis for sulfur and sulfur forms
- ash fusion analysis
- geiseler plasticity data
- physical properties (free swelling index, surface
  areas, densities, pore size distribution etc)

- petrographic analysis
- reflectance analysis
- element analysis
- volatile trace element
- gasification data
- carbonization data.

RATE SCHEDULE : (September 1986)

I. SAMPLES

A. Standard Samples :
1) + 2 Cans   1 l b./-20 mesh $ 5.00
2) = 10 Cans   5 l b./-1/4 inch $ 15.00
3) Buckets   35 l b./-1/4 inch $ 35.00
4) Plastic Drums 200 l b./variable $ 150.00

B. Premium Samples :
1) Ampoules   45 g/-60 mes $ 15.00

II. DATA PRINTOUTS

A. PDP 11/23 $0.50/page or $4.50/Standard
                     Printout (10 pages)

B. IBM 370 $5.00/standard Printout
                     (3 pages)

LIST OF PARTICIPANTS

Prof. P. ANGLESIO

Dipartimento di Energetica Politecnico di Torino
c.so Duca degli Abruzzi 24 - 10129 TORINO -Italy

Prof. J.M. BEÉR

Department of Chemical Engineering
Massachusetts Institute of Technology
CAMBRIDGE Mass. 02139 - U.S.A.

Dr. F. BERETTA

Laboratorio di Ricerche sulla Combustione, C.N.R.
Piazzale Tecchio - 80125 NAPOLI - Italy

Dr. C. BERTRAND

Lafarge Coppee Recherche - Laboratoire Central -
B.P. 8 - 97220 VIVIERS s/RHONE - France

Dr. D. BOUCHEZ

Elf France - Centre de Recherche Elf Solaize -
B.P. 22 - 69360 ST-SYMPHORIEN D'OZON - France

Mr. O. CHARON

Centre de Recherches sur la Physico-Chimie des
Surfaces Solides - C.N.R.S.
24, avenue du Président Kennedy
68200 MULHOUSE - France

Prof. R. CYPRES

Université Libre de Bruxelles
Faculté des Sciences Appliquées
Service de Chimie Générale et Carbo-Chimie
Avenue F.D. Roosevelt, 50 (CP 165)
1050 BRUXELLES - Belgium

Dr. G. De SOETE

Institut Français du Pétrole - B.P. 311
92506 RUEIL-MALMAISON - France

Dr. M. DESPREZ

L'Air Liquide - Centre de Recherche Claude-Delorme
Les Loges-en-Josas - B.P. 176
78350 LES LOGES-EN-JOSAS - France

Prof. R.H. ESSENHIGH

Department of Mechanical Engineering
The Ohio State University
206 West 18th Avenue
COLUMBUS Ohio 43210-1107 - U.S.A.

Dr. G. FLAMENT

C.E.R.C.H.A.R.
Plate-Forme Nationale d'Essais des Charbons
Rue Aimé Dubost - B.P. 19 -
62670 MAZINGARBE - France

Dr. A. GARO

Centre de Recherches sur la Physico-Chimie des
Surfaces Solides - C.N.R.S.
24, avenue du Président Kennedy
68200 MULHOUSE - France

Dr. I. GULYURTLU — Laboratorio Nacional de Engenharia e Tecnologia Industrial (LNETI) - Grupo de Combustao R. Alves Redol - 1000 LISBOA - Portugal

Dr. Z. HABIB — Laboratoire de Thermodynamique - U.A. CNRS 230 Faculté des Sciences de Rouen 76130 MONT-SAINT-AIGNAN - France

Dr. D.R. HARDESTY — Combustion Research Laboratory SANDIA National Laboratories LIVERMORE CA 94550 - U.S.A.

Prof. B.S. HAYNES — Department of Chemical Engineering University of Sydney SYDNEY N.S.W. 2006 - Australia

Dr. M. HERTZBERG — U.S. Bureau of Mines - Pittsburgh Research Center P.O. Box 18070 - PITTSBURGH Pa 15236 - U.S.A.

Dr. J.D. HICKERSON — United States Department of Energy Pittsburgh Energy Technology Center - MS 922 P.O. Box 10940 - PITTSBURGH Pa 15236 - U.S.A.

Prof. J.B. HOWARD — Department of Chemical Engineering Massachusetts Institute of Technology CAMBRIDGE Mass. 02139 - U.S.A.

Dr. P.J. JACKSON — Central Electricity Generating Board Technology Planning and Research Division Marchwood Engineering Laboratories - Marchwood SOUTHAMPTON SO4 4ZB - U.K.

Dr. T. JOHNSON — State Electricity Commission of Victoria Howard Street - RICHMOND, Vic - Australia 3121

Prof. Dr. H. JÜNTGEN — Bergbau-Forschung GmbH Franz-Fischer-Weg 61 - Postfach 130140 4300 ESSEN 13 (Kray) - Germany

Dr. F. KAPTEIJN — Institute of Chemical Technology University of Amsterdam Nieuwe Achtergracht 166 1018 WV AMSTERDAM - Netherland

Prof. Dr. H. KREMER — Ruhr-Universität Bochum Universitätsstrasse 150 - 4630 BOCHUM 1 - Germany

Dr. J. LAHAYE — Centre de Recherches sur la Physico-Chimie des Surfaces Solides - C.N.R.S. 24, avenue du Président Kennedy 68200 MULHOUSE - France

Mr. G. LEYENDECKER   Houillères du Bassin de Lorraine
Direction de l'Electricité et de la Carbonisation
Usine de Marienau - 57600 FORBACH - France

Dr. F.C. LOCKWOOD   Department of Mechanical Engineering
Imperial College of Science and Technology
Exhibition Road - LONDON SW7 2BX - U.K.

Dr. R.E. MITCHELL   Combustion Research Division
Sandia National Laboratories
LIVERMORE Ca 94550 - U.S.A.

Dr. M. MORGAN   International Flame Research Foundation
c/o Hoogovens Groep B.V. - Building 3G-25
P.O. Box 10 000 - 1970 CA IJMUIDEN - Netherland

Dr. H.J. MÜHLEN   Bergbau-Forschung GmbH
Franz-Fischer Weg 61
Postfach 130140 - 4300 ESSEN 13 - Germany

Prof. J.F. MULLER   Laboratoire de Spectrométrie de Masse et de
Chimie Laser - Université de Metz
B.P. 794 - 57012 METZ CEDEX 1 - France

Dr. H.P. ODENTHAL   L&C Steinmüller GmbH
Postfach 100855/100865
5270 GUMMERSBACH - Germany

Mr. T. OUTASSOURT   Centre de Recherches sur la Physico-Chimie des
Surfaces Solides - C.N.R.S.
24, avenue du Président Kennedy
68200 MULHOUSE - France

Dr. J.H. POHL   Energy Systems Associates
15991 Red Hill Ave., 110
TUSTIN Ca 92680 - U.S.A.

Prof. G. PRADO   Centre de Recherches sur la Physico-Chimie des
Surfaces Solides - C.N.R.S.
24, avenue du Président Kennedy
68200 MULHOUSE - France

Dr. J.R. RICHARD   Centre de Recherches sur la Chimie de la
Combustion et des Hautes Températures - C.N.R.S.
1C, avenue de la Recherche Scientifique
45071 ORLEANS CEDEX - France

Dr. P. ROBERTS   Thornton Research Centre
P.O. Box n° 1 - CHESTER CH1 3SH - U.K.

Dr. J.L. ROTH   I.R.S.I.D. - Institut de Recherche de la
Sidérurgie - Station d'Essais : Voie Romaine
B.P. 64 - 57210 MAIZIERES-LES-METZ - France

Dr. E. SAATDJIAN	Laboratoire d'Aérothermique - C.N.R.S. 4ter, route des Gardes - 92190 MEUDON - France
Prof. A.F. SAROFIM	Department of Chemical Engineering Massachusetts Institute of Technology CAMBRIDGE Mass. 02139 - U.S.A.
Dr. Wm. R. SEEKER	Energy and Environmental Research Corporation 18 Mason - IRVINE Ca 92718 - U.S.A.
Prof. I.W. SMITH	C.S.I.R.O. - Division of Fossil Fuels P.O. Box 136 - NORTH RYDE NSW 2113 - Australia
Dr. J.G. SMITH	C.S.I.R.O. - Institute of Energy and Earth P.O. Box 136 - NORTH RYDE NSW 2113 - Australia
Dr. P.R. SOLOMON	Advanced Fuel Research P.O. Box 18343 - EAST HARTFORD CT 06118 - U.S.A.
Prof. E.M. SUUBERG	Brown University - Division of Engineering PROVIDENCE RI 02912 - U.S.A.
Dr. W. THIELEN	L&C Steinmüller GmbH Postfach 100855/100865 5270 GUMMERSBACH - Germany
Dr. K.H. Van HEEK	Bergbau-Forschung GmbH Franz-Fischer-Weg 61 Postfach 130140 - 4300 ESSEN 13 - Germany
Dr. J.M. VLEESKENS	Netherlands Energy Research Foundation Chemistry and Materials Department P.O. Box 1 - 1755 ZG PETTEN - Netherland
Prof. T.F. WALL	Department of Chemical Engineering University of Newcastle NEW SOUTH WALES 2308 - Australia
Prof. A. WILLIAMS	University of Leeds Department of Fuel and Energy LEEDS LS2 9JT - U.K.
Dr. W. ZINSER	Institut für Verfahrenstechnik und Dampfkesselwesen (IVD) Pfaffenwaldring 23 7000 STUTTGART 80 - Germany

Index

Optical microscopy 71, 165, 166, 169, 269

Organic matter 4, 12, 250, 269, 272

Oxidation 11, 104, 106, 109, 117, 225, 245, 305, 306, 363, 440, 454

Oxygen (contents, functionalities) 11, 180, 185, 247, 306, 307, 315, 326, 377, 384 t/m 388, 433, 434

Particle(s) (single, number density, size) 4, 9, 13, 22, 95, 104, 105, 106, 110, 111, 114, 119, 121, 126, 130 t/m 140, 144, 146, 153, 155, 179, 183, 184, 189, 191, 193, 249, 272, 275, 279, 290, 293, 294, 306, 321, 322, 327, 332-334, 347 t/m 357, 365, 366, 369, 370, 377 t/m 380, 386 t/m 389, 437, 454

Particulate 321, 345, 347, 348, 397, 437

Petrographic components 6

Phenolic hydroxyl (groups) 11

Peat 6, 59

Photographic 336

Photography heat speed 273

Plasma (temperature) 63, 342, 344

Plasticity (plastic) 77, 88, 92, 95, 96, 97, 98, 114

Pollutant (formation) 245, 330, 398, 424, 442

Polarization 330

Polymerization 94, 96

Pore (diffusion, structure) 9, 17, 117, 136, 159 t/m 173, 219 t/m 221, 258, 355, 400, 458

Porosity 7, 9, 105, 126, 159 t/m 172, 179, 220, 356

Potassium (ions, chloride) 67, 255, 272, 274, 277, 322, 328, 343

Pressure 105, 130, 219, 223, 224, 255, 349, 357 t/m 366

Power station 250, 403, 404, 464

Pulse (lenght, energy) 106, 130, 323, 324, 329, 330

Pyrometer 130, 140, 144, 156, 338, 353, 388

Pyrite 18, 22, 67, 271, 272, 295

Pyrrhotite 272

Pyrolysis 5, 11, 13, 19, 59, 60, 92, 96, 97, 104 t/m 107, 110, 114, 126 t/m 131, 134 t/m 146, 168, 170, 178, 181, 184, 186, 187, 190 t/m 193, 245, 272, 305, 352, 359, 363, 371, 445, 454

Quartz 270, 273, 288

Radical Species (OH-C^2-O-NH-) 249, 305, 307, 313, 323 t/m 330, 446

Radiation 106, 137, 179, 186, 246, 328, 348, 351, 352, 36, 368, 371, 377, 378, 395, 401, 424 t/m 431, 439

Raman (scattering, spectroscopy, stockes, anti-stockes, resonance, spontaneous Raman scattering) 323 t/m 329

Rayleigh 335, 338

Rate 6, 16, 77, 78, 80, 82, 83, 96, 104, 105, 106, 110, 118, 120, 121, 126 t/m 131, 136, 140 t/m 146, 152, 155, 156, 172, 185, 186, 220, 221, 224, 226, 249, 278 t/m 280, 295, 399, 400, 403, 427, 440, 457

Reaction (primary, secondary, parallel and multiple parallel, single, reaction mode, mechanism) 13, 77, 78, 82, 83, 88, 96, 97, 104, 105, 136, 193, 194, 195, 219, 220, 221, 224, 228, 305, 306, 424, 438, 455

Reactivity 9, 17, 22, 59, 126, 159, 180, 184, 186, 219, 220, 222, 225, 226, 229, 252, 233, 309, 310, 315, 316, 400, 454

Recirculation 250, 396

Refinery 396

Reflectance 4, 6, 7, 59, 118, 120